D. Stoyan · W. S. Kendall · J. Mecke

Stochastic Geometry and Its Applications

Mathematische Lehrbücher und Monographien
Herausgegeben von der Akademie der Wissenschaften der DDR
Karl-Weierstraß-Institut für Mathematik

II. Abteilung
Mathematische Monographien
Band 69

Stochastic Geometry and Its Applications

von D. Stoyan, W. S. Kendall und J. Mecke

Stochastic Geometry and Its Applications

by Dietrich Stoyan, Wilfrid S. Kendall, and Joseph Mecke
with a foreword by D. G. Kendall

with 83 figures and 18 tables

Akademie-Verlag Berlin
1987

Verfasser:

Dr. Dietrich Stoyan
Bergakademie Freiberg, DDR
Sektion Mathematik

Dr. Wilfrid S. Kendall
University of Strathclyde, Glasgow, UK
Department of Mathematics

Prof. Dr. Joseph Mecke
Friedrich-Schiller-Universität Jena, DDR
Sektion Mathematik

ISBN 3-05-500032-3
ISSN 0076-5430

Erschienen im Akademie-Verlag Berlin, DDR - 1086 Berlin, Leipziger Straße 3—4

Printed in the German Democratic Republic
Lizenznummer: 202 · 100/402/86.
Gesamtherstellung: VEB Druckhaus „Maxim Gorki" 7400 Altenburg
Lektor: Dr. Reinhard Höppner
LSV 1085
Bestellnummer: 763 535 3 (6920)
05000

Foreword

My good friends the authors of this book have kindly invited me to add a few words describing 'how it all began'. At present one can write only an anecdotal history of stochastic geometry, and we must recognize that the anecdotes of others will certainly much extend the account given here, and will supply fresh perspectives. A serious attempt to write a history would be premature. The historian of mathematics looks always to the future rather than to the past. He hopes to find early instances of general concepts later seen to be of fundamental significance, and so if he works at a time of rapid development (such as the present) he will overlook many clues in the early record which point to a future not yet revealed.

My own first contact with *classical geometrical probability* occurred during the war, when the Superintendent of my group (Louis Rosenhead) asked me to investigate the following problem; for the sake of clarity I formulate it in the modern terminology. A euclidean plane with a marked origin O carries a Poisson field of unsensed lines with a uniform intensity. Almost surely the point O will lie in the interior of a unique Crofton cell C, with (unlabelled) shape $\sigma(C)$ and area $a(C)$. What probability statements can be made about C which convey information about the strength of a fabric ('a sheet of paper') consisting of the field of lines ('fibres')? Thus it would be useful to be able to calculate the rate of occurrence of splinter-shaped cells C, and the rate of occurrence of cells with large area $a(C)$. A few moments of the $a(C)$-distribution were already known, and I managed to add one more to these. One would have preferred to be able to say something about the asymptotics of the marginal $a(C)$-distribution valid for large areas, and to throw light one way or the other on my conjecture that the conditional law for $\sigma(C) \mid a(C)$ converges weakly, as $a(C) \to \infty$, to the degenerate law concentrated at the circular shape. Unfortunately nothing substantial is known about either of these questions even today, apart from the limited information that can be derived from a massive series of simulations carried out in Stanford by E. I. George (1982) in association with Herbert Solomon.

In 1961 Roger Miles wrote his Cambridge PhD thesis on a generalized version of this problem under the direction of Dennis Lindley and Peter Whittle and in consultation with paper technology experts from Wiggins Teape Research and Development Ltd. This initiated a long series of famous papers by Miles which provide a huge volume of information about the problem as a whole without, however, bringing us any nearer to the answer to the two questions above.

All this work was within the classical Croftonian framework, not however without hints that a statistical theory of shape could play a useful role. What we now call *stochastic geometry* began for me with three papers by Maurice Bartlett: 'The spectral analysis of point processes' (*JRSS* B 1963), 'The spectral analysis of two-dimensional point processes' (*Biometrika* 1964), and especially 'The spectral analysis of line processes' (*5th Berkeley Symp.* 1965). These became available just when Rollo Davidson joined me as a research student in October 1965, and the impact of Bartlett's third paper can be felt in the Smith's Prize Essay which he wrote in 1967. As Bartlett's work had focused on the empirical spectral analysis of point and line processes, I encouraged Davidson to set up an appropriate theoretical framework underlying such an empirical analysis, and the second half of his PhD thesis was concerned with this. see Chapter 2.1 of *Stochastic Geometry* (ed. Harding and Kendall) for a reprint of it. The following two quotations give the flavour of the approach:

'... we can talk of flat- and line-processes, meaning the point-processes that they induce on the appropriate manifolds',
'... my results showed that it is profitable, when considering point-processes, to observe simply whether certain sets contain points of the process or not; the usual approach is, of course, to look at the number of points in these sets'.

The first quotation, coupled with a general specification for a point process on a manifold M, takes us away from Croftonian geometric probability to the consideration of arbitrary random fields of geometric objects, while the second hints at a theory of random sets of very general character to replace the point processes on the representation-manifold. Two such (closely related) theories of random sets very shortly afterwards became available; both were strongly influenced by earlier ideas due to Gustave Choquet. Stochastic Geometry thus became a reality.

When was the phrase 'stochastic geometry' first used? Klaus Krickeberg, who was to play a leading role in its development, thinks that perhaps he and I may have used the phrase informally in the Spring of 1969 when he was in Cambridge and we were planning the Oberwolfach meeting (for June 1969) on *Integral Geometry and Geometrical Probability*. This is very possible, because 'stochastic analysis' was already in common use in the UK as part of the title of the Stochastic Analysis Group set up in December 1961 under the chairmanship of Harry Reuter, and one phrase naturally suggested the other. Certainly there was obviously no other choice for the titles of the two memorial books produced after Davidson's death in 1970. So 1969 was perhaps the year of coining. The initial group of enthusiasts could readily be identified by inspecting the *Tagungsbuch* at Oberwolfach. From the first Ruben Ambartzumian played a very important role and he has continued to influence the development of the subject in characteristic ways.

To some extent and to my great satisfaction there has also been a close association between those interested in stochastic geometry, and those interested in geometrical statistics, so that we now have a broad and lively subject area with abstract and empirical edges to it. I trust that this will continue, and the balance of the present book makes that seem likely.

Shape-theory is generally viewed as part of stochastic geometry and I think that this is as it should be. I have already mentioned one early hint at the need for a theory of shape, and here is a much earlier one. *The Ladies' Diary* (1706—1840), *The Gentleman's Diary* (1741—1840), and *The Lady's and Gentleman's Diary* (1841—1871) are mathematical periodicals known now perhaps only to a few specialists, but are worth very serious study because of the frequency with which important ideas first found explicit mention in their pages. In *TLGD* (1861) there is the following challenge by the London-based mathematician Wesley Stoker Barker Woolhouse (1809—1893):

'Problem 1987. Three points being taken at random in space as the corners of a plane triangle, determine the probability that it shall be acute.'

A solution by Stephen Watson of Haydonbridge, Northumberland, appeared in the 1862 edition of the diary. It begins with the comment

"'Space' is equivalent to a sphere of infinite radius, and it is obvious that the chance will be the same whatever be the radius of the sphere within which the three points may lie; hence we may suppose them to always lie within a sphere of radius unity."

He then continues with an integration argument yielding the probability 33/70, and the probability $4/\pi^2 - 1/8$ for three points in a plane is added in a comment by Woolhouse.

Watson's solution provoked a strong reaction from Augustus de Morgan, who argued in *Trans. Cambridge Philos. Soc.* ***11*** (1871) 145—189 that 'it is very easily shown that the chance of an acute-angled triangle must be infinitely small'. The controversy raged for many years in the journal *Mathematical Questions with their Solutions from the 'Educational Times'*, where Woolhouse's challenge had been reprinted as Question 1333. The article on it by M. W. Crofton in 1867 is particularly interesting. In the pages of these journals we can see classical geometrical probability taking shape under our eyes.

But from our present point of view a most interesting comment is that Watson could very well have approximated to 'infinite space' by expanding an arbitrary compact convex set K about its centroid, and he would then have obtained a different answer based on the shape measure induced on $\sum_3^3$ (a hemisphere) by IID-uniform sampling from the interior of K, so that the resulting probability would depend on the shape of K itself, but not at all upon its size. Of course Watson's choice was a very natural one, because one feels that one is required to respect the isotropy of 'infinite space' — though this is a delusion, for the isotropy is only maintained at the centre of the sphere. Meaningless though Woolhouse's problem is, without further specification, any reader of this book will probably find it instructive as well as amusing to read through these old polemics. As the same protagonists occur again and again in the pages of the journals mentioned, and as they express themselves with considerable freedom, one soon becomes familiar with them on a personal basis, and the early history of our subject comes to life in a most vivid way.

It only remains to say that, given the spherical assumption, the numerical answers obtained by Watson and by Woolhouse are identical with those derived from the recent solution to the general problem in n dimensions given by G. R. Hall in *JAP* (1982) ***19***, 712—715.

David Kendall

Preface

Complicated geometrical patterns occur in many areas of s ience and technology and often require statistical analysis. Examples include the structures studied in geology, sections of porous media, solid bodies, biological tissues, and patterns formed by the distinction between wood and field in a landscape. Analysis of such data sets require suitable mathematical models and appropriate statistical methods. The area of mathematical research that seeks to provide such models and methods is called *stochastic geometry*. The oldest part of this subject considers problems concerning a finite number of geometrical objects of fixed form, whose positions are completely random and (in some sense) uniformly distributed. The famous question of Buffon's needle is the prototype of these problems, which form the subject of *Geometrical Probability*. The modern theory of stochastic geometry (initiated by D. G. Kendall, K. Krickeberg, and R. E. Miles) considers random geometrical patterns (which may be infinite in extent) of more complicated distribution. *Stereology* is that branch of stochastic geometry which studies the problem of recovering information on three-dimensional structures when the only information avaiable is two- or one-dimensional, obtained by planar or linear section.

This monograph grew out of a book originally published in German (Stoyan and Mecke, 1983), but has undergone considerable expansion and reorganization. Its aim is to make the results and methods of stochastic geometry more generally accessible to applied scientists, but also to provide an exposition which is mathematically exact and general, and which takes into account the current state of research in order to serve as an introduction to stochastic geometry for mathematicians. Of course these aims conflict and the resulting compromises have strongly influenced the form of the book. In most parts of the monograph proofs are omitted. The level of exposition is uneven and the subjects are treated with varying thoroughness: some topics are illustrated by numerical examples, some results are stated without much comment, others are accompanied by heuristic arguments, and sometimes substantial issues are dismissed with only a few remarks and a few references to the literature. Throughout the text attempts are made to explain the plausible nature and the underlying ideas of mathematical concepts, in order to facilitate the reader's understanding and to pave the way for a deeper study of the literature. Our hope is that those readers who do not wish to invest much effort in following mathematical arguments will nevertheless be able to interpret and to use most of the formulae.

There is some redundancy in the exposition but we believe this is in most readers' interest. At some points it has been appropriate to use formulae and notation in anticipation of their introduction. In any case readers may prefer to turn directly to the chapters concerning the topics that interest them the most, rather than to read through the book consecutively. Generally we have not sought to use the most elegant possible mathematical style but rather to strike a balance between generality and concrete special cases. For example we use the theory of marked point processes, and this frees us from the need to consider point processes in abstract spaces.

At places we refer to ideas of mathematical physics that are related to the techniques of stochastic geometry. A closer collaboration between mathematical physicists and stochastic geometers might be very fruitful; the two subjects meet at several points but use different languages.

Mathematical terminology is used throughout the book. This exhibits some peculiarities due to historical accident. The word 'process' as in 'point process' and 'line process' does not imply any dependence on time (with the possible exception of point processes on the real line). A more logical terminology would use the phrases 'random point field' and 'random line field'.

The bibliography is extensive, comprising ca. 600 references, but is still far from complete. An important source of additional information is to be found in the series of *Notes on Recent Research in Geometric Probability* (for example BADDELEY; 1977).

The examples in the book are for the most part concerned with the analysis and description of images by numbers and functions, and are drawn from various branches of science. Generally the theoretical basis of the statistical methods is not discussed. In some cases statistical methods are given and these enable the fitting of models to empirical data. Much work remains to be done on statistical theory for geometrical structures. For example little is known of the distribution theory for most estimators appearing in this book.

A brief summary of the contents of the book will illustrate the way in which theory and practice are intertwined. The first chapter briefly introduces areas of mathematics with which most scientists and engineers are not familiar. We recommend the books of HADWIGER (1957), HALMOS (1974), KINGMAN and TAYLOR (1966), LEICHTWEISS (1980), MATHERON (1975), MENDELSON (1975), SANTALO (1976), and SERRA (1982) for a thorough study. We assume a basic knowledge of probability theory and statistics.

In the remaining chapters the development of the exposition does not proceed from the general to the particular but rather in the reverse direction. Thus chapters 2 and 3 discuss the Poisson process and the Boolean model, which are simple cases of the random structures to be discussed in the remainder of the text. Chapter 4 and 5 continue the subject of point processes and give a general discussion; Chapter 6 expounds the general theory of random sets. Chapter 7 briefly introduces the important concept of a random measure, which arises throughout the subject at a more theoretical level. The theory of random processes on geometrical objects is introduced in Chapter 8, which leads on to the discussion of fibre processes in Chapter 9 and tessellations in Chapter 10. The final Chapter is on stereology, which is of great importance in practice and uses results and ideas from all of the preceding discussion.

The book is based on a rough English translation of the German text (but including additional material) which was produced by D. Stoyan. A second draft was made by W. S. Kendall who polished the language, added some more material, and occasionally recast the argument. In particular W. S. Kendall added the new Chapter 8. The text was then edited by J. Mecke and D. Stoyan and brought to its final form by W. S. Kendall after discussion by mail.

During the work on the book we benefited from the help of a number of people. We are most grateful to our colleagues T. Fiksel, K. H. Hanisch, L. Heinrich, D. König, W. Nagel, J. Ohser, P. Schatte, A. Schwandtke, H. Stoyan, K. Voss, and B. D. Ripley, and also the many colleagues who sent us offprints and preprints; in particular P. J. Diggle, H. J. G. Gundersen, B. Matern, and J. P. Serra and his colleagues. Thanks are also due to R. V. Ambartzumian and H. Sukiasian (who translated the German text into Russian) for pointing out a number of errors in the original text, to H. J. G. Gundersen, E. B. Jensen, D. G. Kendall, D. König and B. D. Ripley for their valuable remarks on drafts of various chapters, and to B. D. Ripley for allowing the use of his graphics package in the production of some of the figures. Finally we thank very much D. G. Kendall for supporting the publication of this book in the Wiley series (and also for his kindness in contributing a brief account on the history of stochastic geometry) and R. Höppner of Akademie-Verlag Berlin for his co-operation. Last and not least we thank our families for their support, encouragement, and tolerance.

June 1985 The authors

Contents

1. Mathematical Foundations

Introduction

This chapter summarizes the mathematical concepts to be employed throughout the book and introduces various notation and conventions.

1.1 Set theory

The language of naive set theory is ubiquituous in geometry and even more so in stochastic geometry. The reader will find a thorough introduction in specialist textbooks. Here the notation is briefly summarized and mention is made of important sets to be employed later.

A *set* is a collection of mathematical objects taken from a suitable domain of discourse. If x is an element of a set S this is written as $x \in S$. For our purposes there are two fundamental sets (from which are constructed all others to appear in this book). These are the set of the natural numbers $\{1, 2, \ldots\}$ and the set of the real numbers (the *real line*) $\mathbb{R} = (-\infty, \infty)$. All our constructions being suitably regular, we ignore the profounder aspects of mathematical logic and the theory of sets.

The notation for the sets of natural and real numbers illustrate two useful conventions for the description of sets. The braces $\{\,\}$ above enclose a description of the set of natural numbers by (implicit, infinite) enumeration. The notation (a, b) for two real numbers or infinities a and b describes the set of all real numbers x such that $a < x < b$. This set (a, b) is an *open interval* of the real line $\mathbb{R}$. *Closed* and *half-open intervals* are given by

$$\begin{aligned}
(a, b] &= \{x \in \mathbb{R} : a < x \leqq b\} \quad \text{(half-open)}\\
[a, b) &= \{x \in \mathbb{R} : a \leqq x < b\} \quad \text{(half-open)}\\
[a, b] &= \{x \in \mathbb{R} : a \leqq x \leqq b\} \quad \text{(closed).}
\end{aligned}$$

Here, the braces $\{\,\}$ enclose a description of a set as the collection of elements of another set satisfying some property. A contraction of this notation is often used:

$$\begin{aligned}
&\{x \in \mathbb{R} : x = y + z \quad \text{with} \quad 0 < y < 1 \quad \text{and} \quad 0 < z < 1\}\\
&\quad = \{x : x = y + z \quad \text{with} \quad 0 < y < 1 \quad \text{and} \quad 0 < z < 1\}\\
&\quad = \{y + z : 0 < y < 1 \quad \text{and} \quad 0 < z < 1\}.
\end{aligned}$$

Note that this set is actually the open interval $(0, 2)$.

A *subset* A of a set S is a set all of whose members happen to lie in S. This is written $A \subset S$. If $A, B \subset S$ for some set S then we can construct their *union*, *intersection*, and *difference* as

Union	$A \cup B = \{x \in S : x \in A \text{ or } x \in B\}$
Intersection	$A \cap B = \{x \in S : x \in A \text{ and } x \in B\}$
Difference	$A \setminus B = \{x \in S : x \in A \text{ but not } x \in B\}$.

We also define the *complement* A^c as

$$A^c = \{x \in S : \text{it is not the case that } x \in A\}$$
$$= S \setminus A$$

but in this case the *superset* S must be clearly understood from the context. The empty set $\emptyset$ is the set that contains no elements. Special collections of sets (σ-algebras) are considered in Section 1.8.

Two sets A and B can be used to form the *Cartesian product* $A \times B$ of ordered pairs:

$$A \times B = \{(a, b) : a \in A \text{ and } b \in B\}.$$

More generally the Cartesian product of n sets $A_1, \ldots, A_n$ is

$$A_1 \times \ldots \times A_n = \{(a_1, \ldots, a_n) : a_1 \in A_1, \ldots, a_n \in A_n\}.$$

A very important example is given by

$$\mathbb{R}^2 = \mathbb{R} \times \mathbb{R} = \{(x_1, x_2) : x_1, x_2 \in \mathbb{R}\}$$

which is the co-ordinatization of the plane familiar to map-readers. More generally one obtains

$$\mathbb{R}^3 = \mathbb{R} \times \mathbb{R} \times \mathbb{R}$$

and

$$\mathbb{R}^d = \{(x_1, \ldots, x_d) : x_1, \ldots, x_d \in \mathbb{R}\}.$$

The spaces $\mathbb{R}^2$, $\mathbb{R}^3$, $\mathbb{R}^d$ are often referred to as the *plane*, *space*, and *d-dimensional (Euclidean) space* respectively. An element $x = (x_1, x_2, \ldots, x_d) \in \mathbb{R}^d$ is referred to as a *point* in geometry. However, in stochastic geometry a distinction must be drawn. The study of stochastic geometry frequently concerns random collections of points, referred to as *point processes*. It is convenient to refer to members of such random collections as *points of the process*, or even just as plain *points*. Thus this book refers to points that are merely locations in $\mathbb{R}^d$ (with no membership presumed of some special collection of points) as *location points* or *locations*. A particular case of a location point is the *origin*

$$o = (0, \ldots, 0).$$

The two- and three-dimensional cases of Euclidean space, those of the plane and of space, are the most important for practical purposes. These will be referred to as the cases $d = 2$ and $d = 3$, or as the *planar* and *spatial cases*.

1.2 Topology in $\mathbb{R}^d$

To Euclidean space of whatever dimension is associated a concept of distance. The *Euclidean metric* measures the distance between two location points $x = (x_1, x_2, \ldots, x_d)$ and $y = (y_1, y_2, \ldots, y_d)$ as

$$\|x - y\| = \sqrt{[(x_1 - y_1)^2 + \cdots + (x_d - y_d)^2]}.$$

This distance allows for the definition of the *closed ball* $b(a, r)$ *of radius* r *and centre* a

$$b(a, r) = \{x \in \mathbb{R}^d : \|x - a\| \leqq r\}.$$

Here r is a positive real number and a is a location point. The *open ball* $b^{\text{int}}(a, r)$ is defined similarly but with strict rather than weak inequality.

The metric (or equivalently the closed and open balls) can be used to define special properties for subsets A of Euclidean space. The set A is said to be *bounded* if there is a ball $b(a, r)$ with

$$A \subset b(a, r).$$

A sequence $\{x_1, x_2, \ldots\}$ is said to *converge* to x if

$$\lim_{n\to\infty} \|x_n - x\| = 0.$$

A set A is said to be *open* if for each $x \in A$ a positive number ε can be found (depending on x) such that $b(x, \varepsilon) \subset A$. The system of open sets of $\mathbb{R}^d$ is denoted by $\mathbb{O}$. In the linear case $d = 1$ examples of open sets are the open intervals (a, b). In the general case of $\mathbb{R}^d$ examples include the open balls $b^{\text{int}}(a, r)$ described above, and the open hypercubes

$$(a_1, b_1) \times \cdots \times (a_d, b_d).$$

Finite intersections and arbitrary unions of open sets are again open.

A set A is said to be *closed* if its complement A^c is open. The system of closed sets of $\mathbb{R}^d$ is denoted by $\mathbb{F}$. In the case $d = 1$ examples of closed sets are the closed intervals $[a, b]$. In the general case of $\mathbb{R}^d$ examples include the closed balls $b(a, r)$ and the closed hypercubes

$$[a_1, b_1] \times \cdots \times [a_d, b_d]$$

and the hyperplanes

$$\{x = (x_1, \ldots, x_d) \in \mathbb{R}^d : \sum a_i x_i = b\}$$

for some constants b and $a_1, \ldots, a_d$ with $a_1^2 + \cdots + a_d^2 > 0$. Finite unions of closed sets and arbitrary intersections of closed sets are again closed.

The *interior* A^{int} of a general set A is the union of all open sets contained in A. The *closure* A^{cl} of A is the intersection of all closed sets containing A. Thus A^{int} is the largest open set contained in A, while A^{cl} is the smallest closed set containing A. Hence

$$A^{\text{int}} \subset A \subset A^{\text{cl}}.$$

Moreover, $A^{\text{int}} = ((A^c)^{\text{cl}})^c$. A set A is open precisely when $A^{\text{int}} = A$, and closed precisely when $A^{\text{cl}} = A$.

The difference $\partial A = A^{\text{cl}} \setminus A^{\text{int}}$ is the *boundary* of A. An important example is the *sphere* of centre a and radius r, which is the boundary of $b(a, r)$ and equals the set

$$\partial b(a, r) = \{x \in \mathbb{R}^d : \|a - x\| = r\}.$$

The particular case $\partial b(o, 1)$ is the *unit sphere* of R^d, and is denoted by S^{d-1}.

A set $K \subset \mathbb{R}^d$ is said to be *compact* if it is both closed and bounded. The system of all compact sets is denoted by $\mathbb{K}$. Moreover it is convenient to refer to the system of non-empty compact sets as

$$\mathbb{K} = \mathbb{K}' \setminus \{\emptyset\}.$$

Examples of compact sets include the closed balls $b(a, r)$ and the closed hypercubes.

1.3 Operations on subsets of Euclidean space

Euclidean space is a *vector space* so it allows the vector space operations of

$$\text{addition} \quad x + y = (x_1 + y_1, \ldots, x_d + y_d)$$

and

$$\text{scalar multiplication} \quad c \cdot x = cx = (cx_1, \ldots, cx_d)$$

for location points $x = (x_1, \ldots, x_d)$, $y = (y_1, \ldots, y_d)$ of $\mathbb{R}^d$ and real numbers c. This vector structure allows for the definition of new set operations special to Euclidean space.

Multiplication by real numbers

$$cA = \{c \cdot x : x \in A\}$$

for real numbers c and $A \subset \mathbb{R}^d$. The case $c = -1$ leads to the particular case of *reflection*

$$\check{A} = -A = \{-x : x \in A\} \text{ for } A \subset \mathbb{R}^d.$$

If $A = \check{A}$ then A is said to be *symmetric*.

Translation

$$A_x = A + x = \{y + x : y \in A\} \quad \text{for } x \text{ in } \mathbb{R}^d \text{ and } A \subset \mathbb{R}^d.$$

Minkowski-addition

$$A \oplus B = \{x + y : x \in A, y \in B\} \quad \text{for } A, B \text{ subsets of } \mathbb{R}^d.$$

The operation of Minkowski-addition by a set B both enlarges, translates, and deforms a set A; see Figure 1.1 on page 22 Minkowski-addition is both associative and commutative. Other properties are summarised in the following formulae.

$$A_x = A \oplus \{x\} \tag{1.3.1}$$

$$A \oplus B = \bigcup_{y \in B} A_y = \bigcup_{x \in A} B_x \tag{1.3.2}$$

$$A \oplus B = \{x : A \cap (\check{B})_x \text{ is not empty}\} \tag{1.3.3}$$

$$A \oplus (B_1 \cup B_2) = (A \oplus B_1) \cup (A \oplus B_2). \tag{1.3.4}$$

If $A_1 \subset A_2$ then $A_1 \oplus B \subset A_2 \oplus B$. Formula (1.3.2) shows that Minkowski-addition can be represented as the union of the translates B_x as x runs through A. In the special case of $B = b^{\text{int}}(o, r)$ the Minkowski-sum $A \oplus B$ is the union of all location points that are of distance no more than r from A.

Minkowski-subtraction

$$A \ominus B = \bigcap_{y \in B} A_y. \tag{1.3.5}$$

or

$$A \ominus B = (A^c \oplus B)^c.$$

In general, the Minkowski-substraction is not an inverse for Minkowski-addition, but we do have

$$(A \ominus \check{B}) \oplus B \subseteq A \subseteq (A \oplus \check{B}) \ominus B.$$

If B is the open ball $b^{\text{int}}(o, r)$ then $A \ominus B$ is the union of all location points lying so far within A that balls centred on them and of radius r are completely contained in A.

The theory of image analysis (see Section 1.4) studies the *dilatation* operation

$$A \to A \oplus \check{B}$$

and the *erosion* operation

$$A \to A \ominus \check{B}.$$

These formulae define the sets of location points x fo: which $B + x$ respectively overlaps with or is contained in A. The application of these operations in quantitative image analysis is further described in the next section.

1.4 Some remarks on quantitative image processing and mathematical morphology

In many fields of technology and science there is a growing need for methods of analysis for large quantities of data coming in the form of *images* (which is to say planar sets). Examples are: satellite photographs, geological maps, and microscope images of sections of metals, minerals, and cellular tissue. The sheer quantity of data requires the use of automatic and quantitative methods. In the following, some of the basic ideas are described. (See HUANG (1975) and ROSENFELD and KAK (1982) for the general theory of image processing.) In particular, it will be seen how statistical procedures on random sets (to be described throughout the book) can be performed automatically.

The first step in the automatic analysis of an image is to digitize it. This is performed by means of a grid and can be carried out most conveniently by using a television camera. The process of digitization is often (here we discuss this case alone) a binary procedure yielding values of 0 and 1 for each location of the grid. These values form a two-dimensional array of *bits* or *pixels*, each holding one of the values 0 or 1 (corresponding to white and black). This requires a calibration choice, fixing the value of the grey scale at which the division between white and black is to be made. Such a choice is often readily made because in many applications the components of the image (the *phases*) are defined by the grey-scale intensity that they produce in the image.

When the array of bits given by the digitized image is obtained then it is possible with suitable equipment to determine important quantities automatically. These quantities are effectively numbers of pixels in specific subsets of the image. Examples are the areas of the white and black phases, the length of the boundary between the two phases, and the number of components of the black phase. If one conceives of the image as a realization of a random structure then these measurements serve as statistical estimates of various characteristics of the random structure. For example, the area fraction p is estimated by the proportion of area covered by the black phase (see 3.3 and 6.3), while the specific length of boundary is estimated by the proportion of length of phase boundary to unit area in the image (see 7.3 and 9.3).

So how can such measurements be obtained by modern quantitative image analysis? The central idea is to subject images to repeated transformations and to perform measurements on the transformed images (SERRA, 1982). On the one hand, automatic operations can be applied to the image to free it from the inevitable image defects and artefacts of the image-processing procedure. This is an important preliminary step to clear the ground for human intuition. On the other hand, important functions can be measured quickly and easily. Examples of these functions are the set covariance, granulometry, and contact distribution functions.

The subject of mathematical morphology as developed by MATHERON and SERRA uses transformations employing 'structuring elements' to perform automated image analysis. This theory is a direct application of the Euclidean set operations discussed in the previous section and it is an useful technical aid in the analysis of images. In the following brief survey, A will denote the set which is of interest (also called the *image*, and in practical applications often the set of dark pixels or points on a photographic image of some kind). The structuring element B will in practice be either a disc (often approximated by a hexagon), a line segment, or a two-point set.

Dilatation

This is the operation

$$A \to A \oplus \check{B}.$$

The set A is enlarged and, at least in the case where the structuring element B is a ball, smoothed. In particular, the action of dilatation fills in cavities, repairs fissures, and joins together a fragmented image. If A is a realization of a random closed set (see Chapters 3 and 6) then dilatation by rB followed by measurement of the area of $A \oplus rB$ allows for the estimation of the contact distribution $H_B(r)$ of the random set.

Erosion

This is dual to the operation of dilatation, and is given by

$$A \to A \ominus \check{B}.$$

This shrinks the set A, tending to produce smaller fragments, even separating connected sets into several subsets. This can be helpful in the estimation of the number of particles composing an image. The erosion operation is of great importance for the quantitative estimation of various quantitative characteristics of random sets. Suppose one takes for the structuring element B a two-point set $\{x, y\}$ with $\|x - y\| = r$. Then the area of the eroded set $A \ominus \check{B}$ is an estimate for the set covariance $C(r)$ (see 6.2).

Opening

The opening of A by B is the result of attempting to reverse an erosion by a dilatation. It is given by

$$A \to A_B = (A \ominus \check{B}) \oplus B.$$

The opening A_B of a set A by B has an appearance similar to that of the original set A, but is built only on the portions of the image that survive the initial erosion. Therefore small disconnected fragments of the image disappear under opening and this is useful in systematically eliminating possible image defects or noise.

Closing

This is dual to opening and attempts to reverse a dilatation by an erosion. It is given by

$$A \to A^B = (A \oplus \check{B}) \ominus B.$$

Again A^B bears an approximate resemblance to A. Together with opening, the closing operation is useful in cleaning up an image. The action of closing tends to close up small holes, to join up close but separated subsets, and to smooth out the boundaries of an image.

Figure 1.1 displays typical results of the application of these transformations. The following example shows how repeated application of the transformations leads to composite image transformations of great importance (see also Serra, 1982).

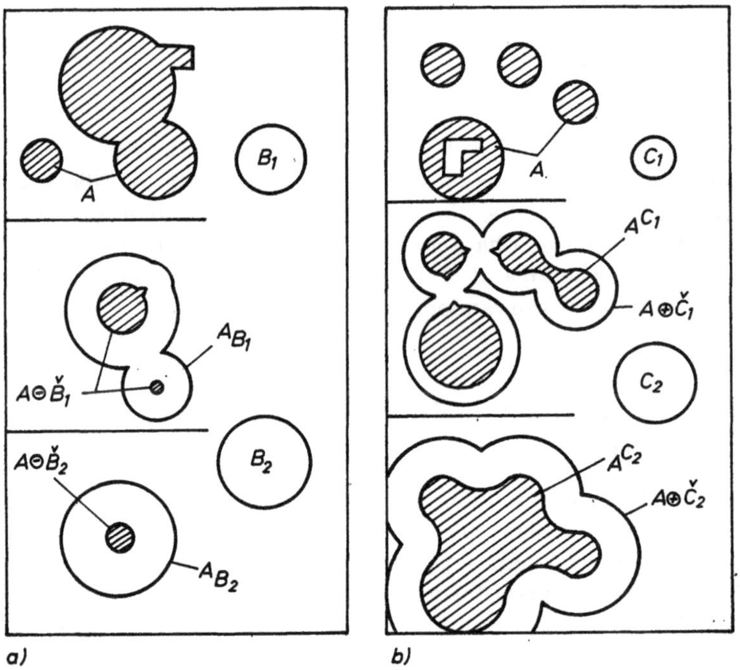

Fig. 1.1(a). The operations of erosion and opening applied to a set. Components that overlap or touch are separated while small components and roughnesses vanish or are reduced. (b) The operations of dilatation and closing applied to a set. Gaps are closed up, concavities vanish or are reduced, and clusters of small particles are merged

Skeleterisation

Consider an image of the plane divided into polygons. The set A is the set of points lying on boundaries between these polygons. These boundaries are thin but have a positive thickness, as indicated in Figure 1.2. It is desired to count the number of polygons neighbouring each polygon. This is trivial but tedious if done by visual inspection. It can be automated as follows. Each individual polygon C is considered in turn. The set $C \oplus b(o, r)$ is constructed, where r is

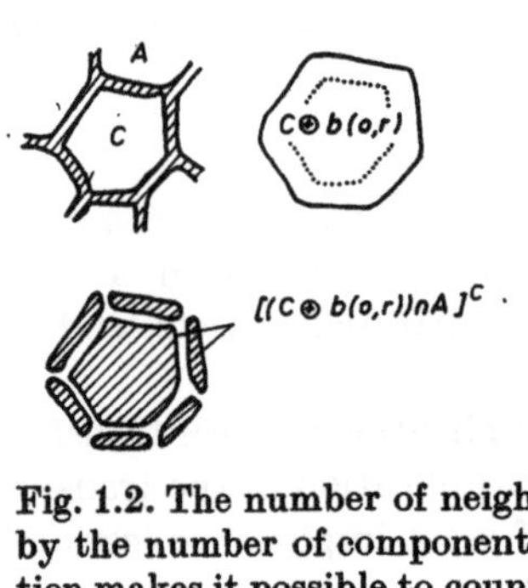

Fig. 1.2. The number of neighbouring polygons to the polygon C is measured by the number of components of $[(C \oplus b(o, r)) \cap A]^c$. This image transformation makes it possible to count the number of sides of a polygon automatically

taken to be approximately twice the thickness of the boundaries of the polygon (this value of r is chosen to eliminate problems due to varying thickness of the boundary). Then the intersection $(C \oplus b(o, r)) \cap A$ has as complement a set made up of n_C connected components. One connected component corresponds to the polygon C itself and the others correspond to the neighbouring polygons (see Figure 1.2).

This means of automatic neighbour counting will break down if the polygons are too irregular, or if the boundary varies too greatly in thickness. However, under only a moderate degree of regularity of the image the procedure still works well, and this is all that could reasonably be expected of an automatic means of image analysis.

1.5 Euclidean isometries

A transformation $\mathfrak{m}: x \to x'$ is said to be a Euclidean *isometry* if it leaves invariant the distance between points x and y. That is to say,

$$\|x - y\| = \|x' - y'\| = \|\mathfrak{m}x - \mathfrak{m}y\|.$$

It can be shown that every isometry of Euclidean space is represented in the form

$$(\mathfrak{m}x)_k = x_k' = b_k + \sum_{l=1}^{d} a_{kl} \cdot x_l \tag{1.5.1}$$

for $k = 1, \ldots, d$, for $b = (b_1, \ldots, b_d) \in \mathbb{R}^d$, and for $A = (a_{kl})$ an orthogonal matrix. In particular the determinant of the matrix $\det A = \pm 1$. The isometry is said to be a *proper isometry* if the determinant is $+1$.

If A is the unit matrix then the isometry is said to be a *translation*; the notation

$$T_y x = x - y$$

is sometimes used. Recalling (1.3.1) we see that the set A_x is the *translated set* of A and can be written

$$T_{-x} A = A_x.$$

Note the composition formula

$$T_x T_y = T_{x+y}. \tag{1.5.2}$$

Rotations about the origin are proper isometries given by (1.5.1) for which $b = (b_1, \ldots, b_d) = 0$. We reserve for them the notation

$$\mathfrak{r}x = Ax$$

A further composition formula is

$$\mathfrak{r} T_x y = T_{\mathfrak{r}x} \mathfrak{r} y \tag{1.5.3}$$

Formula (1.5.1) makes it plain that every proper isometry is the composition of a rotation about the origin with a translation.

An alternative term for isometry is *motion*.

1.6 Convex sets in Euclidean spaces

A subset K of $\mathbb{R}^d$ is called *convex* if for every pair of points x, y in K the intervening line segment also lies in K. That is to say

$$c \cdot x + (1 - c) \cdot y \in K \quad \text{whenever } x, y \in K \text{ and } 0 < c < 1.$$

Important examples of convex sets include the *affine linear subspaces*; that is to say subsets L of $\mathbb{R}^d$ with the property that for every x, y in L the whole line $\{c \cdot x + (1 - c) \cdot y : c \in \mathbb{R}\}$ through x and y also lies in L. The *dimension* of such an affine linear subspace is the smallest whole number s such that L can be given by the formula

$$L = \{c_1 z_1 + \dots c_{s+1} z_{s+1} : c_1, \dots, c_{s+1} \in \mathbb{R}, \sum c_i = 1\}$$

for some points $z_1, \dots, z_{s+1}$ in $\mathbb{R}^d$. Affine linear subspaces of dimension r are called *r-flats* or *r-planes*; if the *r-flat* contains o then it is called a r-subspace. The $(d - 1)$-flats are called *hyperflats* or *hyperplanes*. The 1-flats are called *lines*. In $\mathbb{R}^3$ the 2-flats are called simply *flats*, the 2-planes are called *planes*. Note that the hyperflats of the plane $\mathbb{R}^2$ are precisely its lines.

The r-flats are closed sets but are not bounded and therefore not compact. Convex sets which are also compact are called *ovoids*. Examples are the closed balls, the closed and bounded hypercubes, and the closed discs (intersections of closed balls with 2-flats). The system of all ovoids is denoted by $C(\mathbb{K})$. In the case $d = 1$ the system $C(\mathbb{K})$ coincides with the system of all closed and bounded intervals. In general ovoids can also be called *proper bodies*. Open balls and open hypercubes are not ovoids, though they are convex. The spheres

$$\{x \in \mathbb{R}^d : \|x - a\| = r\}$$

and the torus are not even convex.

An important characteristic for an ovoid K that contains the origin o is the *support function*

$$r_K{}^\circ : \partial b(o, 1) \to \mathbb{R}$$

defined by

$$r_K{}^\circ(u) = \sup_{x \in K} \langle u, x \rangle$$

where $\langle u, x \rangle = u_1 x_1 + \dots + u_d x_d$ is the scalar product of $u = (u_1, \dots, u_d)$ and $x = (x_1, \dots, x_d)$. The support function $r_K{}^\circ$ is continuous on $\partial b(o, 1)$.

If K is symmetric (so $\check{K} = K$) then the support function is uniquely determined by its values on one hemisphere of the sphere $\partial b(o, 1)$. In this case the *modified support function* r_K is defined on the set L_1 of all lines through the origin. For $l \in L_1$ let $e(l)$ be the point on $l \cap \partial b(o, 1)$ in the upper hemisphere $(x_d \geqq 0)$. Then

$$r_K(l) = r_K{}^\circ\big(e(l)\big) \quad \text{for } l \in L_1.$$

In the planar case $(d = 2)$ the line l can be replaced by the angle α formed by l and the x_1-axis in the upper half-plane. Then r_k becomes a function defined on $(0, \pi]$.

The modified support function is given the simpler notation here because the unmodified support function will not be used in this book.

Some set operations preserve the class $C(\mathbb{K})$. In particular if K_1, K_2 belong to $C(\mathbb{K})$ then so do $c \cdot K_1$ for real c, $\check{K}_1$, $K_1 \cap K_2$, and $K_1 \oplus K_2$.

An *ovoid functional* h is a real-valued functional defined on $C(\mathbb{K})$ assigning a value $h(K)$ to every $K \in C(\mathbb{K})$. We are particularly interested in those non-negative ovoid functionals which possess the following properties:

Isometry invariance $h(\mathfrak{m}K) = h(K)$ if $K \in C(\mathbb{K})$ and $\mathfrak{m}$ is an isometry.
Monotonicity if $K_1 \subset K_2$ then $h(K_1) \leq h(K_2)$.
C-additivity $h(K_1) + h(K_2) = h(K_1 \cup K_2) + h(K_1 \cap K_2)$, for
$K_1, K_2 \in C(\mathbb{K})$ if $K_1 \cup K_2 \in C(\mathbb{K})$.

Important examples of ovoid functionals for the cases $d = 1, 2, 3$ are

the *length* $l(K)$ if $d = 1$
the *area* $A(K)$ if $d = 2$
the *volume* $V(K)$ if $d = 3$
the *boundary length* $U(K)$ if $d = 2$

and

the *surface area* $S(K)$ if $d = 3$.

Note that in the case $d = 2$ if K can be contained in a line (that is to say if K is a line segment) then $U(K)$ is defined as twice the length of K. Likewise in the case $d = 3$ if K can be contained in a plane then $S(K)$ is twice the area of K.

The *parallel set* of distance r of a set $A \subset \mathbb{R}^d$ is the set $A \oplus b(o, r)$. The operation of taking a parallel set preserves the properties of convexity, of compactness, of being an ovoid, and of being a ball. We are interested in the dependence on the distance r of the content of a parallel set in the three cases $d = 1, 2, 3$. For the case $d = 1$ this is given simply by

$$l(K \oplus b(o, r)) = l(K) + 2r.$$

In the cases $d = 2$ and $d = 3$ Steiner's formula gives the results

$$A(K \oplus b(o, r)) = A(K) + U(K)\, r + \pi r^2 \tag{1.6.1}$$

$$V(K \oplus b(o, r)) = V(K) + S(K)\, r + 2\pi \bar{b}(K)\, r^2 + 4\pi r^3/3. \tag{1.6.2}$$

Here $\bar{b}$ is yet another ovoid functional, called the *mean breadth*. The mean breadth can be defined explicitly as follows. For each line g through the origin let $b_g(K)$ be the least distance between two parallel flats perpendicular to g and enclosing K entirely between them. Then $\bar{b}(K)$ is defined to be the mean value of $b_g(K)$ averaging over all lines g through the origin using invariant measure. This can be given in a formula as

$$\bar{b}(K) = \frac{1}{2\pi} \int_0^{\pi/2} \int_0^{2\pi} l(\mathfrak{p}_{\beta,\lambda} K) \sin \beta \, d\lambda \, d\beta$$

where $l(\mathfrak{p}_{\beta,\lambda} K)$ is the projected length of K on the line through the origin and through the point $(\sin \beta \cos \lambda, \sin \beta \sin \lambda, \cos \beta)$.

In the special case of polyhedra one can derive the following formula (see SANTALÓ, 1976, p. 226):

$$\bar{b}(K) = \frac{1}{4\pi} \sum_i l_i \chi_i \tag{1.6.3}$$

where l_i is the length of the ith edge and χ_i is the angle between the normals of the faces which meet at the ith side (here $0 < \chi_i < \pi$). Table 1.1 displays mean breadths, together with volumes and surface areas, for various ovoids $K \subset \mathbb{R}^3$. Other formulae for mean breadths can be found in HADWIGER (1957, p. 215) and SANTALO (1976, p. 226).

Table 1.1. Volumes, surface areas, and mean breadths for ovoids K in $\mathbb{R}^3$

K	V	S	$\bar{b}(K)$
Ball of radius r	$(4/3)\,\pi r^3$	$4\pi r^2$	$2r$
Cylinder, radius r, height h	$\pi r^2 h$	$2\pi r(r + h)$	$\frac{1}{2}(h + \pi r)$
Disc, radius r	0	$2\pi r^2$	$(\pi/2)\,r$
Segment, length l	0	0	$l/2$
Prolate spheroid, major axis L, minor axis l; $L/l = 2$	$8.38l^3$	$21.4l^2$	$2.76l$
Oblate spheroid, $L/l = 2$	$2.09L^3$	$8.67L^2$	$1.71L$
Cube, side a	a^3	$6a^2$	$(3/2)\,a$
Rectangular parallelepiped sides a, b, c	abc	$2(ab + bc + ca)$	$(a + b + c)/2$
Square plate, side a	0	$2a^2$	a
Tetrahedron, side a	$0.1179a^3$	$1.732a^2$	$0.9123a$
Octahedron, side a	$0.4714a^3$	$3.464a^2$	$1.175a$

The average breadth $\bar{b}$ is closely related to the integral of mean curvature M by

$$M = 2\pi\bar{b}. \tag{1.6.4}$$

The integral of mean curvature M is the surface integral over ∂K of the mean of the two principal curvatures of the surface. For M to be well defined by this integral the surface ∂K must satisfy suitable regularity conditions, though of course the mean breadth makes sense for any ovoid.

The formulae (1.6.1) and (1.6.2) given above generalize to the case of general dimension d. The volume $\nu_d(K \oplus b(o, r))$ of the parallel bodies of K for K in $C(\mathbf{K})$ are given by the *Steiner formula*:

$$\nu_d(K \oplus b(o, r)) = \sum_{k=0}^{d} \binom{d}{k} W_k(K)\, r^k. \tag{1.6.5}$$

This formula introduces the important *quermassintegrals* or *Minkowski functionals*. They are isometry-invariant, monotone, C-additive ovoid functionals, defined directly by the formula

$$W_k(K) = (\omega_d/\omega_{d-k}) \int_{L_k} \nu_{d-k}(\mathfrak{p}_{S\perp} K) \; U_k(\mathrm{d}S) \tag{1.6.6}$$

in which the symbols have the following meaning:

$$\omega_l = \sqrt{\pi^l}/\Gamma(1 + l/2)$$

is the volume of the unit ball in $\mathbb{R}^l$: note the special cases $\omega_0 = 1$, $\omega_1 = 2$, $\omega_2 = \pi$, and $\omega_3 = 4\pi/3$,

ν_l	is l-dimensional Lebesgue measure.
L_l	is the set of all l-subspaces.
$\mathfrak{p}_{S\perp} K$	is the orthogonal projection of K on the $(d - l)$-subspace orthogonal to $S \in L_l$.
U_l	is the uniform probability distribution on L_l.

A brief introduction to general measure theory is given in section 1.8.

The quantity $W_0(K)$ is equal to the volume $\nu_d(K)$; the quantity $d \cdot W_1(K)$ is the $(d - 1)$-*dimensional measure of area of the boundary* ∂K; W_d is constant and equal to ω_d, and $(2/\omega_d) W_{d-1}$ is the mean breadth functional. In the particular cases of $d = 1, 2, 3$ this gives

$$d = 1: W_0(K) = l(K), \; W_1(K) = 2 \tag{1.6.7}$$

$$d = 2: W_0(K) = A(K), \; W_1(K) = \frac{1}{2} U(K)$$

$$W_2(K) = \pi \tag{1.6.8}$$

$$d = 3: W_0(K) = V(K), \; W_1(K) = \frac{1}{3} S(K), \; W_2(K) = \frac{2}{3} \pi \bar{b}(K)$$

$$W_2(K) = \frac{2}{3} \pi \bar{b}(K), \; W_3(K) = \frac{4}{3} \pi. \tag{1.6.9}$$

The *Hadwiger characterization theorem* states that every non-negative, isometry-invariant, monotone, C-additive ovoid functional h can be written in the form

$$h = \sum_{k=0}^{d} a_k W_k$$

where the a_k are non-negative constants depending on h.

The volume itself is a quermassintegral as noted above. Formula (1.6.5) can be generalized to apply not only to the volume $\nu_d \doteq W_0$ but to the other quermass integrals. The generalized formula is

$$W_j\big(K \oplus b(o, r)\big) = \sum_{k=0}^{d-j} \binom{d-j}{k} W_{k+j}(K) \, r^k \tag{1.6.10}$$

Formulae (1.6.1) and (1.6.2) are special cases of (1.6.5) for $d = 2$ and $d = 3$. In the case $d = 2$ the formula (1.6.10) gives

$$U\big(K \oplus b(o, r)\big) = U(K) + 2\pi r \tag{1.6.11}$$

and in the case $d = 3$ it gives

$$S(K \oplus b(o, r)) = S(K) + 4\pi\bar{b}(K)\, r + 4\pi r^2 \tag{1.6.12}$$

$$\bar{b}(K \oplus b(o, r)) = \bar{b}(K) + 2r. \tag{1.6.13}$$

Matheron (1978) and Weil (1982a) give similar formulae for $V(K \ominus b(o, r))$ under suitable smoothness conditions on ∂K.

The Minkowski functionals for intersections of convex sets with flats satisfy the *Crofton formula*

$$\int\limits_{L_k} \int\limits_{S^\perp} W_j^k(K \cap S_s)\, \nu_{d-k}(\mathrm{d}s)\, U_k(\mathrm{d}S) = \frac{\omega_k \omega_{d-j}}{\omega_d \omega_{k-j}} W_j^d(K) \tag{1.6.14}$$

for $0 \leqq j \leqq k \leqq d - 1$, and $K \in C(\mathbb{K})$.

We set $W_j^k(\emptyset) = 0$. $S^\perp$ is the $(d-k)$-subspace perpendicular to the k-subspace S, ν_{d-k} is the Lebesgue measure on S (considering S as a Euclidean space $\mathbb{R}^{d-k}$) and W_j^k is the jth Minkowski functional on S_s (considering S_s as a Euclidean space $\mathbb{R}^k$). If $j = k$ then (1.6.14) is equivalent to (1.6.6). Further integral-geometric formulae related to projections are given in Hadwiger (1957) and Weil (1983).

1.7 The convex ring

Some of the results concerning convex sets can be generalized for sets in a wider class known as the *convex ring* $\mathcal{R}$, see Hadwiger (1957), Matheron (1975) and Weil (1983). The convex ring $\mathcal{R}$ is the system of all subsets A of $\mathbb{R}^d$ which can be expressed as finite unions of ovoids:

$$A = \bigcup_{i=1}^{n} K_i \quad \text{for } K_i \in C(\mathbb{K}).$$

If A_1 and A_2 both belong to $\mathcal{R}$ then so do $A_1 \cup A_2$ and $A_1 \cap A_2$.

An *additive functional* h on $\mathcal{R}$ is a map $h: \mathcal{R} \to \mathbb{R}$ with the properties

$$h(\emptyset) = 0$$

and

$$h(A_1 \cup A_2) + h(A_1 \cap A_2) = h(A_1) + h(A_2)$$

for $A_1, A_2 \in \mathcal{R}$.

Motion invariance of functionals on $\mathcal{R}$ is defined as for ovoid functionals.

A very important example of an additive and motion invariant functional on $\mathcal{R}$ is given by the *connectivity number* or *Euler-Poincare characteristic* χ. For convex nonempty K it is

$$\chi(K) = 1$$

and this together with the additivity property defines it for general A in $\mathcal{R}$. For suppose that A is represented by

$$A = \bigcup_{i=1}^{n} K_i \quad \text{for } K_i \in C(\mathbb{K}).$$

Then the additivity property gives

$$\chi(A) = \sum_i \chi(K_i) - \sum_{i_1<i_2} \chi(K_{i_1} \cap K_{i_2}) + \cdots + (-1)^n \chi(K_1 \cap \cdots \cap K_n). \tag{1.7.1}$$

It can be shown that $\chi(A)$ is independent of the representation of A as a finite union of ovoids; see HADWIGER (1957) for this fundamental result. Figure 1.3 shows three planar sets of connectivity numbers $+1$, 0, and -1 respectively. In this two-dimensional case $\chi(A)$ equals the number of outer boundary components minus the number of inner boundary components.

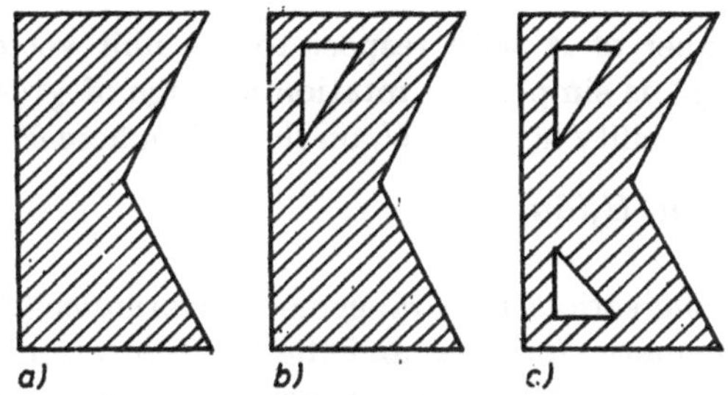

Fig. 1.3. Planar figures with various connectivity numbers. (a) Connectivity number 1; (b) Connectivity number 0; (c) Connectivity number -1

The connectivity number provides a basis for generalizing the Minkowski functionals to act on $\mathcal{R}$. The term $\nu_{d-k}(\mathfrak{p}_{S^\perp} K)$ in the formula (1.6.6) must be replaced by

$$\int_{S^\perp} \chi(K \cap S_s)\, \nu_{d-k}(\mathrm{d}s)$$

where $S^\perp$ is the $(d-k)$-plane through the origin of $\mathbb{R}^d$ that is orthogonal to S and $S_s = S + s$. If W_k also denotes the corresponding *generalized Minkowski functional* then

$$W_k(A) = \frac{\omega_d}{\omega_{d-k}} \int_{L_k} \int_{S^\perp} \chi(A \cap S_s)\, \nu_{d-k}(\mathrm{d}s)\, U_k(\mathrm{d}S) \tag{1.7.2}$$

for $k = 1, \ldots, d$. These functionals are additive. The quantity $W_0(A)$ is the d-dimensional volume of A, $d \cdot W_1(A)$ is the surface of area A, and $W_d(A)/\omega_d$ is the connectivity number of A.

The class of the Minkowski functionals can be generalized in another way, to the *positive extension* introduced by MATHERON (1975) and referred to in 7.3. This generalization yields non-negative functionals. A quantity that arises by this route of generalization is $M^+(A)$, the *integral of mean positive curvature*, defined for $A \in \mathcal{R}$. A further positive characteristic which is a counterpart to the connectivity number is the *convexity number* χ^+. We give an account of its definition after MATHERON (1975, p. 122–123), but using the notation developed in this chapter.

First we define the *convexity number of $A \in \mathcal{R}$ with respect to the vector $\boldsymbol{u}$.* Here $\boldsymbol{u}$ is a unit vector. Let $S_{\boldsymbol{u},r}$ be the hyperplane through $r\boldsymbol{u}$ and perpendicular to $\boldsymbol{u}$. For A in $\mathcal{R}$ let $K_1, K_2, \ldots, K_n$ be the connected components of $A \cap S_{\boldsymbol{u},r}$. The connected component K_i is said to be an *entering set* for A (with respect to $\boldsymbol{u}$ and $S_{\boldsymbol{u},r}$) if there is an open and connected set G containing K_i such that $G \cap K_j$ is empty for $j \neq i$ and $A \cap G \cap S_{\boldsymbol{u},r-\varepsilon}$ is empty for all sufficiently small positive ε. Let $f(r)$ denote the number of entering sets associated with $S_{\boldsymbol{u},r}$. If $\boldsymbol{u}$ is a fixed unit vector then this number $f(r)$ is non-zero for only a finite number of real numbers r, so that we may put

$$\chi^+(A, \boldsymbol{u}) = \sum_r f(r).$$

This is the convexity number of A with respect to the unit vector $\boldsymbol{u}$. The (mean) convexity number χ^+ is simply the rotation average of $\chi^+(A, \boldsymbol{u})$ with respect to the invariant probability measure U_1 on the unit sphere $\partial b(o, 1)$:

$$\chi^+(A) = \int_{\partial b(o,1)} \chi^+(A, \boldsymbol{u}) \, U_1(\mathrm{d}\boldsymbol{u}).$$

One might ask whether the various integral-geometric formulae (the Steiner formula, the principal kinematic formula, and the Crofton formula) hold in the convex ring if the standard Minkowski functionals are systematically replaced by their generalizations. This is indeed the case; see Hadwiger (1957), Davy and Miles (1977), Matheron (1975), and Weil (1983).

1.8 A brief introduction to measure and integration theory

In modern probability theory a central role is played by the concept of a measure. Measures are real-valued functions defined in families of sets enjoying properties of additivity and positivity.

To introduce the idea of a measure we give three simple and linked examples. Suppose that we are concerned to study the contents of a body K in space that is made up of some material. Suppose that the mass of this body is not necessarily homogeneously distributed in space. Associated with each portion A of K are the quantities $\alpha_1(A)$, the volume of A, and $\alpha_2(A)$, the mass of K that is contained in A. Suppose that in this body K there are also small grains of another material. Let the number of these grains lying in A be given by $\alpha_3(A)$. Thus there are three set functions $\alpha_1, \alpha_2, \alpha_3$ defined on the portions of K. If a portion A is divided into parts A_1, A_2 then

$$\alpha_k(A) = \alpha_k(A_1 \cup A_2) = \alpha_k(A_1) + \alpha_k(A_2) \quad \text{for } k = 1, 2, 3.$$

Such set functions are said to be additive.

For mathematical reasons, it is useful to stipulate that such functions should have the property of *σ-additivity*. That is to say, that if a set can be divided into a countable disjoint union of subsets then the value of the set function on the whole set should equal the sum of the values of the set function on the subsets. This consideration together with various mathematical technicalities leads to the realization that such set functions as given above are naturally defined on systems of sets that are closed under the basic set operations of union, intersection, and complementation, and are also closed under the operations of taking countably infinite unions and intersections.

Such systems of sets are called *σ-algebras*. Technically a σ-algebra is a system $\mathcal{X}$ of subsets of some ground set X satisfying three conditions:

(S1) $X \in \mathcal{X}$

(S2) if $A \in \mathcal{X}$ then $A^c \in \mathcal{X}$

(S3) if $A_1, A_2, \ldots \in \mathcal{X}$ then $\bigcup_{k=1}^{\infty} A_k \in \mathcal{X}$.

From these properties others follow immediately:

$\emptyset \in \mathcal{X}$

if $A_1, \ldots, A_n \in \mathcal{X}$ then $A_1 \cap \cdots \cap A_n \in \mathcal{X}$, $A_1 \cup \cdots \cup A_n \in \mathcal{X}$,

if $A, B \in \mathcal{X}$ then $A \setminus B \in \mathcal{X}$,

if $A_1, A_2, \ldots \in \mathcal{X}$ then $\bigcap_{k=1}^{\infty} A_k \in \mathcal{X}$.

Trivial examples of σ-algebras are given by the minimal σ-algebra $\{\emptyset, X\}$ and by the *power set* $\mathcal{P}(X)$, i.e. the system of all subset of X.

A very important example of a σ-algebra is given by the family $\mathcal{B}^d$ of *Borel sets of* $\mathbb{R}^d$. This is the smallest σ-algebra on $\mathbb{R}^d$ that contains all the open subsets of $\mathbb{R}^d$. It is genuinely smaller than the power set $\mathcal{P}(\mathbb{R}^d)$. Loosely speaking it contains all the subsets of $\mathbb{R}^d$ that can be constructed from the open subsets by the basic set operations and by limits. Since these operations can be iterated this is an exceedingly large class of sets. It certainly includes all closed sets, and thus all compact sets, and thus all elementary geometrical bodies discussed in this book. The subclass of all bounded Borel sets of $\mathbb{R}^d$ is occasionally denoted by $\mathcal{B}_0{}^d$.

If a space X comes equipped with a σ-algebra $\mathcal{X}$ then $[X, \mathcal{X}]$ is called a *measurable space*.

A function $f: X \to \mathbb{R}$ is said to be *$\mathcal{X}$-measurable* if for all Borel sets $B \in \mathcal{B}^1$ the inverse image $f^{-1}(B) = \{x \in X : f(x) \in B\}$ belongs to the σ-algebra $\mathcal{X}$ associated with X. Elementary algebraic operations (such as the taking of modulus signs, and the taking of limits) when applied to measurable functions all yield measurable functions. A particular example of a measurable function is the *indicator function* $\mathbf{1}_A$ of a measurable set A:

$$\mathbf{1}_A(x) = \begin{cases} 1 \text{ for } x \in A \\ 0 \text{ otherwise.} \end{cases}$$

The co-ordinate functions are special examples of Borel-measurable functions

$$x = (x_1, \ldots, x_d) \to x_i \quad \text{from } \mathbb{R}^d \text{ to } \mathbb{R}.$$

All continuous functions and hence all differentiable functions on $\mathbb{R}^d$ are Borel-measurable.

If in the definition of a measurable function $\mathbb{R}$ is replaced by a general set Y with associated σ-algebra $\mathcal{Y}$ then a transformation

$$f: X \to Y$$

is said to be $(\mathcal{X}, \mathcal{Y})$-measurable if $f^{-1}(B) \in \mathcal{X}$ for each $B \in \mathcal{Y}$.

Suppose that $[X, \mathcal{X}]$ is a measurable space. A *measure* on $[X, \mathcal{X}]$ is a function $\mu: \mathcal{X} \to [0, \infty]$ with the two properties

(M1) $\mu(\emptyset) = 0$

(M2) $\mu\left(\bigcup_{k=1}^{\infty} A_k\right) = \sum_{k=1}^{\infty} \mu(A_k)$.

for all $A_1, A_2, \ldots \in \mathcal{X}$ with $A_i \cap A_j = \emptyset$ if $i \neq j$. Note that $\mu(A)$ may equal ∞. The property (M2) is referred to as the property of *σ-additivity*. A particular case of this property is finite additivity;

$$\mu(A_1 \cup \cdots \cup A_n) = \mu(A_1) + \cdots + \mu(A_n)$$

if $A_1, \ldots, A_n$ are pairwise disjoint. Another important consequence is that if $A, B \in \mathcal{X}$ and $B \subset A$ then

$$\mu(A \setminus B) = \mu(A) - \mu(B).$$

The measures which occur in this book generally have two other important properties. They are defined on the Borel σ-algebra of $\mathbb{R}^d$ for some d, and they are finite on bounded sets; such measures are referred to as *Radon measures*. The relationship between measure theory and the topology of the ground space is an important and extensive theory: however we only require the two definitions above.

Simple examples of measures on measurable spaces are given by the *Dirac measures* δ_x for x in X. They are defined by

$$\delta_x(A) = \begin{cases} 1 & \text{if } x \in A \\ 0 & \text{otherwise.} \end{cases}$$

On $[\mathbb{R}^d, \mathcal{B}^d]$ the Dirac measures are Radon measures.

A further and most important example is that of *Lebesgue measure* ν_d on $[\mathbb{R}^d, \mathcal{B}^d]$. A fundamental theorem of measure theory has as a special case the result that Radon measures on $[\mathbb{R}^d, \mathcal{B}^d]$ are unequivocably defined by their values on hypercubes. Lebesgue measure is defined by

$$\nu_d(Q) = (b_1 - a_1) \cdots (b_d - a_d)$$

if $Q = [a_1, b_1] \times \cdots \times [a_d, b_d]$. In particular the values of the ν_d-measure can be deduced from this for sets given by elementary geometrical constructions (such as balls, cylinders, toruses, and so forth). These values are given in Table 1.1 since in the case $d = 3$ the Lebesgue measure ν_3 is equal to the volume measure V. In the cases $d = 1$ and $d = 2$ we have $\nu_1 = l$ the length measure and $\nu_2 = A$ the area measure.

If B is a bounded Borel set then $\nu_d(B) < \infty$. This follows from the positivity of ν_d and its finiteness on all hypercubes (including those which contain the bounded set A). Thus ν_d is indeed a Radon measure.

Lebesgue measure has the property of being *isometry-invariant*:

$$\nu_d(\mathfrak{m}A) = \nu_d(A)$$

for all isometries $\mathfrak{m}$ and Borel sets A. There is another important theorem that states that every isometry-invariant Radon measure μ (indeed every translation-invariant Radon measure) on $[\mathbb{R}^d, \mathcal{B}^d]$ is a constant multiple of the Lebesgue measure

$$\mu = c\nu_d, \quad c \geqq 0. \tag{1.8.1}$$

A significant generalization of this theorem asserts that for a wide class of measurable spaces with sufficiently rich symmetry groups (but not too rich) there is essentially only one invariant measure; all invariant Radon measures are constant multiples of each other.

Dirac measures and Lebesgue measure are quite different in that Dirac measures are concentrated on single points while Lebesgue measure gives zero mass to every point. Measures μ on $[\mathbb{R}^d, \mathcal{B}^d]$ that are concentrated on a countably infinite collection of points $x_1, x_2, \ldots$ (so that $\mu(\mathbb{R}^d \setminus \{x_1, x_2, \ldots\}) = 0$) are called *atomic measures*, and are called *counting measures* if in addition they give each point a mass of zero or one. A measure is called a *diffuse measure* if like Lebesgue measure it gives zero mass to each point.

A measure μ on a measurable space $[X, \mathcal{X}]$ defines a triple $[X, \mathcal{X}, \mu]$ which is called a *measure space*. Associated with each measure space is a functional defined on the system of $\mathcal{X}$-measurable functions $f\colon X \to [0, \infty]$ which assigns to each function f a value $\int f \, d\mu$ in $[0, \infty]$. It is called the *integral of f with respect to* μ, or simply the *integral* of f. It can be extended to all $\mathcal{X}$-measurable functions $f\colon X \to \mathbb{R}$ with $\int |f| \, d\mu < \infty$. The extension is given by

$$\int |f| \, d\mu = \int f \mathbf{1}_A \, d\mu - \int (-f) \mathbf{1}_{A^c} \, d\mu$$

where

$$A = \{x \in X : f(x) > 0\}.$$

The integral has the properties

$$\int \mathbf{1}_A \, d\mu = \mu(A) \quad \text{for } A \in \mathcal{X}$$

$$\int (c_1 f_1 + c_2 f_2) \, d\mu = c_1 \int f_1 \, d\mu + c_2 \int f_2 \, d\mu \quad \text{for } c_1, c_2 \in \mathbb{R}$$

$$\int \sum_{k=1}^{\infty} f_k \, d\mu = \sum_{k=1}^{\infty} \int f_k \, d\mu \quad \text{if } f_1, f_2, \ldots \geqq 0$$

and, moreover, these properties define it uniquely. In particular a measure μ is uniquely defined by the system of values

$$\int f \, d\mu \quad \text{for all measurable } f\colon X \to [0, \infty).$$

Occasionally such integrals are written as

$$\int f \, d\mu = \int f(x) \, \mu(dx) = \int f(t) \, \mu(dt)$$

where the *dummy variables* x, t of integration can be chosen arbitrarily from previously undefined symbols.

The integral derived from the measure space $[\mathbb{R}^d, \mathcal{B}^d, \nu_d]$ is called the *Lebesgue integral*. In the case $d = 2$ this integral can be interpreted graphically. A non-negative function $f: \mathbb{R}^2 \to [0, \infty)$ can be considered as giving the height $f(x)$ at each point x of a surface lying over the plane $(x_3 = 0)$ in $\mathbb{R}^3$. Then the integral $\int f \, d\nu_2$ can be interpreted as the volume lying between the surface and the plane $(x_3 = 0)$. A similar interpretation is possible for the case $d = 1$.

A frequent contraction of notation involves removing indicator functions from the integrand and inserting the corresponding sets as *range of integration*. In the case $d = 1$

$$\int 1_{[a,b]} f \, d\nu_1 = \int_{[a,b]} f \, d\nu_1$$

and in the general case

$$\int 1_A f \, d\mu = \int_A f \, d\mu .$$

For elementary functions f on $\mathbb{R}$ the expression $\int_{[a,b]} f \, d\nu_1$ equals the well-known elementary integral $\int_a^b f(x) \, dx$ (when $a < b$).

If two measures μ_1, μ_2 are given on the same measurable space $[X, \mathcal{X}]$ then sometimes one can be given in terms of the other by an integral formula involving a *density* $g: X \to [0, \infty)$:

$$\mu_2(A) = \int_A g \, d\mu_1 \quad \text{for all } A \text{ in } \mathcal{X} .$$

In this case we say μ_2 is *absolutely continuous with respect to* μ_1, and write

$$\mu_2 \ll \mu_1 .$$

The Radon-Nikodym theorem gives a necessary and sufficient condition for one Radon measure μ_2 to be absolutely continuous with respect to another Radon measure μ_1: if for all A in $\mathcal{X}$

$$\mu_1(A) = 0 \quad \text{implies} \quad \mu_2(A) = 0$$

then

$$\mu_2 \ll \mu_1 .$$

If a measure space $[\Omega, \mathcal{A}, \mathsf{P}]$ is such that $\mathsf{P}(\Omega) = 1$ then it may be called a *probability (measure) space* and the measure P is called a *probability measure*. In this case real-valued $\mathcal{A}$-measurable functions defined on Ω are called *random variables*, as are (less commonly) general measurable transformations from $[\Omega, \mathcal{A}]$ to another measurable space.

Thorough introductions to the theory of measure and integration can be found in many textbooks; for example, HALMOS (1974) and KINGMAN and TAYLOR (1966).

2. Point processes I – The Poisson point process

2.1 Introduction

The basic ingredients of practical geometry are points. Accordingly, random point patterns (or *point processes* in mathematical terminology) play a fundamental role in stochastic geometry and they arise directly as the results of investigations into nature and technology. Figures in Chapters 2, 4, and 5 display various point patterns that arise in this fashion. In this context they must be studied by mathematical methods and compared to suitable mathematical models. Frequently they occur as systems of points connected with particles, fibres, or other geometrical objects. For example, a system of particles gives rise to a point pattern generated by the centroids of the particles. This leads on to a second, more indirect, fashion in which random point patterns arise. In many problems of stochastic geometry it is helpful to interpret patterns of geometrical figures as systems of 'points' in suitable 'respresentation spaces'; a 'point' from the system represents a particular geometrical figure in the original pattern. Chapter 8 discusses this in detail, with particular reference to patterns of lines and patterns of triangles.

In the current chapter the simplest and most important random point pattern is studied; the Poisson point process. This study will be carried out in a heuristic fashion, generally avoiding use of the abstract theory of point processes. It is hoped that readers of less mathematical training will thus be encouraged to read further and study the general theory as presented in Chapter 4 and the discussion in Chapter 5 of further models for random point patterns.

An appreciation of the possible varieties of random point patterns leads to an understanding of the central position that the Poisson point process holds among theoretical models. A given random point pattern may well exhibit various kinds of interaction between its constituent points. Thus the points may occur in clusters (see Figures 5.4 and 5.6) or may exhibit great regularity (Figure 5.7). There may be a *hard-core distance* (a minimal interpoint distance). These extreme features may occur together in the same pattern. The subject of point process statistics aims to detect and to quantify such interactions. Models from the theory of point processes can be used both in comparison to the original point pattern and in representation of it.

In the absence of any of the above interactions the point pattern is completely random. A theoretical model of such a pattern gains great importance as a basis for comparison. Calculations can be made of the extent of probable fluctuations in such a model. These fluctuations provide objective grounds for ignoring sufficiently small empirically observed interactions, on the basis that they are insignificant compared with features that might well be observed in a completely random point pattern. If intuitively appealing axioms are imposed (concerning stationarity and lack of interaction) then completely random processes can be described as Poisson point processes. Thus one part of the central role of the Poisson point process is in its service as null hypothesis for statistical tests of interaction.

The other part lies in the service that the Poisson process gives as basic building block for other more complicated models. Theoretical definitions of varieties of point processes frequently make reference to Poisson point processes. Simulation procedures often include the construction of a Poisson point process, which is then modified into the form required.

A history of the concept of a Poisson process can be found in DALEY and VERE-JONES (1986), who report that the first recorded use of the Poisson process in spatial statistics appears to be that of ABBÉ (1878).

2.2 The binomial point process

The most elementary example of a point process is one that contains one point only. A random point ζ *uniformly distributed* in a compact set $W \subset \mathbb{R}^d$ is one which falls in subsets of W with probability proportional to the volume of W. So

$$\mathsf{P}(\zeta \in A) = \nu_d(A)/\nu_d(W) \tag{2.2.1}$$

for all Borel sets A contained in W. Many problems in geometric probability arise from the study of uniformly distributed random points and the determination of the volume ratio in (2.2.1) when A is defined by some geometrical construction. The progress of modern stochastic geometry can to some extent be summarized as the replacement of uniformly distributed random points by more general random point patterns.

A uniformly distributed random point is a trivial random pattern. However, n independent uniformly distributed random points can be superposed to form a new random point pattern, a *binomial point process of n points*. Formally such a process is formed by n independent points $\zeta_1, \ldots, \zeta_n$ uniformly distributed in the same compact set W. From formula (2.2.1)

$$\begin{aligned}\mathsf{P}(\zeta_1 \in A_1, \ldots, \zeta_n \in A_n) &= \mathsf{P}(\zeta_1 \in A_1) \cdot \cdots \cdot \mathsf{P}(\zeta_n \in A_n) \\ &= \nu_d(A_1) \cdot \cdots \cdot \nu_d(A_n)/\nu_d(W)^n \end{aligned} \tag{2.2.2}$$

for $A_1, \ldots, A_n$ Borel subsets of W. This leads to an equivalent definition; the points $\zeta_1, \ldots, \zeta_n$ form a binomial point process in W a subset of $\mathbb{R}^d$ if the random vector $(\zeta_1, \ldots, \zeta_n)$ is uniformly distributed in W^n. Figure 2.1 (on page 39) shows the result of a simulation of a binomial point process in $W = [0, 1]^2$ with $n = 50$ points.

If $\zeta_1, \ldots, \zeta_n$ form a binomial point process in W then the random pattern constituted by these points is denoted by $\Phi_W^{(n)}$. For most purposes the ordering of the points is irrelevant and $\Phi_W^{(n)}$ can be regarded as a random set. It can also be thought of as a random counting measure; for a Borel set A let $\Phi_W^{(n)}(A)$ denote the number of points of $\Phi_W^{(n)}$ falling in A. Thus

$$\Phi_W^{(n)}(\emptyset) = 0, \qquad \Phi_W^{(n)}(W) = n$$

and

$$\Phi_W^{(n)}(A_1 \cup A_2) = \Phi_W^{(n)}(A_1) + \Phi_W^{(n)}(A_2)$$

when A_1, A_2 are disjoint subsets of W. The σ-additivity property also holds. Hence $\Phi_W^{(n)}$ is indeed a random measure.

The binomial point process $\Phi_W^{(n)}$ is our first non-trivial example of a point process. It plays the typical dual role of a point process (as random set and as random measure) which is an important feature of all (simple) random point processes.

Basic properties

The binomial point process $\Phi_W^{(n)}$ earns its name by a distributional property. If A is a Borel subset of W then $\Phi_W^{(n)}(A)$ is binomial of parameters $n = \Phi_W^{(n)}(W)$ and $p = p(A) = \nu(A)/\nu_d(W)$. So the mean number of points per unit volume, or *intensity* of the binomial point process is given by

$$\lambda = n \cdot p(A)/\nu_d(A) = \Phi_W^{(n)}(W)/\nu_d(W).$$

Numbers of points in different subsets of W are *not* independent even if the subsets are disjoint. This is clear since $\Phi_W^{(n)}(A) = m$ necessarily implies $\Phi_W^{(n)}(W \setminus A) = n - m$. The distribution of $\Phi_W^{(n)}$ as a point process is given completely by the *finite-dimensional distributions* discussed in 4.1

$$\mathsf{P}\left(\Phi_W^{(n)}(A_1) = n_1, \ldots, \Phi_W^{(n)}(A_k) = n_k\right)$$

for $k = 1, 2, \ldots$ and arbitrary Borel sets $A_1, \ldots, A_k$ and integers $n_1, \ldots, n_k$ with $n_1 + \cdots + n_k \leqq n$.

The distribution of a point process is determined by its so-called *void-probabilities* $v_K = \mathsf{P}(\Phi_W^{(n)}(K) = 0)$ (here K is an arbitrary compact subset of W). Section 6.1 discusses these ways of specifying distributions in the much more general context of random sets. See also the remarks on void-probabilities in Section 2.4.1.

The void-probabilities for the binomial point process are given by

$$\mathsf{P}\left(\Phi_W^{(n)}(K) = 0\right) = \left(\nu_d(W) - \nu_d(K)\right)^n/\nu_d(W)^n \tag{2.2.3}$$

and, when $A_1, \ldots, A_k$ are disjoint Borel sets with $A_1 \cup \cdots \cup A_k = W$ and $n_1 + \cdots + n_k = n$, the finite-dimensional distributions are given by the multinomial probabilities

$$\mathsf{P}\left(\Phi_W^{(n)}(A_1) = n_1, \ldots, \Phi_W^{(n)}(A_k) = n_k\right) = \frac{n!}{n_1! \cdots n_k!} \cdot \frac{\nu_d(A_1)^{n_1} \cdots \nu_d(A_k)^{n_k}}{\nu_d(W)^n}. \tag{2.2.4}$$

As noted above $\Phi_W^{(n)}(A)$ is of binomial distribution with parameters n and $p(A)$. If the total number of points n tends to infinity and the second parameter $p(A)$ tends to zero in such a way that the product $\lambda \cdot \nu_d(A) = n \cdot p(A)$ remains fixed then $\Phi_W^{(n)}(A)$ is asymptotically of a Poisson distribution of mean $\lambda \cdot \nu_d(A)$. This limit can be obtained if the region W is enlarged to fill out all of R^d in step with n tending to infinity. If the ratio $n/\nu_d(W)$ remains fixed as n increases and W is enlarged then the Poisson limit will hold for $\Phi_W^{(n)}(A)$ for any fixed bounded Borel set A. So if there is in some sense an asymptotic point process Φ under this limit then we may expect that it will satisfy the distributional property:

$$\Phi(A) \text{ is Poisson of mean } \lambda \cdot \nu_d(A) \quad \text{for each bounded Borel set } A.$$

A close analysis of (2.2.4) for $k = 2$ indicates that such a limiting process should be 'independently scattered'. So it will have the property:

if $A_1, \ldots, A_k$ are disjoint bounded Borel sets

then

$\Phi(A_1), \ldots, \Phi(A_k)$ will be independent.

We shall return to these points in Section 2.4.

Simulation

The simulation of a binomial point process follows easily by a superposition procedure once one knows how to simulate a random point uniformly distributed in the required region; the binomial point process can be obtained by independent replication of the simulated random point.

It is straightforward to simulate a random point uniformly distributed in $[0, 1]^2$. If $\{z_n\}$ is a sequence of independent random numbers uniformly distributed in $[0, 1]$ then the points

$$x_i = (z_{2i-1}, z_{2i}) \quad \text{for } i = 1, 2, \ldots \tag{2.2.5}$$

form a sequence of independent random points uniformly distributed in $[0, 1]^2$. Figure 2.1 shows a sample of 50 points obtained in this way. The numbers $\{z_n\}$ will in practice be produced by a random number generator. Careful attention must be paid to the quality of the generator. All random number generators are 'defective' in some way or another, and it must be considered whether or not the defects of the chosen random number generator will lead to undesirable features in the simulated sequence of random points. Ripley (1983; 1986) discusses this and other matters in the use of random number generators for simulations occurring in spatial problems.

A sequence of random points uniformly distributed in the hypercube $[0, 1]^d$ is produced by

$$x_i = (z_{(i-1)d+1}, \ldots, z_{id}) \quad \text{for } i = 1, 2, \ldots. \tag{2.2.6}$$

Translation and scale changes can be used to produce a sequence of points uniformly distributed in any fixed hypercube.

Such simulation procedures cover a large number of cases in practice, as frequently the binomial point process to be simulated will take place in a square or cube. However, other regions do arise and then further work must be done. Simulation of a uniform random point in a general bounded region W is tackled using one of three main techniques. To be definite we discuss only the common planar case.

(*a*) ***Rejection sampling*** A small square or rectangle containing W is found and uniform random points are simulated in this region. The procedure is continued

Fig. 2.1. A simulation of fifty random points uniformly distributed in a square

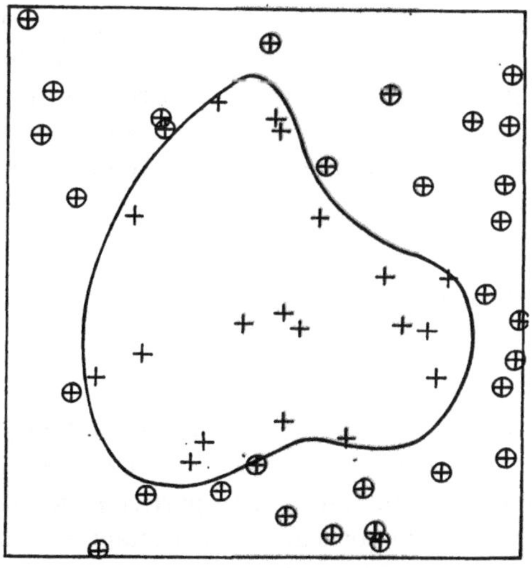

Fig. 2.2. A simulation of nineteen random points (+) uniformly distributed in the set W. The points (⊕) are those points of Figure 2.1 that do not lie in W

until a point first falls in W. This point will be uniformly distributed in W. To obtain a binomial point process the procedure is continued until n points have fallen in W, and these n points constitute the binomial point process. Figure 2.2 illustrates the process.

(b) *Approximation* The region W is replaced by a disjoint union of k open squares approximating W. A random point distributed uniformly over this union is simulated by choosing a square with probability proportional to its area and then simulating a random point uniformly distributed over that square.

For complicated regions a judicious combination of this technique with rejection sampling can prove effective.

(c) *Transformation of coordinates* If the region W has substantial symmetry then some transformation of coordinates may be useful. For example, if W is the unit disc $b(o, 1)$ then a uniform random point can be described in polar coordinates

$$x = (r, \theta) \quad \text{for } r \text{ in } [0, 1] \text{ and } \theta \text{ in } (0, 2\pi].$$

The random variables r and θ are independent, with θ uniform over $(0, 2\pi]$ and r satisfying the law

$$\mathsf{P}(r \leqq t) = t^2 \quad \text{if } 0 \leqq t \leqq 1.$$

Thus if z_1, z_2 are independent random numbers uniform over [0, 1] then the formulae

$$r = \sqrt{z_1} \quad \text{and} \quad \theta = 2\pi z_2$$

provide a method of simulating $x = (r, \theta)$.

As explained in Section 2.4.1(d) the binomial point process arises as a result of conditioning a portion of a stationary Poisson point process. Consequently, it is not necessary to give statistical methods for assessing the hypothesis that a given point pattern is a realization of a binomial point process. One can use the methods for stationary Poisson point processes described in Section 2.7. On the other hand, the simulation procedures for stationary Poisson point processes and general Poisson point processes depend heavily on the simulation methods for binomial point processes described here.

2.3 The Bernoulli lattice process

Introduction

Before turning to the Poisson point process it is worth considering a simple discretized point process which we will call a *Bernoulli lattice process*. This is a discretized analogue of the Poisson point process. In particular the discretized points of a Bernoulli lattice process do not suffer any interaction. The lattices to be considered are Cartesian lattices based at the origin

$$L_\delta = \{x = (m_1\delta, \ldots, m_d\delta) : m_1, \ldots, m_d \text{ integers}\}$$

Here δ is a positive number, the *mesh* of the lattice. A Bernoulli lattice process $\Phi_{p,\delta}$ on L_δ is a random subset of the lattice; each point of L_δ is contained in $\Phi_{p,\delta}$ with probability p independently of all the others. Formally if $\Phi_{p,\delta}(A)$ is the number of points of $\Phi_{p,\delta}$ falling in A then:

$\Phi_{p,\delta}(A)$ is binomially distributed with parameters

$n = \#\{x \in L_\delta : x \in A\}$ and p,

and

if $A_1, \ldots, A_k$ are disjoint

then

$\Phi_{p,\delta}(A_1), \ldots, \Phi_{p,\delta}(A_k)$ are independent.

Basic properties

It is immediate from the definition that if A is a bounded Borel set then

$$\mathsf{E}\Phi_{p,\delta}(A) = p \cdot \#\{x \in L_\delta : x \in A\}. \tag{2.3.1}$$

This means that in some approximate sense $\Phi_{p,\delta}$ has an 'intensity' $p \cdot \delta^{-d}$.

The most important feature of $\Phi_{p,\delta}$ for our purposes is its asymptotic behaviour. Let δ and p tend to zero together in such a way that $p \cdot \delta^{-d} = \lambda$ is held constant. Then for bounded and open A the random variable $\Phi_{p,\delta}(A)$ has a distribution that is asymptotically Poisson of mean $\lambda \cdot \nu_d(A)$. The limit process must be randomly scattered since each individual $\Phi_{p,\delta}$ has this property. So the limit should be the same as for the binomial point process.

More can be said than this. Consider the sequence of interleaved lattices defined by the meshes $\delta_n = 2^{-n}$. They fit one within the other

$$L_{\delta_1} \subset L_{\delta_2} \subset \cdots.$$

Given $x \in L_{\delta_n}$ let the *cell* of x be

$$C_{n+1}(x) = \{y \in L_{\delta_{n+1}} : x - y \in [-2^{-n-1}, 0]^d\}.$$

Then $C_{n+1}(x)$ contains 2^d locations of $L_{\delta_{n+1}}$, moreover

$$L_{\delta_{n+1}} = \cup \{C_{n+1}(x) : x \in L_{\delta_n}\}$$

as a disjoint union. Let $p_n = 2^{-nd}\lambda$. It is possible to define a sequence of Bernoulli lattice process Φ_{p_n,δ_n} on the same probability space such that

$$\Phi_{p_n,\delta_n}(\{x\}) = 0 \text{ implies that } \Phi_{p_{n+1},\delta_{n+1}}\big(C_{n+1}(x)\big) = 0. \tag{2.3.2}$$

The sequence of Bernoulli lattice process Φ_{p_n,δ_n} share the same approximate 'intensity' since $p_n \cdot \delta_n{}^{-d} = \lambda$. So it is intuitively clear and easy to prove that they converge in some sense to a limiting point process Φ. This limiting process must have the distributional properties:

$\Phi(A)$ is Poisson of mean $\lambda\nu_d(A)$ if A is a bounded Borel set,

$\Phi(A_1), \ldots, \Phi(A_k)$ are independent if $A_1, \ldots, A_k$

are disjoint Borel sets. This is because of the asymptotic properties of Bernoulli lattice process described above. This identifies the limit Φ as a stationary Poisson point process of intensity λ as described in Section 2.4.

Simulation

The Bernoulli lattice process can be simulated in a straightforward manner. First, one must generate one random variable z_x for each x in L_δ so that the z_x are independent and uniformly distributed over [0, 1]. The required Bernoulli lattice process can be simulated by the union of all points x for which $z_x < p$.

2.4 The stationary Poisson point process

2.4.1 Definition and defining properties

Since a general treatment of point processes appears in Chapter 4, only a brief explanation of terminology is given here. As seen above, a random point pattern or *point process* Φ can be regarded either as a random sequence (more strictly a random set) $\Phi = \{x_1, x_2, \ldots\}$ or else as a random measure. As a random measure Φ can be applied to sets and for each Borel set B the measure $\Phi(B)$ is the number of points of Φ that fall in the set B. Because our random point patterns are all locally finite the set function Φ is a locally finite σ-additive Borel measure.

As a random set Φ can be intersected with other sets; if B is a Borel set then $B \cap \Phi$ is the random set of points of Φ that also belong to B. Again because our random point patterns are all locally finite the random set Φ will always be closed and $B \cap \Phi$ will be finite whenever B is bounded.

A stationary Poisson point process Φ is characterized by two fundamental properties which have already appeared as asymptotic properties in Sections 2.2 and 2.3. They are: (1) Poisson distribution of point counts, the number of points of Φ in a bounded Borel set B has a Poisson distribution of mean $\lambda \nu_d(B)$ for some constant λ; (2) Independent scattering, the numbers of points of Φ in k disjoint Borel sets form k independent random variables.

Property (2) is also known as the 'completely random' or 'purely random' property.

The number λ occurring in (1) is the characteristic parameter of the stationary Poisson point process. It gives the mean number of points to be found in a unit volume

$$\lambda \nu_d(B) = \mathsf{E}\Phi(B) \quad \text{for all bounded Borel } B.$$

It is called the *intensity* or *density* of the stationary Poisson point process Φ. In the following we assume that λ is positive and finite (if $\lambda = 0$ then the point pattern contains no points while an infinite λ corresponds to a pathological case).

The intensity can also be interpreted in infinitesimal terms. If B is a set of small Lebesgue measure then by property (1) the following asymptotics hold

$$\mathsf{P}\big(\Phi(B) = 0\big) = 1 - \lambda\nu_d(B) + o\big(\nu_d(B)\big),$$
$$\mathsf{P}\big(\Phi(B) = 1\big) = \lambda\nu_d(B) + o\big(\nu_d(B)\big),$$
$$\mathsf{P}\big(\Phi(B) > 1\big) = o\big(\nu_d(B)\big).$$

Basic properties

Let Φ be a stationary Poisson point process of intensity λ. From properties (1) and (2) the whole distribution of the stationary Poisson point process can be determined once the intensity λ is known.

(a) *Finite-dimensional distributions.* It can be shown directly from (1) and (2) that if $B_1, \ldots, B_k$ are disjoint bounded Borel sets then $\Phi(B_1), \ldots, \Phi(B_k)$ are independent Poisson random variables with means $\lambda\nu_d(B_1), \ldots, \lambda\nu_d(B_k)$. Thus

$$\mathsf{P}\big(\Phi(B_1) = n_1, \ldots, \Phi(B_k) = n_k\big)$$
$$= \frac{\lambda^{n_1+\cdots+n_k}[\nu_d(B_1)]^{n_1} \cdots [\nu_d(B_k)]^{n_k}}{n_1! \cdots n_k!} \exp\left(-\sum_{i=1}^{k} \lambda\nu_d(B_i)\right). \quad (2.4.1)$$

From this formula the joint probabilities $\mathsf{P}\big(\Phi(B_1 = n_1, \ldots, \Phi(B_k = n_k)\big)$ can be evaluated for general (overlapping) $B_1, \ldots, B_k$.

(b) *Stationarity and isotropy.* For a point process $\Phi = \{x_n\}$ to be stationary the translated process $\Phi_x = \{x_n + x\}$ must have the same distribution for all x in $\mathbb{R}^d$. A point process is isotropic if the same is true for all rotated processes $\mathfrak{r}\Phi = \{\mathfrak{r}x_n\}$ where $\mathfrak{r}$ is a rotation about the origin. A process is motion-invariant if it possesses both of these properties. The stationary Poisson point process Φ is defined by properties (1) and (2) above and the specification of the intensity λ. These properties (and the characteristic λ) are clearly invariant under rotation and translation. Therefore the stationary Poisson point process Φ must be stationary and isotropic that is to say motion-invariant.

This may be verified directly by noting that the finite-dimensional distributions above remain the same whether one uses a stationary Poisson point process Φ or its translation Φ_x or its rotation $\mathfrak{r}\Phi$.

(c) *Void-probabilities.* The *void-probabilities* of a point process are the probabilities of there being no point of the process in given test sets B

$$v_B = \mathsf{P}\big(\Phi(B) = 0\big).$$

In the case of Φ being a stationary Poisson point process

$$v_B = \exp\big(-\lambda\nu_d(B)\big). \quad (2.4.2)$$

The contact distributions are closely associated with the void probabilities. If B is a Borel set with $\nu_d(B) > 0$ and $o \in B$ then the *contact distribution function* $H_B(r)$ (*with respect to* B) is given by

$$H_B(r) = 1 - v_{rB} = 1 - \mathsf{P}\big(\Phi(rB) = 0\big) \quad \text{for } r \geq 0.$$

A particularly important case uses $B = b(o, 1)$ the unit sphere, yielding the *spherical contact distribution* or *first contact distribution function*

$$H_s(r) = 1 - \mathsf{P}\big(\Phi\big(b(o, r) = 0\big)\big) = 1 - \exp\left(-\lambda\omega_d r^d\right) \quad (2.4.3)$$

for $r \geqq 0$. This is the distribution function of the distance from o to the nearest point of Φ.

(d) *Conditioning and binomial point processes.* If Φ is a stationary Poisson point process then one can consider the restriction of Φ to a compact set W under the condition that $\Phi(W) = n$. This conditioning yields a new point process which is actually the binomial point process in W with n points.

This assertion can be proved by showing that the finite-dimensional distributions of the two processes coincide. In fact, it is enough to consider only the void-probabilities. If K is a compact subset of W then the void-probability for K of the conditioned stationary Poisson point process is given by

$$\begin{aligned} \mathsf{P}\big(\Phi(K) = 0 \mid \Phi(W) = n\big) &= \mathsf{P}\big(\Phi(K) = 0,\ \ \Phi(W) = n\big)/\mathsf{P}\big(\Phi(W) = n\big) \\ &= \mathsf{P}\big(\Phi(K) = 0\big)\ \mathsf{P}\big(\Phi(W \setminus K) = n\big)/\mathsf{P}\big(\Phi(W) = n\big) \\ &= \frac{\big(\nu_d(W) - \nu_d(K)\big)^n}{\nu_d(W)^n} \end{aligned}$$

after substitution and cancellation. This formula coincides with formula (2.2.3) for the void-probabilities of the binomial point process.

The general theory of simple point processes contains a theorem asserting equality of the distributions of point processes if their system of void-probabilities is equal. However in this case it is straightforward to directly compute the finite-dimensional distributions of the conditioned process (in a manner similar to the above) and to check that these are equal to the finite-dimensional distributions of the binomial point process as given by formula (2.2.4).

Characterizations of the stationary Poisson point process

The properties (1) and (2) of the stationary Poisson point process are not logically independent. Rényi (1967) has shown that (1) implies (2); the Poisson distribution property forces the point process to have the independently scattered property (this follows from the remark at the end of (d) above). However it is not enough for property (1) to hold only for the class of all convex subsets of $\mathbb{R}^d$. Moreover, Moran (1976) has shown that it is not enough to assume the independence and Poissonian distribution of counts of points in k disjoint convex sets for some fixed k. Thus if property (1) is weakened then property (2) may be necessary for characterization of the stationary Poisson point process.

Another set of properties also characterizes the Poisson process, and should be borne in mind as it frequently provides a *prima facie* case for assuming that an empirical point pattern is a realization of a stationary Poisson point process. This characterization asserts that a process must be a stationary Poisson point process if the following three properties are satisfied:

(I) *Simplicity.* No two points coincide so that the process is a *simple* point process. That is to say there are no multiple points.

(II) *Stationarity.* This is as defined above.

(III) *Independent scattering.* This is property (2) as given at the beginning of the section.

Note that properties (I) and (II) together imply the property that the point process is *ordinary*, a property normally required instead of simplicity in the point process literature.

In stochastic geometry the property of simplicity generally holds by definition since objects in the same location are geometrically equivalent. Usually it is evident from the nature of the point pattern whether or not it is simple.

The stationarity assumption has greater significance. It implies that the statistical properties of the point process do not depend on the location of the observer. In some cases, stationarity can be tested statistically but often it must be assumed on a tentative basis in order to make progress in the analysis of the point pattern.

The most important assumption is that of independent scattering. In effect it asserts that there is no interaction between the points of the pattern. It can be tested statistically. In some cases, it is suggested by underlying biological or physical theories; Example 2.2 illustrates this. But plausible or not it frequently provides the starting point for statistical analysis, even if only as a hypothesis to be subjected to attack.

2.4.2 Moments and moment measures

Just as the moments of random variables are important characteristics, so are corresponding entities for the Poisson process and for general point processes. However, the step of increasing complexity from a random variable to a random point process is matched in the nature of the corresponding moments. While random variables have moments that are real numbers a point process has moments that are measures. In this section the moment measures are described for the specific case of the stationary Poisson point process. In this case the explicit formulae have very simple forms. Chapter 4 discusses moment measures for general point processes.

Consider the stationary Poisson point process Φ regarded as a random measure on the Borel sets in $\mathbb{R}^d$. If B is a Borel set then $\Phi(B)$ is a random variable with a first moment

$$\Lambda(B) = \mathsf{E}\Phi(B).$$

If B_1, B_2 are two Borel sets then $\Phi(B_1)$, $\Phi(B_2)$ are two random variables with a non-centred covariance

$$\mu^{(2)}(B_1 \times B_2) = \mathsf{E}\Phi(B_1)\,\Phi(B_2).$$

Both Λ and $\mu^{(2)}$ are moments but depend on particular Borel sets or products of Borel sets. The nature of this dependence is σ-additive (this follows from Φ being a random measure) and so they can both be represented as measures.

The first-order quantity Λ has a straightforward form. By property (1) the random variable $\Phi(B)$ is Poisson of mean $\lambda \cdot \nu_d(B)$ and so

$$\Lambda(B) = \mathsf{E}\Phi(B) = \lambda \cdot \nu_d(B) \quad \text{for all Borel sets } B. \tag{2.4.4}$$

So in the case of the stationary Poisson point process the *first-order moment measure* or *intensity measure* Λ is a constant multiple of Lebesgue measure and the multiple is given by the intensity of the Poisson process.

The second-order quantity $\mu^{(2)}$ is evaluated by using both properties (1) and (2). To evaluate $\mu^{(2)}(B_1 \times B_2)$ one notes that both B_1 and B_2 can be decomposed into disjoint unions

$$B_1 = B_1 \cap B_2 \cup B_1 \setminus B_2,$$
$$B_2 = B_1 \cap B_2 \cup B_2 \setminus B_1.$$

Using the fact that Φ is a random measure and property (2) one can establish

$$\begin{aligned}\mu^{(2)}(B_1 \times B_2) &= \mathsf{E}\Phi(B_1)\,\Phi(B_2)\\ &= \mathsf{E}\Phi(B_1 \setminus B_2)\,\mathsf{E}\Phi(B_2 \setminus B_1) + \mathsf{E}\Phi(B_1 \cap B_2)\,\mathsf{E}\Phi(B_2 \setminus B_1)\\ &\quad + \mathsf{E}\Phi(B_1 \setminus B_2)\,\mathsf{E}\Phi(B_1 \cap B_2) + \mathsf{E}\big(\Phi(B_1 \cap B_2)\big)^2\\ &= \mathsf{E}\Phi(B_1)\,\mathsf{E}\Phi(B_2) + \mathsf{E}\big(\Phi(B_1 \cap B_2)\big)^2\\ &\quad - \big(\mathsf{E}\Phi(B_1 \cap B_2)\big)^2.\end{aligned}$$

Property (1) shows that $\Phi(B_1 \cap B_2)$ is Poisson of mean $\lambda \cdot \nu_d(B_1 \cap B_2)$ and so has second non-centred moment of

$$\lambda \cdot \nu_d(B_1 \cap B_2) \cdot \big(1 + \lambda \cdot \nu_d(B_1 \cap B_2)\big).$$

So with the aid of (2.4.4) the final formula can be derived

$$\begin{aligned}\mu^{(2)}(B_1 \times B_2) &= \Lambda(B_1)\,\Lambda(B_2) + \lambda \cdot \nu_d(B_1 \cap B_2)\\ &= \lambda^2 \cdot \nu_d(B_1) \cdot \nu_d(B_2) + \lambda \cdot \nu_d(B_1 \cap B_2).\end{aligned} \tag{2.4.6}$$

So the *second-order moment measure* $\mu^{(2)}$ can be expressed in terms of Lebesgue measure. It should be noted that $\mu^{(2)}$ is a measure on $\mathbb{R}^d \times \mathbb{R}^d$.

Variances and covariances can be calculated directly from the second-order moment measure (this is also true for general point processes). The relevant formulae are

$$\operatorname{var}\big(\Phi(B)\big) = \mu^{(2)}(B \times B) - \big(\Lambda(B)\big)^2,$$
$$\operatorname{cov}\big(\Phi(B_1), \Phi(B_2)\big) = \mu^{(2)}(B_1 \times B_2) - \Lambda(B_1) \cdot \Lambda(B_2)$$

for all Borel sets B, B_1, and B_2. These formulae follow immediately from the definitions of variance and covariance. In the case of the stationary Poisson point process one can use (2.4.6) (or calculate directly from properties (1) and (2)) that

$$\operatorname{cov}\big(\Phi(B_1), \Phi(B_2)\big) = \lambda \cdot \nu_d(B_1 \cap B_2). \tag{2.4.7}$$

As a random measure, Φ counts the number of points falling in the set that serves as argument. Therefore $\mu^{(2)}$ can also be expressed as the expectation of this sum:

$$\begin{aligned}\mu^{(2)}(B_1 \times B_2) &= \mathsf{E}\,\#\,\{(x, y) : x \in \Phi \cap B_1, y \in \Phi \cap B_2\}\\ &= \mathsf{E} \sum_{x,y \in \Phi} \mathbf{1}_{B_1}(x)\,\mathbf{1}_{B_2}(y).\end{aligned}$$

The two terms in (2.4.6) correspond to the dissection of this sum into the sum over distinct pairs of points $x, y \in \Phi$ and the sum over equal points $x = y \in \Phi$.

For some purpose it is convenient to subtract out the second of these terms. The result is the *second-order factorial moment measure* $\alpha^{(2)}$

$$\alpha^{(2)}(B_1 \times B_2) = \mathsf{E}\,\# \{(x, y) : x \in \Phi \cap B_1, y \in \Phi \cap B_2, x \neq y\}$$
$$= \mathsf{E} \sum_{x,y \in \Phi}^{\#} \mathbf{1}_{B_1}(x)\,\mathbf{1}_{B_2}(y)\,.$$

Here $\sum^{\#}$ stands for the summation over all (x, y) such that $x \neq y$.

The difference between $\mu^{(2)}(B_1 \times B_2)$ and $\alpha^{(2)}(B_1 \times B_2)$ lies in the expectation of the sum

$$\sum_{x,y \in \Phi : x = y} \mathbf{1}_{B_1}(x)\,\mathbf{1}_{B_2}(y) = \sum_{x \in \Phi} \mathbf{1}_{B_1 \cap B_2}(x)\,.$$

Hence

$$\mu^{(2)}(B_1 \times B_2) = \alpha^{(2)}(B_1 \times B_2) + \Lambda(B_1 \cap B_2)\,.$$

In the case of the stationary Poisson point process,

$$\alpha^{(2)}(B_1 \times B_2) = \lambda^2 \cdot \nu_d(B_1) \cdot \nu_d(B_2) = \Lambda(B_1) \cdot \Lambda(B_2)\,.$$

So in this case the second-order factorial moment measure $\alpha^{(2)}$ is a constant multiple of Lebesgue measure on $\mathbb{R}^d \times \mathbb{R}^d$. The density of $\alpha^{(2)}$ with respect to Lebesgue measure is known as the *second-order product density* $\varrho^{(2)}$ and the above shows that

$$\varrho^{(2)}(x_1, x_2) = \lambda^2 \quad \text{for } x_1, x_2 \text{ in } \mathbb{R}^d.$$

The second-order product density has an infinitesimal interpretation. If B_1, B_2 are two infinitesimally small disjoint Borel sets of volumes $\mathrm{d}V_1$, $\mathrm{d}V_2$ and if $x_1 \in B_1$, $x_2 \in B_2$ then

$$\varrho^{(2)}(x_1, x_2)\,\mathrm{d}V_1\,\mathrm{d}V_2 = \lambda^2\,\mathrm{d}V_1\,\mathrm{d}V_2$$

is the probability that Φ places a point in each of B_1 and B_2.

The 'first-order product density' has already been defined. By analogy with the above it is none other than the intensity λ.

Just as high-order moments can be defined for random variables so high-order moment measures can be defined for point processes. The *nth-order moment measure* $\mu^{(n)}$ is a measure on $\mathbb{R}^{dn}$ defined by

$$\mu^{(n)}(B_1 \times \cdots \times B_2) = \mathsf{E}\Phi(B_1) \ldots \Phi(B_n)$$
$$= \mathsf{E} \sum_{x_1,\ldots,x_n \in \Phi} \mathbf{1}_{B_1}(x_1) \cdots \mathbf{1}_{B_n}(x_n)$$

for Borel sets $B_1, \ldots, B_n$. The *nth order factorial moment measure* $\alpha^{(n)}$ is a measure on $\mathbb{R}^{dn}$ given by

$$\alpha^{(n)}(B_1 \times \ldots \times B_n) = \mathsf{E} \sum_{x_1,\ldots,x_n \in \Phi}^{\#} \mathbf{1}_{B_1}(x_1) \cdots \mathbf{1}_{B_n}(x_n)$$

where here $\sum^{\#}$ stands for summation over n-tuples $(x_1, \ldots, x_n)$ of n distinct points. The *nth-order product density* $\varrho^{(n)}$ is the density of $\alpha^{(n)}$ with respect to (nd)-

dimensional Lebesgue measure. In the case of stationary Poisson point processes these product densities are given by simple formulae

$$\varrho^{(n)}(x_1, \ldots, x_n) = \lambda^n \tag{2.4.8}$$

and $\alpha^{(n)}$ has the form

$$\alpha^{(n)}(B_1 \times \cdots \times B_n) = \lambda^n \cdot \nu_d(B_1) \cdots \nu_d(B_n). \tag{2.4.9}$$

For large n the relationships between $\alpha^{(n)}$ and $\mu^{(n)}$ are complicated. For example,

$$\begin{aligned}\mu^{(3)}(B_1 \times B_2 \times B_3) = \lambda^3 \cdot \nu_d(B_1) \cdot \nu_d(B_2) \cdot \nu_d(B_3) \\ + \lambda^2 \cdot \nu_d(B_1) \cdot \nu_d(B_2 \cap B_3) + \lambda^2 \cdot \nu_d(B_2) \cdot \nu_d(B_1 \cap B_3) \\ + \lambda^2 \cdot \nu_d(B_1 \cap B_2) \cdot \nu_d(B_3) + \lambda \cdot \nu_d(B_1 \cap B_2 \cap B_3)\end{aligned}$$

for Borel sets B_1, B_2, B_3.

There is an infinitesimal interpretation for $\varrho^{(n)}(x_1, \ldots, x_n)$ which follows that of $\varrho^{(2)}(x_1, x_2)$.

2.4.3 The Palm distribution of a stationary Poisson point process

Many problems in point process theory require the consideration of an arbitrary 'typical' point of a point process Φ. For example, the nearest neighbour distance distribution function D describes the distribution of the distance from a typical point x of Φ to the nearest point in $\Phi \setminus \{x\}$. The idea of a typical point is clear enough but needs to be made mathematically precise. Points sampled by some systematic method (such as by choosing the nearest to a given origin) are *not* typical because they *have* been so sampled. Palm distribution theory makes the notion of a typical point precise. In intuitive terms the Palm distribution probabilities are the conditional probabilities of point process events given that a point (the typical point) is observed at a specific location.

Two heuristic approaches to the Palm distribution are given in this section, and it is calculated for the case of a stationary Poisson point process. One approach is essentially local and involves conditioning the distribution of the point process on there being a point at a given position x. The other approach is more global and applicable only to stationary point processes. It is related to statistical methods of estimating Palm probabilities.

To discuss the local approach some notation is necessary. Probabilities will be discussed that involve the conditioning of Φ on the event that Φ contains a point at x. If Y is some point process property (perhaps the property that $\Phi(B) = n$ for some fixed Borel set B) then

$$\mathsf{P}(\Phi \text{ has property } Y \parallel x) = \mathsf{P}(\Phi \text{ has property } Y \mid x \in \Phi).$$

Further notation introduced in Chapter 4 enables us to write the event 'Φ has property Y' as '$\Phi \in Y$', interpreting Y as a member of a σ-algebra of sets of realizations. In any case stationarity makes it clear that

$$\mathsf{P}(\Phi \text{ has property } Y \parallel x) = \mathsf{P}(\Phi_x \text{ has property } Y \parallel o).$$

Here Φ_x denotes the shifted point process $\{x_1 + x, x_2 + x, \ldots\}$, where Φ is the set $\{x_1, x_2, \ldots\}$.

In this notation the nearest neighbour distance distribution function D for a stationary point process is given by

$$D(r) = \mathsf{P}\big(\Phi(b(o, r)) > 1 \parallel o\big) = 1 - \mathsf{P}\big(\Phi(b(o, r)) = 1 \parallel o\big).$$

The conditioning event $\{o \in \Phi\}$ has probability zero if Φ is stationary. This means that the conditioning must be carefully defined to avoid possible ambiguity. Chapter 4 uses the Radon-Nikodym theorem, which frequently illuminates the meaning of conditional probability constructions in general probability theory. However, in the case of the stationary Poisson point process the Palm probabilities can be calculated by a limit procedure. This is exemplified by the following.

Calculation of the nearest neighbour distance distribution function for the stationary Poisson point process. To begin with a similar conditional distribution function is evaluated, in which the conditioning is on a point being *near* to o. The conditional probability

$$D_\varepsilon(r) = 1 - \mathsf{P}\big(\Phi[b(o, r) \setminus b(o, \varepsilon)] = 0 \mid \Phi[b(o, \varepsilon)] = 1\big)$$

is well defined for positive ε smaller than r, since $\mathsf{P}\big(\Phi[b(o, \varepsilon)] = 1\big) = \lambda \cdot \omega_d \varepsilon^d \times \exp(-\lambda\omega_d\varepsilon^d) > 0$. Using the definition of conditional probability and property (2) of the stationary Poisson point process

$$\begin{aligned} D_\varepsilon(r) &= 1 - \frac{\mathsf{P}\big(\Phi[b(o, r) \setminus b(o, \varepsilon)] = 0\big)\, \mathsf{P}\big(\Phi[b(o, \varepsilon)] = 1\big)}{\mathsf{P}\big(\Phi[b(o, \varepsilon)] = 1\big)} \\ &= 1 - \mathsf{P}\big(\Phi[b(o, r) \setminus b(o, \varepsilon)] = 0\big) \\ &= 1 - \exp\big[-\lambda\big(\nu_d[b(o, r)] - \nu_d[b(o, \varepsilon)]\big)\big] \end{aligned}$$

It is appealing to consider the nearest neighbour distance distribution function to be the limit of the above as $\varepsilon \to 0$. Setting $D(r) = \lim_{\varepsilon \to 0} D_\varepsilon(r)$ yields the result

$$D(r) = 1 - \exp\big(-\lambda \cdot \nu_d[b(o, r)]\big) = 1 - \exp(-\lambda \cdot \omega_d r^d) \qquad (2.4.10)$$

for $r \geqq 0$. The same result is established in Chapter 4 using the rigorous definition of Palm probabilities.

The right-hand sides of (2.4.10) and (2.4.3) are equal. This leads to a relationship between the spherical contact distribution and nearest neighbour distance distribution for the stationary Poisson point process

$$D(r) = H_s(r). \qquad (2.4.11)$$

This plays a role in the statistical analysis of Poisson processes.

In the case $d = 2$ the mean and variance of D are given by

$$m_D = 1/\big(2\sqrt{\lambda}\big)$$

$$\sigma_D^2 = 1/(\pi\lambda) - 1/(4\lambda).$$

Arguing as for (2.4.10) leads to the result

$$\mathsf{P}\big(\Phi(B) = n \parallel o\big) = \mathsf{P}\big(\Phi(B) = n\big) \quad \text{for } n = 0, 1, 2, \ldots.$$

and for any compact set B not containing o. This suggests that the Palm distribution for the stationary Poisson point process is essentially the distribution of the original stationary Poisson point process together with a point adjoined at o; the content of the theorem below.

Slivnyak's theorem. P (Φ has property $Y \parallel o) = \mathsf{P}(\Phi \cup \{o\}$ has property Y). In the σ-algebra language of Chapter 4,

$$\mathsf{P}(\Phi \in Y \parallel o) = \mathsf{P}(\Phi \cup \{o\} \in Y). \tag{2.4.12}$$

A proof of Slivnyak's theorem is given in Section 4.4.

The second more global approach to Palm probabilities examines all points of Φ that fall within some fixed Borel test set B with $\nu_d(B) > 0$. Suppose $N(\Phi, B, Y)$ of these are such that Φ_{-x} has property Y. Then one can put

$$\mathsf{P}\ (\Phi \text{ has property } Y \parallel o) = \mathsf{P}(\Phi \in Y \parallel o) = \mathsf{E}N(\Phi, B, Y)/\lambda\nu_d(B).$$

Since $\lambda\nu_d(B)$ is the expected number of points falling in B this equation exhibits the Palm probability for the property Y as the fraction of points x expected to fall in B such that Φ_x has the property Y. This definition does not depend on B (by the stationarity of Φ).

It can be shown that the two approaches yield the same Palm distribution.

The *reduced second-order moment measure function* K is related to Palm distribution theory. This will be examined at greater length in Section 4.5. A brief indication of the nature of the relationship follows from noting that

$$\lambda K(r) = \mathsf{E}\big(\Phi[b(o, r)] \parallel o\big) - 1$$

so that $\lambda K(r)$ is the mean number of points in a ball of radius r and centre a typical point, if the typical point is itself not counted in the mean. Formula (2.4.12) yields

$$\lambda K(r) = \mathsf{E}\big(\Phi[b(o, r)] + 1\big) - 1 \tag{2.4.13}$$

so

$$K(r) = \omega_d r^d \quad \text{for } r \geqq 0.$$

2.5 The general Poisson point process

The stationary Poisson point process has a constant intensity measure; the mean number of points per unit area does not vary over space. Many point patterns arising in applications exhibit fluctuations that make implausible such a lack of spatial variation. For example, they may display an apparent trend of increase in a certain direction, as can be observed in the pattern of locations of trees in some forests. Suppose that the forest is in an area for which height above sea-level increases, or soil becomes progressively more dry, in a given direction. Then the mean number of trees per unit area will decrease in line with the trend. Figures 2.3 and 2.4 show point patterns which clearly exhibit such trends.

The general Poisson point process Φ provides a more general stochastic model appropriate for such point patterns. It is characterized completely by a

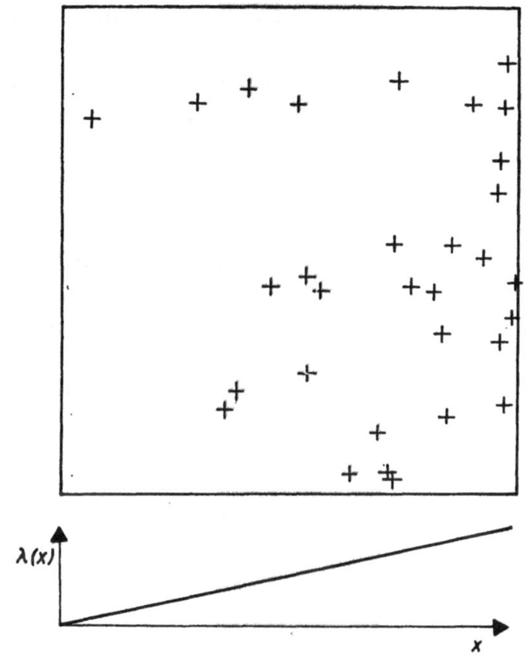

Fig. 2.3. A simulation of a sample of a Poisson process which has a linearly varying intensity function $\lambda(\cdot)$. The pattern has been generated from that of Figure 2.1 by the thinning procedure described in the text.

diffuse Radon measure Λ on $\mathbb{R}^d$ which is called the *intensity measure* of Φ and corresponds to the intensity measure previously introduced for the stationary Poisson point process. *A general Poisson point process Φ with intensity measure Λ is a point process with the two properties following:*

(1′) *Poisson distribution of point counts*; the number of points in a bounded Borel set B has a Poisson distribution with mean $\Lambda(B)$

$$\mathsf{P}\big(\Phi(B) = m\big) = \big[\Lambda(B)\big]^m \cdot \exp\big(-\Lambda(B)\big)/m! \text{ for } m = 0, 1, 2, \ldots \quad (2.5.1)$$

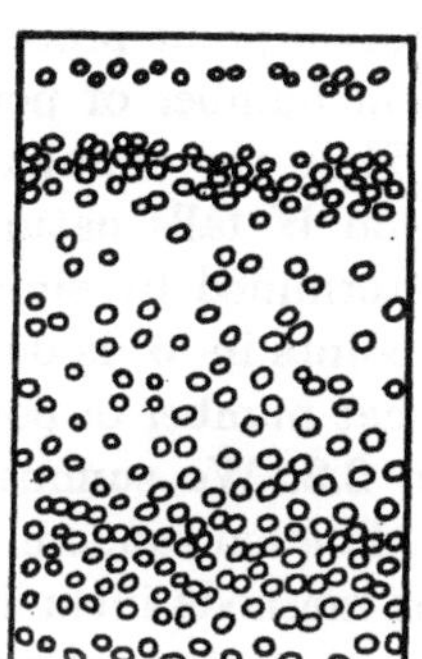

Fig. 2. 4. Cell nuclei developing in the pyramidal layer of the cerebral cortex of an embryo. The diagram is taken from Economo and Koskinas (1925). See also Stephan (1975)

(2') *Independent scattering*; the numbers of points in k disjoint Borel sets form k independent random variables.

It is clear from property (1') that such a process Φ is in general *not* stationary.

If the Radon measure Λ has a density with respect to Lebesgue measure then it can be written as

$$\Lambda(B) = \int_B \lambda(x)\,\mathrm{d}x \quad \text{for Borel sets } B.$$

The density $\lambda(x)$ is called the *intensity function* of the general Poisson point process. It has an appealing and intuitive infinitesimal interpretation; $\lambda(x)\,\mathrm{d}V$ is the infinitesimal probability that there is a point of Φ in a region of infinitesimal volume $\mathrm{d}V$ situated at x.

As for the stationary Poisson point process in Section 2.4.2, so here one can define *moment measures*

$$\mu^{(n)}(B_1 \times \cdots \times B_n) = \mathsf{E}\Phi(B_1) \cdots \Phi(B_n) = \mathsf{E} \sum_{x_1,\ldots,x_k \in \Phi} \mathbf{1}_{B_1}(x_1) \cdots \mathbf{1}_{B_k}(x_k)$$

and *factorial moment measures*

$$\alpha^{(n)}(B_1 \times \cdots \times B_n) = \mathsf{E} \sum_{x_1,\ldots,x_k \in \Phi}^{\neq} \mathbf{1}_{B_1}(x_1) \cdots \mathbf{1}_{B_k}(x_k).$$

The factorial moment measures are given by

$$\alpha^{(n)}(B_1 \times \cdots \times B_n) = \Lambda(B_1) \cdots \Lambda(B_n).$$

Consequently in the case of Λ having a density λ the *product densities* $\varrho^{(n)}$ are given by

$$\varrho^{(n)}(x_1, \ldots, x_n) = \lambda(x_1) \cdots \lambda(x_n).$$

Further generalizations of the Poisson process are considered in Sections 5.2 and 5.3.

2.6 Simulation of Poisson point processes

2.6.1 Simulation of a stationary Poisson point process

The starting point for simulating the stationary Poisson point process is property (d) in Section 2.4.1; conditioning on the total number of points in a compact set produces a binomial point process. Thus the simulation of a stationary Poisson point process in a compact region W falls naturally into two stages. First, the number of points in W is determined by simulating a Poisson random variable, then the position of the points in W is determined by simulating a binomial point process in W with that number of points. The second stage has already been covered in Section 2.2. We summarise two possibilities for the first stage, that of simulating a Poisson random variable. ATKINSON (1979a, b) surveys and compares various methods including those sketched here. It should be borne in mind that for many statistical purposes it may be advisable *not* to simulate the total number of points but to simulate binomial processes of N points for suitably chosen representative numbers N. This corresponds to a decision in the statistical analysis of the pattern to condition on the ancillary statistic that is the total number of points in the pattern.

Which of the methods here is appropriate for simulating a Poisson random variable depends on the mean $\lambda \cdot \nu_d(W)$ of the random variable. If $\lambda \cdot \nu_d(W)$ is small then the standard method for simulating a discrete distribution can be used. This involves calculating the Poisson probabilities

$$p_n = \left(\lambda \cdot \nu_d(W)\right)^n \cdot \exp\left(-\lambda \cdot \nu_d(W)\right)/n!$$

and setting $x_0 = 0$, $x_n = x_{n-1} + p_{n-1}$. If X has a uniform distribution over the interval $[0, 1]$ and Y is set equal to the n for which

$$x_{n-1} \leqq X < x_n$$

then Y has a Poisson distribution of mean $\lambda \cdot \nu_d(W)$. Improvements on this method involve such matters as re-ordering the intervals

$$[x_{n-1}, x_n)$$

so as to speed up the search for Y.

An alternative for small $\lambda \cdot \nu_d(W)$ is to simulate a linear Poisson process, exploiting the fact that then the interpoint distances are independent exponential random variables. Random variables e_i are generated to be independent and exponential of mean 1 (this can be achieved by the transform method; if z_i is uniform on $[0, 1]$ then $e_i = -\ln(z_i)$ is as desired). The desired Poisson random variable Y is the smallest n for which

$$\sum_{i=1}^{n+1} e_i > \lambda \cdot \nu_d(W).$$

An advantage of this method is its flexibility; it is an appropriate method to meet successive requirements to generate Poisson random variables of different means.

In the case when $\lambda \cdot \nu_d(B)$ is large then some form of rejection technique must be used. ATKINSON (1979a, b) gives details.

A final possibility uses a form of the polar co-ordinate method on page 40. This is called 'radial generation'; QUINE and WATSON (1984) exploit this in a simulation study of Poisson-Voronoi polyhedra (see 10.6.1).

2.6.2 Simulation of a general Poisson point process

We consider here the problem of simulating a general Poisson point process for which the intensity measure has a density function $\lambda(x)$ bounded above by λ^*. A thinning procedure can be used, as suggested by LEWIS and SHEDLER (1979); see also OGATA (1981). Section 5.1 discusses the thinning operation for point processes.

The procedure falls into two stages. Firstly, a stationary Poisson point process of intensity λ^* is simulated as above. Secondly, the resulting point pattern is thinned by deleting each point x independently of the others with probability $[1 - \lambda(x)/\lambda^*]$. If the points of the stationary Poisson point process pattern are $\{x_1, x_2, \ldots\}$ then this thinning can be performed with the aid of an independent sequence $\{z_1, z_2, \ldots\}$ of random numbers uniformly distributed over $[0, 1]$. The point x_k is deleted if $z_k > \lambda(x_k)/\lambda^*$. In the terminology of Section 5.1.1 the thinning probability is $p(x) = \lambda(x)/\lambda^*$. Formula (5.1.1) ensures that the point pattern of undeleted points is indeed a general Poisson point process of intensity function $\lambda(x)$. The pattern in Figure 2.3 was generated in this manner.

2.7 Statistics for the stationary Poisson point process

The discussion in Section 2.1 of the central role of the stationary Poisson point process makes clear the importance of statistical methods for deciding whether or not a given point pattern is Poisson. This section is a brief survey of such methods. It confines itself to the case of *planar* point patterns; the methods presented carry over to point patterns on the line, in space, and on a sphere. There is a vast literature on the particular case of Poisson processes on the line; see for example, Brillinger (1978), Cox and Lewis (1966), Snyder (1975), Karr (1986) and Kutoyants (1984).

Estimation of the intensity

The most fundamental statistical question for the stationary Poisson point process concerns the estimation of the intensity λ when the process is observed through a window W. A general unbiased estimator for the intensity λ of a stationary point process is given by

$$\hat{\lambda} = \Phi(W)/\nu_2(W). \tag{2.7.1}$$

Other approaches consider the point process as sampled by *quadrats* (square subregions) or by interpoint distances. If the window W is divided into a large number of quadrats (or other regions) of equal area a^2 then the counts of points falling in the various quadrats can be used. A simple method to obtain a preliminary estimate of λ consists of first the determination of the fraction p_0 of empty quadrats and then the estimation of λ by the employment of the formula for an empty quadrat

$$p_0 = \exp(-\lambda a^2)$$

Shiyomi (1978) discussed other quadrat methods for the estimation of λ.

Methods employing interpoint distances are discussed for example, in Persson (1971), Diggle (1975, 1977, 1983), Byth (1982), and Ripley (1979b, 1981).

Confidence intervals for λ can be based on (2.7.1) since $\hat{\lambda} \cdot \nu_2(W) = \Phi(W)$ has a Poisson distribution. In the case of $\Phi(W)$ being large Crow and Gardner

(1959) and SACHS (1984, p. 182) give a simple approximate $100(1-\alpha)\%$ confidence intervals for λ, employing the normal approximation and a continuity correction. For example,

$$\left[\frac{z_{\alpha/2}}{2} - \sqrt{\Phi(W)}\right]^2 \leqq \lambda\nu_2(W) \leqq \left[\frac{z_{\alpha/2}}{2} + \sqrt{(\Phi(W)+1)}\right]^2. \tag{2.7.2}$$

Here the $z_{\alpha/2}$ are as might be expected from the normal approximation:

$$z_{\alpha/2} = 1.65,\ 1.96,\ 2.58 \quad \text{for } \alpha = 0.10,\ 0.05,\ 0.01$$

respectively.

This confidence interval is useful in simple planning of the size of window required. If δ is the desired breadth of the confidence interval and α is the required confidence level then

$$\delta \cdot \nu_2(W) \simeq \left[\frac{z_{\alpha/2}}{2} + \sqrt{(\lambda\nu_2(W)+1)}\right]^2 - \left[\frac{z_{\alpha/2}}{2} - \sqrt{(\lambda\nu_2(W))}\right]^2.$$

from which follows the approximation

$$\nu_2(W) \simeq \frac{4z_{\alpha/2}^2\lambda}{\delta^2}$$

where λ here must be estimated by a preliminary rough estimate or by using *a priori* information.

A numerical example can be taken from Example 2.2 on page 61. If $\alpha = 0.05$, $\lambda = 0.0325$, and $\delta = 0.01$ then the formula yields a window size of

$$\nu_2(W) \simeq 5000.$$

This suggests that a 70×70 square is sufficient to obtain 95% confidence in accuracy to two decimal places.

The next statistical questions to be considered concern the model assumptions; those of stationarity and of the Poisson process hypothesis.

Testing the hypothesis of stationarity

If the point pattern is a stationary Poisson point process then the quantity F given below has a probability distribution that is approximately an F-distribution on $2n_1 + 1$ and $2n_2 + 1$ degrees of freedom.

$$F = \frac{\nu_2(W_1)\,(2n_2+1)}{\nu_2(W_2)\,(2n_1+1)}.$$

Here n_1 and n_2 are the point numbers in two disjoint subregions W_1, W_2 of the window W. It is assumed that the order of the indices $_1$ and $_2$ is chosen to make F greater than one. COX (1953) and SACHS (1984, page 187) discussed this approximation. The stationarity hypothesis is rejected at a significance level α if

$$F > F_{2n_1+1,2n_2+1;\ \alpha/2}.$$

This test is strictly appropriate only if the subregions W_1 and W_2 are chosen *a priori* to investigate specific possibilities of trend. For example, such a trend might relate the intensity of a Poisson process of tree locations to some measure of soil dryness. It should also be noted that the significance calculations above depend on the Poisson nature of the point process.

Testing the Poisson process hypothesis

There is a vast multiplicity of tests to which one can subject the hypothesis that a given point pattern is Poisson. There can be no single criterion to give the 'best' test to apply (see the instructive discussion in DIGGLE, 1983). Which test is most appropriate must depend on the nature of the alternative hypothesis envisaged and on limitations imposed by measurement methods.

Distance methods

In some circumstances, the most readily available measurements are of distances between points of the pattern, and between points of the pattern and chosen locations. This is the case when point pattern of forest tree locations is being investigated. Distance methods respect such limitations by employing these measurements alone. If the point pattern has been mapped so that counts of the total number of points in each possible region are readily available, then distance measurements are less appropriate than the count methods given later in this section.

The basis of distance methods as applied to testing the Poisson process hypothesis is the equation (2.4.11)

$$D(r) = H_s(r) \quad \text{for } r \geqq 0.$$

This equation asserts that for a stationary Poisson point process the nearest-neighbour distance distribution and the spherical contact distribution are the same. From distance measurements one might hope to obtain empirical distribution functions corresponding to D and H_s above and to test for their equality. However, a difficulty arises. The independence assumptions required by the appropriate methods of classical statistics do not hold for samples of distance measurements.

The remedy for this difficulty is to employ simulation to sample a subset of the possible distance so as to maintain an approximate independence. A further difficulty then arises; in order to obtain nearest-neighbour distances one must sample the points of the pattern at random but without enumerating all points of the pattern (for otherwise distance methods lose their advantage). How can one sample the points of the pattern at random without biasing towards points occurring in particular locations?

BYTH and RIPLEY (1980) gave an example of a method that overcomes both these problems. Semi-systematic sampling is used to chose a number $2m$ of locations in the window W. Half of these are used as bases from which to measure distances to nearest point in the pattern. These distances are denoted by $u_1, \ldots, u_m$. The other half of the locations are used to define subregions within which the point pattern is surveyed exhaustively. The common size of the subregions is chosen so that one would expect about five points of the pattern to fall within each subregion. A point of the pattern is selected at random from each subregion and for each of these points the nearest-neighbour distance is measured. This gives rise to a further m distances $v_1, \ldots, v_m$. Under the Poisson hypothesis (since then $D(r) = H_s(r)$) these two samples of m numbers each are drawn from the probability distribution given by

$$F(r) = 1 - \exp(-\lambda\pi r^2) \quad \text{for } r \geqq 0$$

(neglecting edge-effects). The samples are approximately independent and independent of each other, if m is not too large. BYTH and RIPLEY (1980) suggest that m should be around 5% of the total number of points in the window W.

As a consequence of the Poisson hypothesis the quantities $u_1^2, \ldots, u_m^2, v_1^2, \ldots, v_m^2$ are approximately independent with exponential distribution of parameter $\lambda\pi$. Thus the statistics

$$h_F = \sum_{i=1}^{m} u_i^2 \Big/ \sum_{i=1}^{m} v_i^2$$

and

$$h_N = \frac{1}{m} \sum_{i=1}^{m} u_i^2/(u_i^2 + v_i^2)$$

have distributions that are approximately the F-distribution on $(2m, 2m)$ degrees of freedom and the normal distribution of mean 1/2 and variance $m/12$ respectively. The measurements u_i, v_i are paired for the formula involving h_N; this pairing may be made at random.

These approximate distributional results allow the construction of statistical tests. Excessively large or small values for h_F or h_N indicate departures from the Poisson hypothesis. Which statistic is appropriate depends on the alternative hypothesis in mind. If the alternative involves clustering then h_F is more suitable, while h_N is preferred if the alternative describes regularity in the pattern.

Details of these two tests and comparisons with other tests can be found in BYTH and RIPLEY (1980). Other references to distance methods include RIPLEY (1979b, 1981), DIGGLE, BESAG, and GLEAVES (1976), DIGGLE (1975, 1977, 1983), and HINES and HINES (1979).

Methods using the reduced second-order moment function K

When the point pattern has been mapped exhaustively so that measurements of the locations of all points in the pattern are available then it becomes worth estimating more sophisticated statistics. In particular estimation of the reduced second-order moment function is useful. Formula (2.4.13) gives the reduced second-order moment function of a planar stationary Poisson point process as

$$K(r) = \pi r^2 \quad \text{for } r \geqq 0.$$

If an estimate of $K(r)$ deviates considerably from this simple form then the Poisson hypothesis is thrown in doubt. Moreover, comparison of the theoretical form with an estimate can do much to reveal the structure of the point pattern. Section 4.6 contains a brief discussion of estimators for $\lambda^2 K(r)$ for general point processes.

RIPLEY (1979b, 1982, 1984), OHSER (1983), and OHSER, RIPLEY, and STOYAN (1985) considered the accuracy of estimators for $\lambda^2 K(r)$ in the case of a stationary Poisson point process. They gave approximations connected to the variance of the estimator $\hat{\kappa}_R$ (which is defined by equation (4.6.3)) using as data a stationary Poisson point process observed through a circular window of radius a. For example,

$$\operatorname{var}\left(\hat{\kappa}_R(r)/\hat{\lambda}^2\right) \simeq 2\lambda^{-2}(r/a)^2\,[1 + 0.61r/a + 0.083\lambda r^3/a]. \tag{2.7.3}$$

Here $\hat{\lambda}$ is the estimator for λ given by (2.7.1).

OHSER, RIPLEY, and STOYAN (1985) used numerical integration to calculate the exact variances for the estimator $\hat{\kappa}_i$ in the cases of rectangular and circular windows.

The formula (2.7.3) can be used to calculate approximate confidence intervals for the square root of the reduced second-order moment function,

$$L(r) = \sqrt{(K(r)/\pi)} \quad \text{for } r \geqq 0$$

under the assumption that $\hat{\kappa}_R(r)$ is approximately normally distributed. For example, consider a circular window of radius a. An approximate $100(1-\alpha)\%$ confidence interval can be constructed for the quantity $L(ca)$, using the approximation given by (2.7.3); from the above a confidence interval can be obtained for $\lambda^2 K(r)$ in terms of $\hat{\kappa}_R(r)$, and then square-roots may be taken and λ may be approximated by n/a^2. The resulting interval has endpoints

$$r \cdot \sqrt{\left[1 \pm \frac{z_{\alpha/2}}{n}\sqrt{2\left(\frac{1}{c^2} + \frac{0.61}{c} + 0.026nc\right)}\right]}$$

where $c = r/a$. Strictly speaking, this is valid only for a single value of r that has been fixed *a priori*. In applications it is more natural to consider the behaviour of an estimator $\hat{L}(r)$ of $L(r)$ when r runs over a whole interval $[0, r_{\max}]$. Here $r_{\max}$ is chosen so that the interval covers the range of r that is of interest. Thus information is required on distributions such as that of

$$\sup_{r \leqq r_{\max}} \left| \sqrt{\left[\frac{\hat{\kappa}_R(r)}{\pi\hat{\lambda}^2}\right]} - L(r) \right|.$$

Some asymptotic results are available about this distribution, using the theory of U-statistics. A brief and useful survey is in RIPLEY (1981, p. 161). Alternatively simulation can be employed to provide Monte Carlo tests. An estimate of the function L is calculated for the data and compared to estimates based on simulations of binomial processes, with the same number of points and the same window as the dataset. Significance tests can be carried out with the aid of the upper and lower envelopes of the simulation estimators. A simple example is given in Example 2.2. Further examples and discussion can be found in Ripley (1977 and discussion, 1981).

A particularly useful aspect of the square-root transform is that the upper and lower envelopes of the simulated estimates of L run parallel to the graph of the theoretical function for small r. This is illustrated in Figure 2.7, and is supported by theoretical arguments.

Quadrat count methods

The case when the point pattern is sampled by means of counts of numbers of points falling in several subregions is intermediate between the case where one can usefully estimate the reduced second-order moment function and the case where one is restricted to distance methods. The sample window is divided into several subregions (often *quadrats* or squares) of equal area $\nu_2(Q)$. Under the hypothesis of a stationary Poisson point process the number of points counted in each quadrat is Poisson of mean $\lambda \cdot \nu_2(Q)$ and counts in disjoint quadrats are independent. Statistical tests can be based on these distributional properties.

The simplest test is the index-of-dispersion test, which is based on the equality of mean and variance for a Poisson distribution. The classical χ^2-goodness-of-fit test examines the distribution of the number-counts. The Greig-Smith method investigates the extent to which the independence properties are valid.

Index-of-dispersion test (see Sachs, 1984)

The index-of-dispersion I is defined by

$$I = (n - 1)\, s^2/\bar{x}$$

where n is the number of quadrats, $\bar{x}$ is the mean number of points per quadrat, and s^2 is the sample variance of the number of points per quadrat. Under the hypothesis of a stationary Poisson point process the index I follows a χ^2-distribution of $n - 1$ degrees of freedom provided that $n > 6$ and $\lambda\nu_2(Q) > 1$ (Diggle, 1983). So if I exceeds $\chi^2_{n-1;\alpha/2}$ or is smaller than $\chi^2_{n-1;1-\alpha/2}$ then the test rejects the stationary Poisson point process hypothesis at significance level $100\alpha\%$. If I exceeds the first level then there is evidence in favour of clustering as the variability in the process is greater than that for the Poisson process. If I is lower than the second level then there is tentative evidence for some regularity in the point pattern as the variability is smaller.

χ^2-goodness-of-fit test.

If the number of squares with i points is n_i, for $i = 0, 1, 2, \ldots$, then these numbers can be compared with those predicted on the basis of a Poisson distribution with mean $\hat{\lambda}$ estimated as in the first part of this section. The test statistic is the standard χ^2-statistic. Example 2.2 provides an illustration of this method.

Greig-Smith method (see Mead, 1974; Greig-Smith, 1964; Kershaw, 1964)

The Greig-Smith method uses not only the quadrat counts on their own but also the counts grouped by neighbouring quadrats. This enables the detection of clustering at different scales. Example 2.1 also illustrates this.

It is assumed that the number of quadrats is $n = 2^q$ for some whole number q. The quantities $s_1, s_2, \ldots$ are calculated

$$2^{q-1}s_1^2 = \sum_j \begin{pmatrix}\text{number in}\\ j\text{th quadrat}\end{pmatrix}^2 - \frac{1}{2}\sum_k \begin{pmatrix}\text{number in the } k\text{th}\\ \text{pair of quadrats}\end{pmatrix}^2,$$

$$2^{q-2}s_2^2 = \sum_j \begin{pmatrix}\text{number in } j\text{th}\\ \text{pair of quadrats}\end{pmatrix}^2 - \frac{1}{2}\sum_k \begin{pmatrix}\text{number in the } k\text{th}\\ \text{foursome of quadrats}\end{pmatrix}^2$$

etc.
In the case $q = 6$ the quadrats can be numbered in chess-board notation, since $2^q = 64$. The pairs of quadrats are then the pairs $(a1, a2), (a3, a4), \ldots, (b1, b2), (b3, b4), \ldots$. The foursomes of quadrats are then $(a1, a2, b1, b2), (a3, a4, b3, b4), \ldots$.

The quantities s_i^2 are unbiased estimators for the variance of the random number of points in a small quadrat. Thus the index-of-dispersion formula can be generalized, calculation made of the indices

$$I_j = 2^{q-j}s_j^2/\bar{x}$$

and the χ^2-test used on each index. Here $\bar{x}$ is as in the index-of-dispersion test. The degrees of freedom for the χ^2-statistics $I_1, I_2, \ldots$ are $2^{q-1}, 2^{q-2}, \ldots$.

The results of the tests should be interpreted as follows. The hypothesis of a stationary Poisson point process can be accepted if all the indices I_j lie between two-sided critical values of the appropriate χ^2-statistics. Otherwise, the values of the indices indicate the nature of the deviation from the hypothesis. Thus if $I_3/2^{q-3}$ is significantly greater than one then there is evidence of clustering on the scale of groups of four quadrats etc.

Further tests for the planar Poisson process are discussed in Ripley (1979b, 1981), and Diggle (1983).

Example 2.1. *Midpoints of large Mars craters*

Midpoints of large craters in Mars (where here 'large' is interpreted as of diameter greater than 200 km) form a point pattern on the sphere which is the surface of Mars. Lipskij *et al.* (1977) presented this data. Because this is a point pattern on the sphere rather than the plane, the theory above does not strictly apply; we must speak of the *homogeneous* Poisson point process hypothesis (homogeneous under rotations of the sphere) rather than the stationary Poisson point process hypothesis. However, the modifications to the theory are entirely routine.

The surface of Mars being divided into 32 regions of equal area, the point counts of the regions yielded the numbers in Table 2.1. In this case, the Greig-Smith method can be applied with $q = 5$ and $\bar{x} = 0.97$. Table 2.1 also indicates the means by which the quadrats were grouped for the application of the Greig-Smith method. The variances s_i^2 and the dispersion indices are presented in Table 2.2 and compared with the χ^2-values at $\alpha = 0.05$. For $i = 1, 3$, and 5 the index values exceed the critical values. This implies that the differences between North and South hemispheres ($i = 5$), between quarter-spheres ($i = 3$), and between 1/32nd-spheres ($i = 1$) are greater than is acceptable for a homogeneous Poisson point process, and thus there is clustering at these scales.

The homogeneous Poisson point process hypothesis is also rejected by the index-of-dispersion test. The index-of-dispersion $I = 30 \times 2.17/0.97 = 67.05$, while the corresponding χ^2-value is 43.77 at $\alpha = 0.05$.

Table 2.1. Counts of large Mars craters in 32 segments of the planet's surface

0°	90°	180°	270°	360°		
0	0	0	1		48.5°	Northern hemisphere
0	1	0	0		30°	
0	1	0	2		14.5°	
1	0	0	2			
					0°	
0	0	0	4			Southern hemisphere
0	0	1	0		14.5°	
0	2	1	2		30°	
6	4	2	1		14.5°	

Table 2.2. Test-statistics referred to the stationary Poisson point process hypothesis for the Mars crater data

i	Degrees of freedom	s_i^2	$\chi^2_{0.95}$	I_i	$\chi^2_{0.05}$
1	16	1.97	8.0	32.5	26.3
2	8	0.84	2.7	6.9	15.5
3	4	4.84	0.7	20.0	9.5
4	2	0.16	0.1	0.3	6.0
5	1	7.03	0.0	7.2	3.8

When used to compare North and South hemispheres the stationarity test yields $n_1 = 8$, $n_2 = 23$, $F = 2.76$. Thus the hypothesis of equal intensity on both hemispheres must be rejected at the 5%-level, as $F_{17,47;0.025} \simeq 2.1$.

These deviations from the homogeneous Poisson point process hypothesis presumably result from the fact that the crater pattern originated by cause of showers of meteors or from modifications of the original pattern by geological processes. Marcus (1977) gave a detailed study of point processes on planetary and lunar surfaces.

Example 2.2. *Traces of α-particles on a detector*

Figure 2.5 shows the pattern formed on a particle detector of 117 centres of traces of α-particles generated by decay of radon in gases within the earth. Details of the measurements are to be found in Fleischer and Mogro-Campero (1978). Some of the tests described above are applied to test the hypothesis that the points are produced by a stationary Poisson point process.

An estimate for λ is given by formula (2.7.1)

$$\hat{\lambda} = 117/60^2 = 0.0325$$

using as units the units of the figure. Figure 2.6 shows estimates of the nearest neighbour distance distribution function D and of the distribution function of first contact H_s. These are given in comparison to the expected function

$$1 - \exp(-\hat{\lambda}\pi r^2).$$

Agreement between these functions is disappointing, which may call in question the stationary Poisson point process hypothesis.

Fig. 2.5. Centres of traces of α-particles on a detector in a 60×60 square (1 unit = 2 μm)

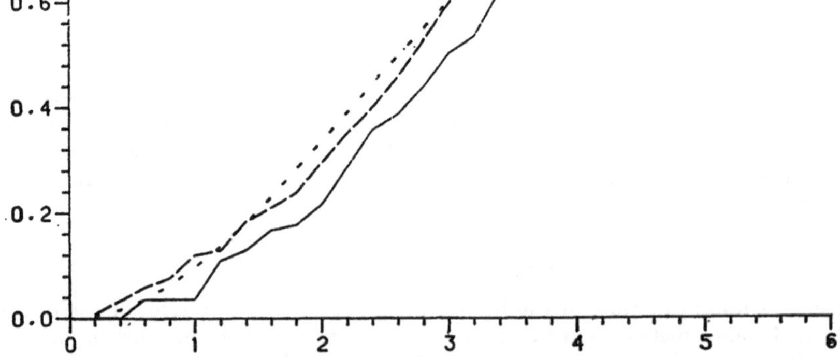

Fig. 2.6. Estimates for D and H_s derived from the data of Example 2.2, and compared with the theoretical distribution. Graph of estimate of D, ———; Graph of estimate of H_s, - - - - - - -; Theoretical curve, · · · · · · · ·.
Length in units of figure 2.5. One unit equals 2 micrometres.

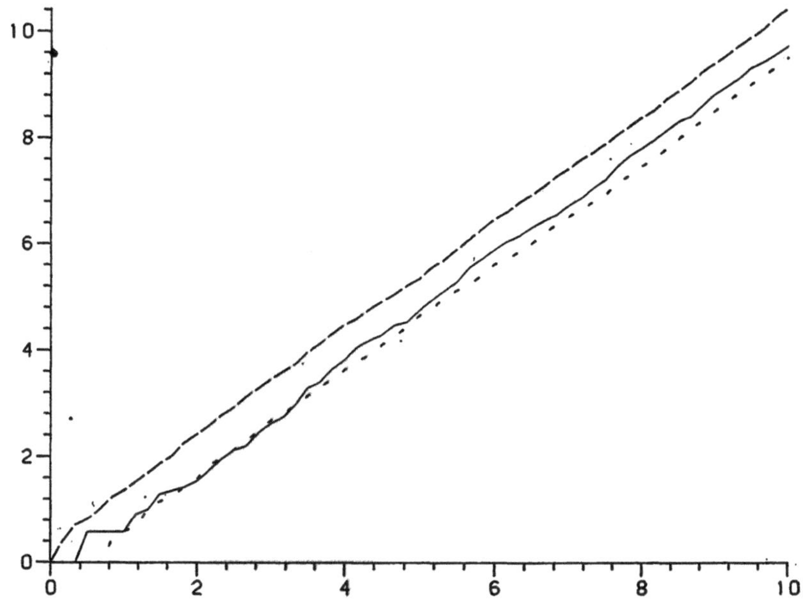

Fig. 2.7. Estimate for L for the dataset illustrated in Figure 2.5, together with upper and lower 5%-envelopes of 199 estimators using simulations of a comparable binomial process. There is evidence of interaction over the range [1, 3], since the dataset estimator touches the lower envelope at points in that range. Estimator of L using data, ———; Upper envelope, - - - - - - - -; Lower envelope, · · · · · · · ·

Lenght in units of figure 2.5. One unit equals 2 micrometres.

The L function was estimated and compared with upper and lower 5%-envelopes of estimates based on 199 simulations of binomial processes, each with 117 points and using the same window as the dataset. The results are graphed in Figure 2.7. The estimate based on the data touches the lower envelope over the region [1, 3]. This is evidence for an interaction of repulsion over that range and evidence against the hypothesis of a stationary Poisson point process.

Quadrat count methods were applied to a 8 × 8 grid. The index-of-dispersion is

$$I = 63 \times 1.76/1.83 = 60.6$$

which lies between the critical χ^2 values at $\alpha = 0.05$. Thus the stationary Poisson point process hypothesis is not rejected by this test either.

Table 2.3. Results for the Greig-Smith method applied to the α-particle trace data.

i	Degress of freedom	s_i^2	$\chi^2_{0.95}$	I_i	$\chi^2_{0.05}$
1	32	2.36	19.0	41.3	46.2
2	16	3.47	8.0	30.4	26.3
3	8	5.06	2.7	22.2	15.5
4	4	7.06	0.7	15.5	9.5
5	2	4.25	0.1	4.6	6.0
6	1	4.50	0.0	2.5	3.8

The χ^2-goodness-of-fit test gives a contrary result. The test statistic yields a value of 0.783 for five classes, if the counts for numbers 4, 5, ... are united into one class. The critical χ^2-value is

$$\chi^2_{3;00.5} = 7.82 .$$

The Greig-Smith method yields results as tabulated in Table 2.3. This test indicates some clustering in the scale of groups of two, four, and eight quadrats. Again, the stationary Poisson point process hypothesis is rejected at the $\alpha = 0.05$ significance level, though the hypothesis is not rejected at the $\alpha = 0.01$ level.

3. Random closed sets I – The Boolean model

3.1 Introduction and basic properties

The *Boolean model* (also known as the *Poisson grain model*) is an important and relatively simple example of a random set. It is both flexible and amenable to calculation.

Two kinds of application can be distinguished. In the first kind, the Boolean model describes a particular empirical random set. In the second kind, it approximates more complicated random sets.

Before a general definition is given it may be helpful to consider one of the simplest possible examples of a Boolean model. Suppose points are scattered in the plane according to a stationary Poisson process of intensity λ. On each of these points a disc of fixed radius r is placed. The union of all of these discs is an example of a Boolean model; a particular realization is displayed in Figure 3.1. In the terminology to be introduced the points of the Poisson process are the *germs* of the model while the discs are the *primary grains*.

Despite its simplicity this random set already has interesting properties that depend on the ratio r^2/λ. In fact, there are unanswered questions; for example, at what critical value of r^2/λ do unbounded components of the Boolean model first occur? (See for example, Gilbert (1961). The problem is part of the subject of 'continuous percolation theory'.)

This construction generalizes to produce the general stationary Boolean model. The discs are replaced by independent realizations of a random compact set. Suppose $\Phi = \{x_1, x_2, \ldots\}$ is a stationary Poisson process in $\mathbb{R}^d$ of intensity λ. Let $\Xi_1, \Xi_2, \ldots$ be a sequence of independent identically distributed random compact sets in $\mathbb{R}^d$ that are independent of the Poisson process Φ and

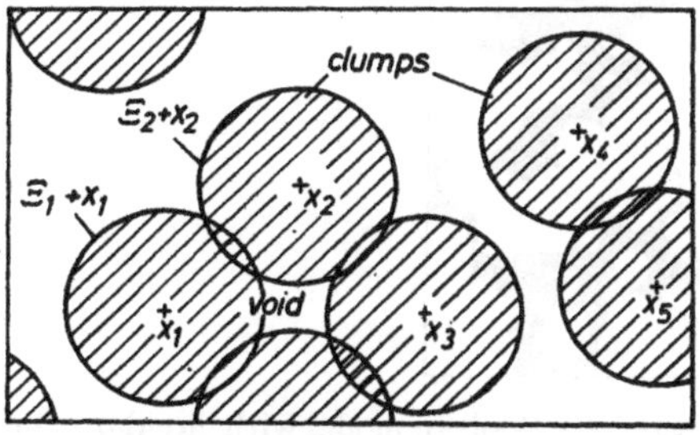

Fig. 3.1. Diagramatic image of a planar Boolean model. The empty space enclosed by three of the primary grains is a *void*.
Key: Germ x_n, $+$; Primary grain, Ξ_n

satisfy a technical condition given below. The *Boolean model* Ξ is constructed by using the *germs* x_n and the *primary grains* Ξ_n

$$\Xi = \bigcup_{n=1}^{\infty} (\Xi_n + x_n) = (\Xi_1 + x_1) \cup (\Xi_2 + x_2) \cup \cdots$$

The technical condition mentioned above is that

$$\mathsf{E}\nu_d(\overset{\circ}{\Xi}_0 \oplus K) < \infty \quad \text{for all compact } K. \tag{3.1.1}$$

Here and in the sequel, Ξ_0 denotes a further random compact set of the same distribution as the Ξ_n but independent both of them and of the germ process Φ. We say that Ξ is a *Boolean model with primary grain* Ξ_0.

The technical condition (3.1.1) ensures that only finitely many of the grains $\Xi_n + x_n$ meet any given compact set. Thus, in particular, it ensures that the property of being a closed set is inherited by Ξ from the primary grains. The logic behind the condition will be made clearer by the determination of the probability of Ξ hitting a compact set; see formula (3.1.2).

A simpler but more restrictive condition than the one given above is as follows; if R_n is the radius of the ball circumscribing Ξ_n then

$$\mathsf{E}R_n{}^d < \infty .$$

The construction of Ξ is illustrated in Figure 3.1 while Figure 3.2 gives an example of a realization of a Boolean model.

In this introduction the notion of a compact random set has been used without the careful treatment that the concept merits. Chapter 6 gives a discussion. In this chapter an intuitive understanding suffices.

A more theoretical and less constructive point of view defines the Boolean model as the union of sets arising from a Poisson process in the family $\mathbb{K}'$ of all non-empty compact subsets of $\mathbb{R}^d$. A treatment using this approach can be

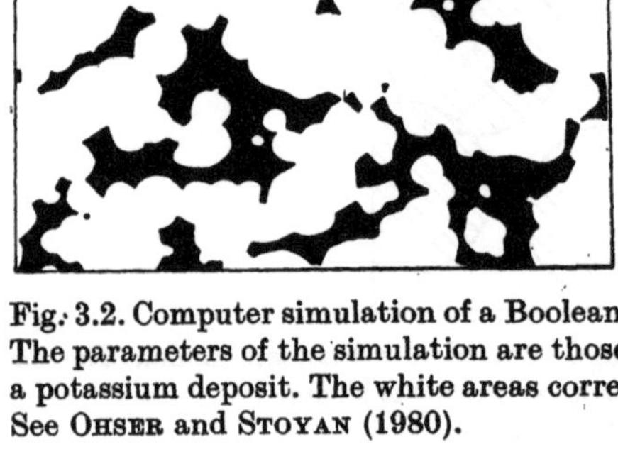

Fig. 3.2. Computer simulation of a Boolean model with circular primary grains. The parameters of the simulation are those fitted to the pattern of a section of a potassium deposit. The white areas correspond to potassium bearing regions. See OHSER and STOYAN (1980).

found in MATHERON (1975). The technical condition mentioned above is then naturally expressed as requiring the governing measure of the Poisson process to be a σ-finite Borel measure on $\mathbb{K}'$, using a suitable topology on $\mathbb{K}'$.

Typical possibilities for the primary grains include: discs of random radius (generalizing the simple example above), segments of random length and orientation, and random finite clusters of points. The primary grains are characterized by their distribution M. As they are random sets this distribution is a probability measure on $\mathbb{K}'$; it is the *mark distribution* of the marked point process $\{[x_n; \Xi_n]\}$; see Section 4.2 and Chapter 6.

Table 3.1

In general Euclidean space $\mathbb{R}^d$	
$\overline{W}_k$	mean of kth Minkowski functional.
For spatial sets, $d = 3$;	
$\overline{V}$	mean value
$\overline{S}$	mean surface area,
$\overline{M}$	mean of integral of mean curvature,
$\overline{\overline{b}}$	mean of mean breadth.
(N.b. $\overline{M} = 2\pi\overline{\overline{b}}$)	
For planar sets, $d = 2$;	
$\overline{A}$	mean area,
$\overline{U}$	mean perimeter length.

Note: The first mean in $\overline{M}$ and in $\overline{\overline{b}}$ is statistical while the second refers to a geometric average taken over all possible directions.

If the primary grains are convex then statistical averages of various numerical measures of convex sets are important. Their notation and explanation are given in Table 3.1. In the case of non-convex grains several of the quantities still have meaning and in such cases the same symbols will be used.

The Boolean model is a basic model in stereology and stochastic geometry. As GIGER and HADWIGER (1968) have shown it also arises from traditional assumptions of independence and uniformity (in a manner similar to some characterizations of the Poisson process). It has been thoroughly studied by Matheron, Serra, and their colleagues; see MATHERON (1967, 1975) and SERRA (1982) and references therein.

There are two possibilities for application of the Boolean model. In the first place, it is a natural model for sparse systems of particles distributed at random. Here the sparse nature of the system is modelled by a low value for the intensity λ of the Poisson process of germs. If λ is small relative to the size of the grains then primary grains will not often overlap and so Ξ will consist in the main of separated particles. Formula (3.1.10) gives the probability that an arbitrary primary grain is overlapped by others.

A typical example of such systems is the set of nodular graphite particles in cast iron (Figure 3.6). A random sparse pattern of plants may also yield such a pattern in the area covered by vegetation.

With increasing λ the number of overlappings increases, as is exemplified in Table 3.2. Of course such overlappings do occur in nature; one has only to consider pores in cheese or areas of weeds in fields.

In the second place the Boolean model may provide a good description for a very irregular pattern observed in nature. An example of such a pattern is given in Figure 3.2. As can be seen from the figure (and in contrast to what is often the case for sparse systems) the primary grains and their locations at germs generally fail to have operational interpretations. The role of the Boolean model in such a case is to provide irregular random sets of a suitable form, and its method of construction need not correspond to any physical reality. This point is returned to in Section 3.5.

The Boolean model has been used, at least in special forms, since about 1940 (and perhaps even before the war). The following list gives some examples drawn mainly from the natural sciences.

Armitage (1949): random clumping of dust or powder particles.
Mack (1954, 1956): corrections for clumping in estimation of particle number in the above problem; see also Kellerer (1983).
Diggle (1981): distribution of heather in a forest.
Jacod and Joathon (1971): form of geological structures generated by sedimentation; see also Ohser and Stoyan (1980): form of geological deposits of potassium (Figure 3.2 is the result of a simulation of a Boolean model using as model parameters those obtained by fitting to the geological data. The white zones in the figure correspond to areas that are rich in Potassium).
Nutting (1913) and Marchant and Dillon (1961): patterns in photographic emulsion; see also Frieden (1983).
Matern (1960): distribution of forests.
Porod (1952): colloids in gel form.
Sonntag, Stoyan and Hermann (1981): structural inhomogeneities in amorphous matter.
Widom and Rowlinson (1970): systems of water droplets in the study of liquid-vapour phase transitions.

The grains of a Boolean model are not required to be connected sets. For example, they may be sets of discrete points. In such a case the Boolean model is a point-process, in fact, a Neyman-Scott point process (see Section 5.3).

In optics and in other fields of physics the shot-noise model is sometimes used (Frieden, 1983). Here the grains Ξ_n are replaced by random disturbance functions and the set-theoretic union by addition of functions. Formulae for this model are in places quite closely analogous to those for the Boolean model.

Bombing problems form a situation where the Boolean model may be usefully applied. Robbins (1944, 1945) and Neyman and Scott (reported in 1972) studied random closed sets with such applications in view. Here the germs are the points of impact of the bombs, and the grains are the areas covered by the splinters from the corresponding bomb explosion. Robbins' account was confined to theoretical issues; Neyman and Scott briefly discussed details of one particular application concerning the clearance of land-mines from the landing beaches in Normandy. Generally in such problems the Boolean model seems plausible for representing the total area covered by splinters while the shot-noise model is more suitable for considerations of intensity of covering.

In description of given samples and situations in nature the Boolean model has a three-fold value. First, a relatively parsimonious description is available for the random set in question (by means of the intensity λ of the germ process and the various characteristics of the primary grains). Second, it is at least possible that the model assumptions may aid understanding of the process of formation of the structure (though it will be emphasized later that this is by no means invariably the case; see Section 3.5). Third, the formulae to be derived for the Boolean model may be used for estimation of quantities not available for direct measurements.

Stationarity and isotropy

From the stationarity of the Poisson process Φ of germs, and the identical distribution of the primary grains, it follows that the Boolean model defined above is stationary. That is to say, its distribution is translation-invariant; see Section 6.1. As a consequence of more general considerations NGUYEN and ZESSIN (1979) show that the Boolean model is *ergodic* (see Section 6.1 for a discussion of this concept). A particular consequence of ergodicity is that

$$\lim_{r\to\infty} \frac{\nu_d[\Xi \cap b(o, r)]}{\nu_d[b(o, r)]} = p$$

where $p = \mathsf{P}(o \in \Xi)$ is the volume fraction of the Boolean model (see below).

If the primary grains have isotropic distribution (distribution invariant under rotations about the origin o) then the Boolean model is also isotropic in distribution. Isotropy and stationarity together imply that the Boolean model has distribution invariant under rigid motions.

Capacity functional

The distribution of a Boolean model is uniquely determined by its capacity functional or hitting distribution T_Ξ;

$$\begin{aligned} T_\Xi(K) &= \mathsf{P}(\Xi \cap K \text{ is not empty}) \\ &= \mathsf{P}(\Xi \cap K \neq \emptyset) \end{aligned}$$

for any compact K.

As given above it can be defined for any random set. Indeed, it determines uniquely the distribution of a general random closed set, as is discussed in Section 6.1.

In the case of the Boolean model, T_Ξ can be expressed in a relatively simple form, in pleasant contrast to the complexity of expressions for many other probabilities connected with this model. By application of the Poisson assumption T_Ξ can be shown to take the form

$$\begin{aligned} T_\Xi(K) &= 1 - \exp\{-\lambda \mathsf{E}\nu_d(\check{\Xi}_0 \oplus K)\} \\ &= 1 - \exp\{-\lambda \mathsf{E}\nu_d(\Xi_0 \oplus \check{K}^\circ)\}. \end{aligned} \qquad (3.1.2)$$

Compare for example the approximate formulae given in MORAN (1973) and KELLERER (1983) for the probability that a given set is contained completely in Ξ.

The proof of (3.1.2) is of methodological interest. Let K be a fixed compact set. From the original germ process Φ a new *thinned* process Φ_p can be produced. The thinning takes place by deleting x_n from Φ if $(\Xi_n + x_n) \cap K$ is empty. Thus

$$\Phi_p = \{x_n \in \Phi : (\Xi_n + x_n) \cap K \text{ is not empty}\}.$$

Whether or not a germ x_n is deleted by this thinning is independent of thinning happening to other germs. This is because of the independence properties of the Boolean model; deletion depends only on the location of the germ and on the corresponding grain. This independence means that Φ_p is again a Poisson process, but nonstationary with intensity measure Λ_p. Here

$$\Lambda_p(B) = \lambda \int_B p(x)\, dx \quad \text{for } B \text{ a Borel set}$$

and $p(x) = \mathsf{P}((\Xi_0 + x) \cap K$ is not empty). Note that $p(x)$ is the probability that a germ placed at x is not deleted. See Section 5.1 for a discussion of such constructions.

The capacity $T_\Xi(K)$ is given by

$$1 - T_\Xi(K) = \mathsf{P}(\Phi_p \text{ has no points}).$$

By the properties of non-stationary Poisson processes the distribution of the total number of points in Φ_p depends only on the total measure

$$\Lambda_p(\mathbb{R}^d) = \lambda \int_{\mathbb{R}^d} p(x)\, dx.$$

If $\Lambda_p(\mathbb{R}^d)$ is finite then the probability of Φ_0 having no points is

$$\exp\left[-\Lambda_p(\mathbb{R}^d)\right] = \exp\left[-\lambda \int_{\mathbb{R}^d} p(x)\, dx\right]$$

So it suffices to evaluate $\Lambda_p(\mathbb{R}^d)$.

The probability $p(x)$ is given by

$$p(x) = \mathsf{P}(x \in \check{\Xi} \oplus K)$$

since $(\Xi_0 + x)$ meets K precisely when $x \in \check{\Xi}_0 \oplus K$. Thus

$$\begin{aligned}\Lambda_p(\mathbb{R}^d) &= \lambda \int_{\mathbb{R}^d} \mathsf{P}(x \in \check{\Xi}_0 \oplus K)\, dx \\ &= \lambda \int_{\mathbb{R}^d} \mathsf{E}\mathbf{1}_{(\check{\Xi} \oplus K)}(x)\, dx \\ &= \lambda \mathsf{E}\nu_d(\check{\Xi}_0 \oplus K)\end{aligned}$$

where in the last step the expectation is interchanged with the integral. This last expectation is finite, by the imposition of the technical condition required in the definition of the Boolean model.

Thus the first equality of (3.1.2) is established:

$$T_\Xi(K) = 1 - \exp\left(-\Lambda_\rho(\mathbb{R}^d)\right)$$
$$= 1 - \exp\left(-\lambda \mathsf{E}\nu_d(\check{\Xi}_0 \oplus K)\right).$$

The second equality follows by noting that

$$(\check{\Xi}_0 \oplus \check{K}) = -(\Xi_0 \oplus \check{K}).$$

Remarks. (1) It is useful to introduce the notation

$$\psi(K) = -\ln \mathsf{P}(\Xi \cap K \text{ is empty}).$$

Then (3.1.2) can be re-expressed as

$$\psi(K) = \lambda \mathsf{E}\nu_d(\check{\Xi}_0 \oplus K)$$
$$= \lambda \mathsf{E}\nu_d(\Xi_0 \oplus \check{K}). \tag{3.1.3}$$

(2) From the proof it is clear that similar formulae hold for random set models constructed as above but using non-stationary Poisson processes as germ processes.

(3) Formula (3.1.3) makes clear the arbitrary element that exists in the choice of location for each primary grain Ξ_n. This can be useful; suppose that z_n are points produced by a construction depending on the corresponding primary grain, so that z_n depends on Ξ_n alone and not on any other grains nor on any other germs. Then a new Boolean model can be constructed using Φ for the grains and $\Xi_n^* = \Xi_n + z_n$ for the primary grains. For example, take $\Xi_n = b(o, X_n)$ where X_n is a random radius. Then pick

$$z_n = (0, 0, \ldots, 0, -X_n).$$

In any case, by the translation-invariance of Lebesgue measure

$$\psi^*(K) = \mathsf{E}\nu_d(\Xi_0^* \oplus \check{K})$$
$$= \mathsf{E}\nu_d[(\Xi_0 \oplus \check{K}) + z_0]$$
$$= \mathsf{E}\nu_d(\Xi_0 \oplus \check{K}) = \psi(K).$$

Hence, the two Boolean models Ξ and Ξ^* have the same capacity functional and so must have the same distribution.

Basic characteristics of the Boolean model

Formula (3.1.2) provides the means of calculating some important characteristics of the Boolean model.

The volume fraction

The volume fraction p is the mean fraction of volume occupied by Ξ in a region of unit volume. This mean fraction does not depend on the choice of region, by virtue of the stationarity of Ξ. A further consequence of stationarity is that

$$p = \mathsf{P}(o \in \Xi)$$

a result that holds for general stationary random sets (see Section 6.2).

Putting $K = \{o\}$ in (3.1.2) it is seen that

$$\mathsf{P}\{o \in \Xi\} = 1 - \exp\left(-\lambda \mathsf{E}\nu_d(\Xi_0)\right).$$

Here, Ξ_0 is as before a typical primary grain of Ξ. Hence

$$p = 1 - \exp\{-\lambda \overline{V}\} \quad \text{where } \overline{V} = \mathsf{E}\nu_d(\Xi_0). \tag{3.1.4}$$

Of course, if the grains of the Boolean model all have zero volume (for example, if they are points or line segments in the plane) then the volume fraction will be zero.

The covariance

If a random set has volume fraction $p > 0$ then its second-order properties are summarized in its covariance, as discussed in Chapter 6.

The covariance is defined as

$$C(\boldsymbol{r}) = \mathsf{P}(o \in \Xi \text{ and } \boldsymbol{r} \in \Xi) \quad \text{for } \boldsymbol{r} \text{ in } \mathbb{R}^d.$$

A simple manipulation reveals that $C(\boldsymbol{r})$ is the volume fraction of $\Xi \cap (\Xi - \boldsymbol{r})$; however this intersection is not in general a Boolean model.

If Ξ is a Boolean model of primary grain Ξ_0 then

$$C(\boldsymbol{r}) = 2p - 1 + (1 - p)^2 \exp\left[\lambda \gamma_{\Xi_0}(\boldsymbol{r})\right] \tag{3.1.5}$$

where $\gamma_{\Xi_0}(\boldsymbol{r}) = \mathsf{E}\nu_d\left(\Xi_0 \cap (\Xi_0 - \boldsymbol{r})\right)$.

Formula (3.1.5) can be proved by considering the following evaluation of $C(\boldsymbol{r})$

$$\begin{aligned} C(\boldsymbol{r}) &= \mathsf{P}\left(o \in \Xi \cap (\Xi - \boldsymbol{r})\right) \\ &= 1 - \{\mathsf{P}(o \text{ not in } \Xi) + \mathsf{P}(o \text{ not in } \Xi - \boldsymbol{r}) - \\ &\qquad \mathsf{P}(o \text{ not in } \Xi \cap (\Xi - \boldsymbol{r}))\} \\ &= 2p - 1 + \mathsf{P}(o \text{ not in } \Xi \cap (\Xi - \boldsymbol{r})). \end{aligned}$$

An application of (3.1.2) shows that

$$\begin{aligned} \mathsf{P}(o \text{ is not in } \Xi \cap (\Xi - \boldsymbol{r})) &= \mathsf{P}(\Xi \cap (\Xi - \boldsymbol{r}) \text{ is empty}) \\ &= \exp\left[-\psi(\{o, \boldsymbol{r}\})\right] \end{aligned}$$

where ψ is defined by (3.1.3). On evaluation of the Minkowski sum

$$\begin{aligned} \psi(\{o, \boldsymbol{r}\}) &= \lambda \mathsf{E}\nu_d[\check{\Xi}_0 \cup (\check{\Xi}_0 + \boldsymbol{r})] \\ &= \lambda\{\mathsf{E}\nu_d\check{\Xi}_0 + \mathsf{E}\nu_d(\check{\Xi}_0 + \boldsymbol{r}) - \mathsf{E}\nu_d[\check{\Xi}_0 \cap (\check{\Xi}_0 + \boldsymbol{r})]\} \\ &= \lambda\{2\mathsf{E}\nu_d(\check{\Xi}_0) - \mathsf{E}\nu_d[\Xi_0 \cap (\Xi_0 - \boldsymbol{r})]\} \end{aligned}$$

and the proof of (3.1.5) is concluded by noting that

$$1 - p = \exp\left[-\lambda \mathsf{E}\nu_d(\Xi_0)\right].$$

As is discussed in Section 6.2 the covariance of a random set is closely related to the concepts of covariance function and correlation function for random fields (but note the covariance function defined above differs from the conventional covariance function in that here the covariance is not centred about the mean).

If the random set Ξ is isotropic as well as stationary then $\gamma_{\Xi_0}(\boldsymbol{r})$ and $C(\boldsymbol{r})$ depend only on the length $\|\boldsymbol{r}\|$ of the vector $\boldsymbol{r}$. In such a case it is convenient to abuse notation by presenting C and γ_{Ξ_0} as functions of length alone rather than of vectors:

$$\gamma_{\Xi_0}(r) = \gamma_{\Xi_0}(r\boldsymbol{u}) \quad \text{and} \quad C(r) = C(r\boldsymbol{u})$$

where $\boldsymbol{u}$ is some unit vector.

In the case of convex grains some properties of the function γ_{Ξ_0} are known, and can be found in MATHERON (1975) and STOYAN (1979). In particular the derivative can be found: if $r > 0$ and $\boldsymbol{u}$ is a unit vector then $\gamma_{\Xi_0}(r\boldsymbol{u})$ has derivative

$$\frac{\mathrm{d}}{\mathrm{d}r}\gamma_{\Xi_0}(r\boldsymbol{u}) = -\mathsf{E}\{\nu_{d-1}[\mathfrak{p}_{\boldsymbol{u}}(\Xi_0 \cap (\Xi_0 + r\boldsymbol{u}))]\}.$$

Here $\mathfrak{p}_{\boldsymbol{u}}(A)$ is the orthogonal projection of A on the hyperplane that has $\boldsymbol{u}$ as normal vector. The geometric configuration is illustrated in Figure 3.3.

It is also the case that $\gamma_{\Xi_0}(r\boldsymbol{u})$ is convex in r. This follows from the observation that $-\mathrm{d}\gamma_{\Xi_0}(r\boldsymbol{u})/\mathrm{d}r$ is decreasing in r.

The function γ_{Ξ_0} goes under the name of 'distance probability function' in physics (it was introduced by POROD, 1951), at least in the case when Ξ_0 is isotropic. A typical instance is that of a deterministic set Ξ_0^* given a uniformly distributed random rotation about the origin o. In this case the corresponding averaged function $\bar{\gamma}_{\Xi_0}$ plays a part in the theory of X-ray analysis of matter, see GLATTER (1979), GLATTER and KRATKY (1982) and HANISCH (1984b).

The quantity $\bar{\gamma}_{\Xi_0}$ is closely related to the so-called *distance distribution* P, which is the density of the distance between two independent random points which are uniformly distributed in Ξ_0^*. The relationship is

$$P(r) = 4\pi r^2 \bar{\gamma}_{\Xi_0}(r)/\nu_d(\Xi_0^*)^2$$

for $r \geq 0$.

There is also a close relationship between $\bar{\gamma}_{\Xi_0}$ and the distribution function L of random chord lengths in Ξ_0^* when the chords are generated by a statio-

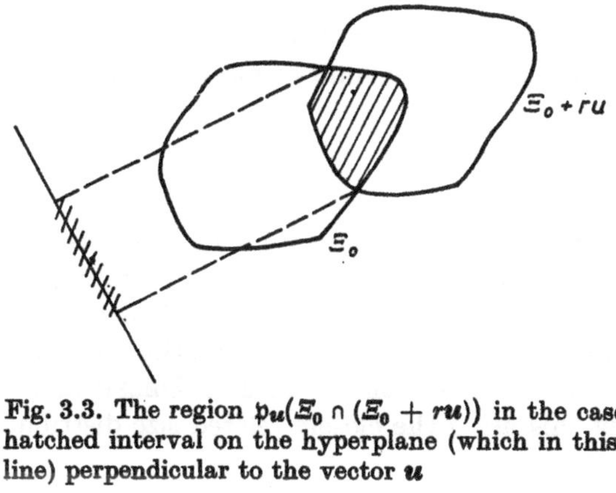

Fig. 3.3. The region $\mathfrak{p}_{\boldsymbol{u}}(\Xi_0 \cap (\Xi_0 + r\boldsymbol{u}))$ in the case $d = 2$. The region is the hatched interval on the hyperplane (which in this two-dimensional case is a line) perpendicular to the vector $\boldsymbol{u}$

nary and isotropic Poisson line process

$$1 - L(r - 0) = -\frac{4}{S(\Xi_0^*)} \frac{\mathrm{d}}{\mathrm{d}r} \bar{\gamma}_{\Xi_0}(r) \quad \text{for } r \geqq 0 .$$

See MATHERON (1975, p. 87), ENNS and EHLERS (1978), KINGMAN (1969).

In general $\bar{\gamma}_{\Xi_0}$ is not an easy function to calculate. For a sphere of random radius, with radius distribution function F_R, it is given by

$$\gamma_{\Xi_0}(r) = \bar{\gamma}_{\Xi_0}(r) = \int\limits_{r/2}^{\infty} \frac{4}{3} \pi x^3 \left(1 - \frac{3r}{4x} + \frac{r^3}{16x^3}\right) \mathrm{d}F_R(x) .$$

Here $\gamma_{\Xi_0}(r) = \gamma_{\Xi_0}(r\boldsymbol{u})$ for $\boldsymbol{u}$ a unit vector, following the abuse of notation mentioned above. In the literature formulae can be found for $\gamma_{\Xi_0}(r)$, $L(r)$, $P(r)$, and $I(s)$ (see page 91) for various bodies Ξ_0^*. Among the bodies for which calculations have been made are: ellipsoids, cylinders (see KELLERER, 1971), prisms (see MITTELBACH, 1964), general parallelepipeds (see page 304), and isotropic random sets such as the typical cells of the VORONOI tessellation (GILBERT, 1962; BRUMBERGER and GOODISMAN, 1983), and typical cells of the Poisson plane tessellation (page 84). The form of Ξ_0^* is often determined by considering $\bar{\gamma}_{\Xi_0}$, L, P, or I and using trial-and-error methods; a form is assumed for Ξ_0^* and the theoretical functions are compared to the empirical data (see KRATKY, 1983).

Other cases are mentioned in GLATTER (1979).

A useful asymptotic form relating C to $\bar{\gamma}_{\Xi_0}$ is given in SONNTAG, STOYAN, and HERMANN (1981). Let $k(r)$ be the correlation function, related to C by

$$k(r) = \frac{C(r) - p^2}{p(1 - p)} \quad \text{for } r \geqq 0 . \tag{3.1.6}$$

As the intensity λ of the process Φ tends to zero, the primary grains remaining fixed, so p tends to zero and

$$\lim_{\lambda \to 0} k(r) = \frac{\bar{\gamma}_{\Xi_0}(r)}{\bar{\gamma}_{\Xi_0}(0)} .$$

Thus for a 'dilute' Boolean model the correlation function k is asymptotically the same as the normalized distance probability function of Ξ_0^*.

Contact distribution functions

In the description and statistical analysis of a random closed set, analysis by means of the contact distribution functions forms an important tool (see Sections 3.3 and 6.2). Even in the case where identification of particles or of single pores is impossible or inappropriate, contact distribution functions permit discussion of the local structure of random closed sets. DELFINER (1972) gave an excellent discussion of this using the closely related size distribution functions.

In passing let it be noted that the relationship between contact distribution and size distribution, while close, is not explicit save in one special case. MATHERON (1975; p. 53) discussed the essential ideas here and established an explicit connection for the special case, which is when the structuring element is a line segment. His discussion concerns granulometries, from which the notion of size distribution is derived. WATSON (1978) also considered this point.

A *contact distribution function* or *hit distribution function* for a random closed set Ξ is defined by using a specified test set or structuring element B, a compact subset of $\mathbb{R}^d$ containing the origin o. Then the contact distribution function is defined as

$$H_B(r) = 1 - \mathsf{P}(\Xi \cap rB \text{ is empty})/(1 - p) \quad \text{for } r \geqq 0$$

where p is the volume fraction of Ξ.

Since B contains the origin, $\Xi \cap rB$ being empty implies that Ξ does not contain the origin. So the value of $H_B(r)$ is the conditional probability that the r-fold enlargement of B intersects (or makes contact with) Ξ, conditional on Ξ not containing o.

By definition H_B is a distribution function: if $R = \inf \{s : \Xi \cap sB \text{ is not empty}\}$ then

$$H_B(r) = \mathsf{P}(R \leqq r \mid R > 0).$$

The shape of the structuring element B plays an important role. Two particularly important cases are the extremes where on the one hand, B is a line segment of unit length (leading to the *linear contact distribution function* H_l), and on the other hand B is a unit sphere (leading to the *spherical contact distribution function* H_s). Of course in non-isotropic cases the orientation of the line segment plays an important part in defining H_l. Section 6.2 discusses briefly the use of the linear contact distribution function in orientation analysis.

The spherical contact distribution function $H_s(r)$ may be interpreted as the distribution function of distance from a point chosen randomly outside Ξ, measured to the nearest point of Ξ. Hence, H_s is often referred to as 'the law of first contact'.

In general a formula follows directly from (3.1.2):

$$H_B(r) = 1 - \exp\{-\lambda[\mathsf{E}\nu_d(\check{\Xi}_0 \oplus rB) - \mathsf{E}\nu_d(\Xi_0)]\} \quad \text{for } r \geqq 0. \tag{3.1.7}$$

Covering and hitting probabilities

Consider a primary grain chosen at random in such a way that every primary grain is equally likely to be chosen. Strictly speaking this is not possible as there are infinitely many grains. However, it can be approached as a limiting case. The relationship between this 'typical' grain and the rest of the Boolean model is of interest. Suppose that Ξ_* is such a typical grain, with corresponding germ x_*, and let

$$\Xi' = \bigcup \{\Xi_i + x_i : x_i \neq x_*\}$$

be the union of the remaining grains. Then the following probabilities are of interest:

$$p_0 = \mathsf{P}(x_* \text{ is covered by } \Xi')$$

$$p_G = \mathsf{P}(\Xi_* + x_* \text{ intersects } \Xi')$$

$$p_S = \text{fraction of surface of } \Xi_* \text{ covered by } \Xi'.$$

(Of course p_S only makes sense if the grains of Ξ have finite nonzero surface area.)

Table 3.2 Probabilities that a randomly chosen grain of a Boolean model with identical circular grains is isolated (is not covered by any other primary grains).

p	$1 - p_G$
0.1	0.656
0.2	0.410
0.3	0.240
0.4	0.130
0.5	0.063
0.6	0.026
0.7	0.008

The exact relationship is easy to work out. The above illustrates explicitly the extent to which individual grains are obscured as the area fraction p increases.

These probabilities satisfy

$$p_0 = p \tag{3.1.8}$$

$$p_S = p \tag{3.1.9}$$

$$p_G = 1 - \mathsf{E} \exp\left[-\psi(\Xi_0)\right]. \tag{3.1.10}$$

In the particular case of the grains being spheres of fixed radius r,

$$p_G = 1 - \exp\left\{-\lambda \nu_d\big(b(o, 2r)\big)\right\}.$$

The dependence of $1 - p_G$ on p is tabulated in Table 3.2 for the two-dimensional case, with grains being circles of fixed radius spheres so that

$$p_G = 1 - (1 - p)^4.$$

Formula (3.1.8) can be established as follows. By an analogue of the theorem of Slivnyak (see 2.4.3 and 4.4) the typical grain $\Xi_* + x_*$ may be taken to be independent of the remainder $\{\Xi_i + x_i : x_i \neq x_*\}$, and the remainder may be taken to form a Boolean model with the same distribution as the original. Furthermore it may be assumed that $x_* = 0$ and $\Xi_* = \Xi_0$. Thus $p_0 = \mathsf{P}(o \in \Xi) = p$.

The proofs of (3.1.9) and (3.1.10) are similar.

3.2 The Boolean model with convex grains

An assumption made frequently in practice is that the Boolean model has isotropic and convex grains. This has the advantage of considerable simplification of various formulae. Moreover, practical experience shows that the resulting structures are sufficiently flexible to find a useful place in many statistical problems.

The simplified formula for the capacity functional

If Ξ_0 is convex then $\psi(K)$ can be calculated for convex K by means of the so-called generalized Steiner formula (see page 173)

$$\psi(K) = \lambda \mathsf{E} \nu_d(\Xi_0 \oplus \check{K}) = \frac{\lambda}{\omega_d} \sum_{k=0}^{d} \binom{d}{k} \overline{W}_k W_{d-k}(K). \tag{3.2.1}$$

Here $\overline{W}_k$ denotes the mean of the Minkowski functional of Ξ_0, so

$$\overline{W}_k = \mathsf{E} W_k(\Xi_0) \quad \text{for } k = 1, \ldots, d$$

and ω_d is the volume of the d-dimensional unit ball.

Thus an important simplification is obtained: for convex K the value of the capacity functional depends only on the means of the Minkowski functionals of the grains.

Example. Let Ξ_0 be a sphere $b(o, R)$ of random radius R. Suppose R has moments

$$\mathsf{E} R^i = m_R^{(i)} \quad \text{for } i = 1, 2, \ldots.$$

Then by the Steiner formula for $K \in C(\mathbf{K})$ in dimension $d = 2$, the planar case,

$$\psi(K) = \lambda[A(K) + 2m_R^{(1)} U(K) + 2\pi m_R^{(2)}],$$

in dimension $d = 3$, the spatial case,

$$\psi(K) = \lambda \left[V(K) + m_R^{(1)} S(K) + 2\pi m_R^{(2)} \bar{b}(K) + \frac{4}{3} \pi m_R^{(3)} \right].$$

Thus in the plane the capacity functional ψ depends only on the first two moments of R, and in three-space only on the first three moments.

Intensities of Minkowski measures

From a Boolean model various random measures can be derived. For example a three-dimensional Boolean model Ξ yields the *surface measure* S_Ξ, defined for a Borel set B by

$$S_\Xi(B) = \nu_2\{(\partial \Xi) \cap B\}$$

the two-dimensional measure of the intersection with B of the boundary $\partial\Xi$ (this is discussed further in Section 7.3.4). The random measure S_Ξ inherits stationarity from the Boolean model Ξ. Therefore the mean measure $\mathsf{E}S_\Xi(B)$ is invariant under translations and so is a multiple of Lebesgue measure:

$$\mathsf{E}S_\Xi(B) = S_V \nu_2(B).$$

The constant S_V is called the *intensity* of the random measure S_Ξ; see page 200.

The intensity S_V can be interpreted as the *mean surface area per unit volume.* An analogous quantity for a planar Boolean model is L_A, the *mean boundary length per unit area.* These quantities are, up to multiplicative factors, particular cases of the intensities w_k of the Minkowski measures described in Section 7.3.

MATHERON (1975) showed that

$$w_0 = p \tag{3.2.2}$$

$$w_k = \lambda(1-p)\,\overline{W}_k : k = 1, 2, \ldots, d \tag{3.2.3}$$

where $\overline{W}_k$ is the mean of the kth Minkowski functional of Ξ_0 as above. The following formulae for the planar and spatial cases may prove useful as they yield particular interpretations for the intensities w_k.

In the planar case ($d = 2$):

$$p \quad = A_A \quad \text{the specific area} \tag{3.2.4}$$

$$w_1 \quad = \lambda(1-p)\,\overline{U}/2 \tag{3.2.5a}$$

$$L_A \quad = \lambda(1-p)\,\overline{U} \quad \text{but see the note on p. 79 following} \tag{3.2.5b}$$

$$w_2 \quad = \lambda(1-p)/\pi \tag{3.2.6a}$$

$$N_A{}^+ = \lambda(1-p). \tag{3.2.6b}$$

Here L_A is the mean boundary length per unit area and $N_A{}^+$ is the *specific convexity number* discussed on page 201.

In the spatial case ($d = 3$):

$$p \quad = V_V \quad \text{the specific volume} \tag{3.2.7}$$

$$w_1 \quad = \lambda(1-p)\,\overline{S}/3 \tag{3.2.8a}$$

$$S_V \quad = \lambda(1-p)\,\overline{S} \text{ but see the note on p. 79 following} \tag{3.2.8b}$$

$$w_2 \quad = \lambda(1-p)\,\overline{M}/3 \tag{3.2.9a}$$

$$M_V{}^+ = \lambda(1-p)\,\overline{M} \tag{3.2.9b}$$

$$w_3 \quad = \lambda(1-p)\,4\pi/3 \tag{3.2.10a}$$

$$N_V{}^+ = \lambda(1-p). \tag{3.2.10b}$$

Here S_V is the mean surface area per unit volume, $M_V{}^+$ is the *specific positive mean curvature*, and $N_V{}^+$ is the *specific convexity number*.

Note: In the case $p = 0$ (for example in the case of a segment process) formula (3.2.5b) should be

$$2L_A = \lambda \overline{U} \text{ and } 2S_V = \lambda \overline{S}$$

Corresponding formulae for the *specific connectivity numbers* N_A and N_V and the *specific mean curvature* M_V are much more complicated (these quantities are discussed in Section 7.3). As shown by MILES (1976) they are given by

$$N_A = \lambda \cdot (1 - p) \cdot [1 - \lambda \overline{U}^2/(4\pi)]$$

$$M_V = \lambda \cdot (1 - p) \cdot [\overline{M} - \pi^2 \lambda \overline{S}^2/32]$$

$$N_V = \lambda \cdot (1 - p) \cdot [1 - \lambda \overline{M}\,\overline{S}/(4\pi) + \pi^2 \lambda^2 \overline{S}^3/96].$$

On the proof of (3.2.8b) and (3.2.6b): Formula (3.2.8b) is a direct consequence of (3.1.9). This is because (3.1.9) implies that for every primary grain the ratio of 'mean surface area not covered by other primary grains' to ,total mean surface area of grain' is $1 - p$. Summing this relation over all primary grains yields

$$S_V = \lambda(1 - p)\,\overline{S}.$$

In the case of (3.2.6b) the definition of N_A^+ must be recalled; see Section 7.3. Using a Cartesian co-ordinate system for $\mathbb{R}^2$ a special point m_n may be defined for each primary grain Ξ_n; m_n is defined to be the point minimal in Ξ_n with respect to the lexicographical ordering using the co-ordinate system. If new grains $\Xi_n^* = \Xi_n + m_n$ are defined then Ξ^* formed by the system of the grains $\{\Xi_n^*\}$ and the original germs $\{x_1, x_2, \ldots\}$ is still a Boolean model, with the same distribution as Ξ. This follows from remark (3) on page 71.

By its definition the quantity N_A^+ is precisely the proportion of primary grains Ξ_n^* whose germs x_n are covered by the union of the remaining grains. But Ξ^* has the same distribution as Ξ, and (3.1.8) indicates that in the case of Ξ the corresponding proportion is p.

Modified versions of formulae (3.2.4) to (3.2.10b), and of the formulae for N_A and N_V above, hold in the case of non-convex grains. The modifications are that $\overline{M}^+$ must replace $\overline{M}$, and in (3.2.6b) and (3.2.10b) the right-hand side must become

$$\lambda(1 - p)\,\overline{\chi}^+.$$

Here $\overline{M}^+$, $\overline{\chi}^+$ are means of the quantities M^+ and χ^+ discussed in section 1.7.

In KELLERER (1983) the connectivity number of $\Xi \cap F$ for a given test set F is given, for the planar case. In this reference simulation is used to determine the number of clumps (connected aggregates of primary grains) and voids (as indicated in Figure 3.1) for circular primary grains in a rectangle. Approximations are given for the mean number N_c of clumps per unit area. (Note that the mean number of voids per unit area is $N_c - N_A$.)

Contact distribution functions

By means of (3.2.1) the formula (3.1.7) can be simplified for convex structuring elements. It becomes

$$H_B(r) = 1 - \exp\left[-\frac{\lambda}{\omega_d}\sum_{k=1}^{d}\binom{d}{k} r^k \overline{W}_k W_{d-k}(B)\right] \quad \text{for } r \geqq 0. \quad (3.2.11)$$

Note that the $k = 0$ term of the sum cancels with the $-\lambda \mathsf{E}\nu_d(\Xi_0)$ term of the exponent in (3.1.7).

The important planar and spatial cases are as follows:

Planar case ($d = 2$),

$$\begin{aligned} H_B(r) &= 1 - \exp\left\{-\lambda\left[\frac{\overline{U}U(B)}{2\pi} r + A(B)\, r^2\right]\right\} \\ &= 1 - \exp\left\{-\frac{1}{1-A_A}\left[\frac{L_A U(B)}{2\pi} r + N_A{}^+ A(B)\, r^2\right]\right\} \quad \text{for } r \geqq 0 \end{aligned}$$

Spatial case ($d = 3$),

$$\begin{aligned} H_B(r) &= 1 - \exp\left\{-\lambda\left[\frac{\overline{S}M(B)}{4\pi} r + \frac{\overline{M}S(B)}{4\pi} r^2 + V(B)\, r^3\right]\right\} \\ &= 1 - \exp\left\{-\frac{1}{1-V_V}\left[\frac{S_V M(B)}{4\pi} r + \frac{M_V{}^+ S(B)}{4\pi} r^2 \right.\right. \\ &\qquad \left.\left. + N_V{}^+ V(B)\, r^3\right]\right\} \quad \text{for } r \geqq 0. \end{aligned}$$

Linear contact distribution function H_l

$$H_l(r) = 1 - \exp\left[-\lambda \frac{\overline{W}_1 \omega_{d-1}}{\omega_d} r\right] = 1 - \exp\left[-\lambda_1{}^{(d)} r\right]. \quad (3.2.12)$$

Note that this is an exponential distribution of parameter $\lambda_1{}^{(d)}$. The parameter $\lambda_1{}^{(d)}$ is defined by its occurrence here, and also by (3.2.14).

Example. Range of vision in a forest (after Polya, 1918) In a forest the set of tree cross sections can be interpreted as a Boolean model realization, with circular primary grains. If the mean radius of a primary grain is m_R then

$$H_l(r) = 1 - \exp\left[-2\lambda m_R r\right] \quad \text{for } r \geqq 0.$$

Taking for example the values $m_R = 0.20$ m and $\lambda = 0.01$ m^{-2} then

$$H_l(r) = 1 - \exp\left(-0.004r\right)$$

and hence $H_l(500) = 0.135$.

That is to say that in a forest which can be described by such a Boolean model, with parameters as above, an observer placed at some 'typical' point and looking in a 'typical' direction would have a probability of 13.5% of being able to see more than 500 m. Further visibility properties are summarized in the *star* which is the set

$$S_x(\Xi) = \{y \in \mathbb{R}^d : [x, y] \cap \Xi \text{ is empty}\}$$

defined for x not in Ξ. See SERRA (1982), WIEACKER (1986), and YADIN and ZACKS (1985).

Spherical contact distribution function H_s

$$H_s(r) = 1 - \exp\left[-\lambda \sum_{k=1}^{d} \binom{d}{k} \overline{W}_k r^k\right] \quad \text{for } r \geqq 0 \tag{3.2.13}$$

Example. Consider a Boolean model with spherical grains in three-space: random spheres whose centres are distributed by a Poisson process in $\mathbb{R}^3$. The contact spherical hit distribution function is the distribution function of the distance to the nearest sphere from a point chosen at random but lying outside all spheres. If $m_R{}^{(k)}$ is the kth moment of the radius distribution function then

$$H_s(r) = 1 - \exp\left\{-\frac{4}{3}\pi\lambda r[3m_R{}^{(2)} + 3m_R{}^{(1)}r + r^2]\right\}.$$

Intersections with linear subspaces

Let L be a linear subspace or linear variety of $\mathbb{R}^d$ (so not necessarily passing through the origin). For Ξ a Boolean model in $\mathbb{R}^d$ consider the intersection $\Xi_L = \Xi \cap L$. Then Ξ_L is again a Boolean model on L, when L is identified with $\mathbb{R}^e$ for e the dimension of L. That Ξ_L is a Boolean model follows from considering its capacity functional. If the primary grains of Ξ are convex then so of course are those of Ξ_L.

In the case of isotropy, when Ξ has convex primary grains, the induced Boolean model Ξ_L has the following characteristics:

the intensity of its germ process is $\lambda_e{}^{(d)}$ and

$$\lambda_e{}^{(d)} = \lambda \frac{\omega_{e-n}}{\omega_d} \overline{W}_e \tag{3.2.14}$$

and the mean value $\overline{W}_k{}^{(e)}$ of the kth Minkowski functional of the primary grain is

$$\overline{W}_k{}^{(e)} \doteq \frac{\omega_e \omega_{d-k}}{\omega_{e-k}\omega_{d-e}} \frac{\overline{W}_k}{\overline{W}_e} \quad \text{for } e \geqq k. \tag{3.2.15}$$

Proofs of these statements may be found in MATHERON (1975, p. 146), where there is also a discussion of second moments and the anisotropic case.

In the particular case of intersection of a planar Boolean model with a line the proof of (3.2.14) and (3.2.15) is as follows. Suppose Ξ is a Boolean model in the plane which is isotropic and has convex grains. Let L be a line and let K be an arbitrary compact convex subset of L (that is, a bounded interval in L) of linear measure $l(K)$.

Clearly $\mathsf{P}(\Xi \cap K$ is empty$) = \mathsf{P}(\Xi_L \cap K$ is empty$)$ thus $\psi(K) = \psi_L(K)$, which relates the capacity functional of Ξ_L to that of Ξ.

Let $\lambda_1^{(2)}$ be the intensity of the germ process of Ξ_L. Then

$$\psi_L(K) = \lambda_1^{(2)}[l(K) + \bar{l}_1]$$

where $\bar{l}_1$ is the mean length of segments of Ξ_L. This follows from (3.1.3) and the convexity both of K and of the primary grains of Ξ_L.

On the other hand

$$\psi(K) = \lambda \mathsf{E} A(\Xi_0 \oplus K)$$
$$= \lambda \left[\bar{A} + \frac{\bar{U}}{\pi} l(K)\right]$$

from the planar case ($d = 2$) of (3.2.1). Equating coefficients of $l(K)$ it follows that

$$\lambda_1^{(2)} = \lambda \frac{\bar{U}}{\pi} = \lambda \frac{\omega_1}{\omega_2} \overline{W}_1 \quad \text{as } \bar{U} = 2\overline{W}_1,$$

$$\bar{l}_1 = \lambda \frac{\bar{A}}{\lambda_1^{(2)}} = \frac{\omega_2}{\omega_1} \cdot \frac{\overline{W}_0}{\overline{W}_1} \quad \text{as } \bar{A} = \overline{W}_0.$$

Distribution function of chord lengths outside Ξ

Let Ξ be a general Boolean model of convex grains. We consider the special case of the previous section, when Ξ is intersected with a line L. The intersection $\Xi_L = \Xi \cap L$ yields two interwoven sequences of chords or intervals on L; one sequence forms Ξ_L and the other the exterior $L \setminus \Xi_L$. The chord lengths of the exterior sequence form a sequence of independent and identically distributed random variables of exponential distribution. If furthermore the Boolean model Ξ is isotropic then the parameter of this exponential distribution is $\lambda_1^{(d)}$, where

$$\lambda_1^{(d)} = \lambda \frac{\omega_{d-1}}{\omega_d} \overline{W}_1.$$

Clearly in the anisotropic case the parameter must depend on the direction of the line.

The interior chord lengths from Ξ_L have a complicated distribution depending on the distribution of Ξ_0 the typical primary grain. The sequence of interior chord lengths are, however, independent and identically distributed as long as the primary grains are convex.

The simple exponential form of the linear contact distribution function (3.2.12) for convex grains can be seen as a direct consequence of the properties described above of the chord lengths of the exterior sequence.

Formulae for some special Boolean models with isotropic convex grains

Two-dimensional models

(1) Primary grains are *random segments* with uniform (isotropic) orientation and of mean length m_1.

In this case the linear contact distribution function can be found from the above paragraphs and

$$\lambda_1^{(2)} = 2\lambda m_1/\pi.$$

The spherical contact distribution function is

$$H_s(r) = 1 - \exp\{-\lambda r(2m_1 + \pi r)\} \quad \text{for } r \geqq 0.$$

(2) Primary grains are *random discs* with radius distribution function F_R and moments $m_R^{(k)}$ for $k = 1, 2, \ldots$.

The area fraction is given by

$$p = 1 - \exp\{-\lambda\pi m_R^{(2)}\}$$

moreover

$$\lambda_1^{(2)} = 2\lambda m_R^{(1)}$$

$$H_s(r) = 1 - \exp\{-\lambda\pi r(2m_R^{(1)} + r)\} \quad \text{for } r \geqq 0$$

and the covariance function is given by

$$C(r) = 2p - 1 + (1-p)^2 \exp\left\{\lambda \int_{r/2}^{\infty} \left[2x^2 \arc\cos \frac{r}{2x} - \frac{r}{2}\sqrt{(4x^2 - r^2)}\right] \mathrm{d}F_R(x)\right\}.$$

Three-dimensional models

(3) Primary grains are *random segments* with uniform (isotropic) orientations of mean length m_1.

$$H_s(r) = 1 - \exp\{-\lambda\pi r^2(m_1 + 4r/3)\} \quad \text{for } r \geqq 0.$$

(4) Primary grains are *random circular discs* with uniform (isotropic) orientation and radius distribution moments $m_R^{(k)}$ for $k = 1, 2, \ldots$.

$$\lambda_1^{(3)} = \lambda\pi m_R^{(2)}/2$$

$$H_s(r) = 1 - \exp\{-\lambda\pi r(2m_R^{(2)} + \pi m_R^{(1)} r + 4r^2/3)\} \quad \text{for } r \geqq 0.$$

(5) Primary grains are *random spheres* with radius distribution function F_R and radius distribution moments $m_R^{(k)}$ for $k = 1, 2, \ldots$.

$$p = 1 - \exp\{-(4/3) \cdot \lambda\pi m_R^{(3)}\}$$

$$\lambda_1^{(3)} = \lambda\pi m_R^{(1)}$$

$$H_s(r) = 1 - \exp\{-\lambda\pi r(4m_R^{(2)} + 4m_R^{(1)}r + 4r^2/3)\} \quad \text{for } r \geqq 0$$

$$C(r) = 2p - 1 + (1-p)^2 \exp\left[\lambda \frac{4}{3}\pi \int_{r/2}^{\infty} x^3 \left(1 - \frac{3r}{4x} + \frac{r^3}{16x^3}\right) \mathrm{d}F_R(x)\right]$$

for $r \geqq 0$.

(6) Primary grains are *Poisson polyhedra* with parameter ϱ (see Section 10.5.2). The expected volume of a primary grain is then

$$\overline{V} = \frac{6}{\pi\varrho^3}.$$

The other formulae are

$$p = 1 - \exp\{-6\lambda/(\pi\varrho^3)\}$$

$$\lambda_1^{(3)} = 6\lambda/(\pi\varrho^2)$$

$$H_s(r) = 1 - \exp\left\{-\lambda r\left[\frac{24}{\pi\varrho^2} + \frac{3\pi}{\varrho} r + \frac{4}{3}\pi r^2\right]\right\} \quad \text{for } r \geqq 0$$

$$C(r) = 2p - 1 + (1-p)^2 \exp\left[\frac{6\lambda \exp(-\varrho r)}{\pi\varrho^3}\right] \quad \text{for } r \geqq 0.$$

Stoyan (1979a) considered some qualitive properties of the covariance function C. For example, C varies continuously with the primary grain Ξ_0, in the sense that C only varies by a small amount if the primary grain is varied by a small amount in the sense of the Hausdorff metric. Moreover, C possesses a kind of monotonicity property; among Boolean models possessing primary grains of equal shape and mean volume, a more variable size distribution leads to a larger C.

3.3 Statistics for the Boolean model

Determination of basic characteristics; general remarks

The first step in statistical analysis for the Boolean model follows that for general random closed sets; estimators $\hat{p}$, $\hat{C}$, $\hat{H}_B$ must be obtained for the basic characteristics of volume or area fraction p, covariance C, and the contact distribution function H_B for suitable test sets (also called *structuring elements*) B. Methods of determining such estimates apply to all random closed sets; therefore they are described not here but in Section 6.3.

A main objective of statistical analysis for the Boolean model is the determination of characteristics associated with the model itself such as λ, $\overline{U}$, $\overline{A}$, $\overline{S}$, and $\overline{V}$. A further and very important problem concerns the decision as to whether or not the Boolean model is appropriate in discussing a particular data-set.

These problems present less difficulty if it is possible to discriminate clearly between primary grains. Such a discrimination can be carried out, albeit with some difficulty, for the empirical set presented in Figure 3.6. The primary germs can then be defined in some systematic fashion, leading to separate studies of the primary grains and the point-pattern of germs. Study of the grains is a problem in random set statistics while study of the germs involves point-process statistics.

Figure 3.2 is a two-component system or 'black-white structure'. In two-component systems it is often not clear which component should be interpreted as the realization of a Boolean model. It may then be necessary to perform some steps of statistical analysis in parallel, comparing results obtained from assuming first that one and then that the other component forms a Boolean model.

A final decision in favour of one component will often result from considering an empirical contact distribution function $\hat{H}_B$.

Testing the model assumptions

Whether or not the Boolean model assumptions are appropriate is evaluated mainly by considering empirical contact distribution functions $\hat{H}_B$; see SERRA (1982), OHSER (1980), OHSER and STOYAN (1980), and DIGGLE (1981). This is because formulae for H_B are relatively simple in the case of convex isotropic primary grains, and this enables comparison between empirical and theoretical contact distribution functions to be made. Furthermore, variation in the shape of B allows different geometrical aspects of the data to be studied and taken into consideration. Finally, the exponential form of H_B, as seen in (3.1.7) and (3.2.11), is typical for the Boolean model and leads to straightforward tests.

However, a strong limitation is inherent in the use of contact distribution functions. Their empirical evaluation is practicable only when Ξ (or B) is planar or linear. Stereological questions intervene when the structure under investigation is three-dimensional. For this reason it is important that the planar or linear intersection of a Boolean model with convex grains is again a Boolean model with convex grains; see p. 81 for a general discussion that includes this as a special case. MILES (1976) shows that the same is true for a projection of a thick slice of a Boolean model.

In the following discussion the Boolean model is always assumed to be planar, with convex primary grains.

As a result of formula (3.2.11) the logarithm of $1 - H_B(r)$ must be a polynomial of the form:

$$\ln\big(1 - H_B(r)\big) = -a(B)\,r - b(B)\,r^2 \quad \text{for } r \geqq 0,\ a(B), b(B) \geqq 0 \tag{3.3.1}$$

for every convex B. Consequently the following test suggests itself. For various shapes of B the empirical contact distribution function $\hat{H}_B(r)$ is computed from the given image. If the logarithm

$$\ln\big(1 - \hat{H}_B(r)\big)$$

is well approximated by quadratic polynomials as in (3.3.1), for each B considered, then the Boolean model can be considered not inappropriate as a model to fit the data. Ripley (1985a) suggested the use of functions similar to H_B but derived from *dilatation*, *opening*, and *closing* the random set (a brief discussion of these operations is given in Section 1.4).

It should be noted that the regression problem implied in the above is by no means trivial. The regression is non-linear; the error structure is different for different r; and indeed the empirical contact distribution function $\hat{H}_B(r)$ clearly exhibits correlation as between different values of r, since evaluations at different r values are still calculated from the same sample.

The use of the covariance rather than the contact distribution function is much more complicated. This is because of the complexity of formulae for the function $\gamma_{\Xi_0}(r)$ that appears in (3.1.5).

With estimates for the parameters of the model having been obtained by some means, simulation-based tests can be carried out of goodness-of-fit. An example of such a procedure can be found in Diggle (1981).

Example 3.1. *A pattern of lichen (genus Placodium) on a stone.*

The pattern is illustrated in Figure 3.4. Estimation of parameters of the pattern of lichen was carried out using a Quantimet QTM 720. Area fraction and mean boundary length per unit area were estimated as

$$\hat{p} = 0.35$$

$$\hat{L}_A = 0.011\ \text{mm/mm}^2.$$

The mean chord length $\bar{l}$ in the area not occupied by the lichen was estimated as

$$\bar{l} = 90.5\ \text{mm}.$$

The question is, whether a Boolean model with convex isotropic grains provides a good fit for these data.

Of course, in a strict sense such a model is implausible because of the irregularity of the boundaries of the regions occupied by the lichen. However, if only the large-scale features of the pattern are considered then the model makes some sense: visual inspection suggests the suitability of a model involving grains.

Figure 3.5 graphs the logarithms of the empirical contact distribution functions based on the unit segment and on the unit square

$$-\frac{1}{r}\ln(1 - \hat{H}_l(r))$$

and

$$-\frac{1}{r}\ln(1 - \hat{H}_Q(r)).$$

The linear contact distribution function has theoretical form derived (via (3.2.12)) from

$$H_l(r) = 1 - \exp(-\lambda_1^{(2)} r)$$

where $\lambda_1^{(2)} = \lambda \bar{U}/\pi$ and $\bar{U}$ is the mean perimeter of the primary grains. (It should be recalled that the form of the distribution of chord lengths outside a Boolean model does not depend on the grain distribution in the case of convex grains.) The data points in Figure 3.5 are fitted well by a horizontal line.

Thus the empirical linear contact distribution function has the form expected of a Boolean model. The fitted value of $1/\lambda_1^{(2)}$ is 125.0; the difference between this and the value 90.5 for mean-chord-length could be a consequence of the roughness of the grain boundaries and the limitations imposed by the restricted window of observation.

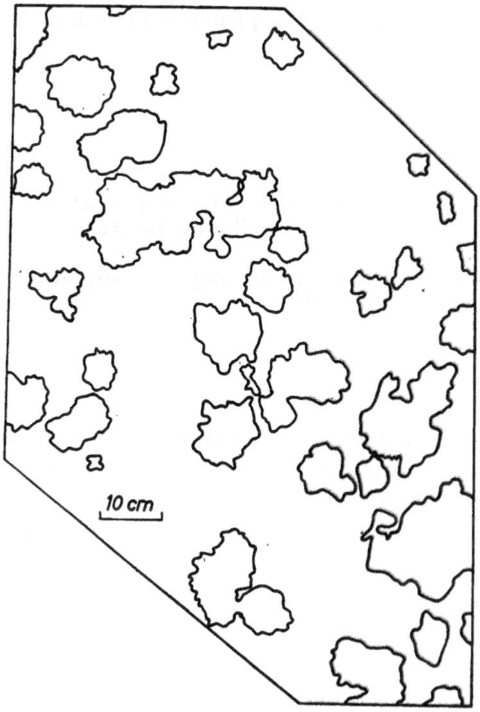

Fig. 3.4. A pattern of lichen on a stone. It is similar to a sample from a Boolean model with convex grains except that the boundaries are very rough

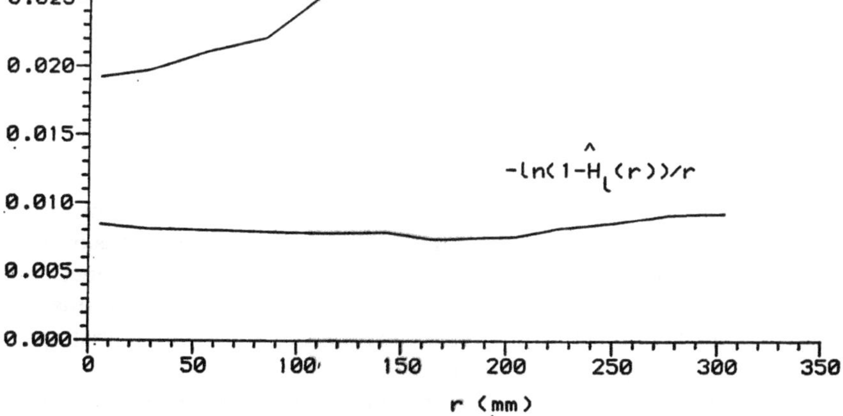

Fig. 3.5. Graphs of the logarithms of the empirical linear and quadrat contact functions for the lichen pattern of Figure 3.4

The empirical contact distribution function $\hat{H}_Q$ using the quadrat is fitted by a line

$$y = c_0 + c_1 r$$

where

$$c_0 = -\hat{a}(Q) = 0.0109$$
$$c_1 = -\hat{b}(Q) = 0.00015.$$

Here the fit is not good. However, further comparison of $\hat{H}_Q$ and $1 - \exp(-c_0 r - c_1 r^2)$ shows that the accuracy of the fit to the actual contact distribution function is more satisfactory.

This test procedure suggests that the pattern in Figure 3.4 may be interpreted, at least on a large-scale approximation, as a Boolean model.

Estimation of model parameters

The parameters of a Boolean model with convex primary grains may be estimated via the analysis of intensities, or of empirical contact distribution functions, or of covariances. Each of these methods is discussed in turn.

The intensities $A_A = p$, L_A, and N_A^+ are connected with the model parameters by three equations:

$$A_A = 1 - \exp(-\lambda \bar{A})$$
$$L_A = \lambda(1 - A_A)\bar{U}$$
$$N_A^+ = \lambda(1 - A_A).$$

These equations correspond to the formulae (3.2.4) to (3.2.6b). By an application of the method of moments they give means of estimating the three parameters λ, $\bar{U}$, and $\bar{A}$ see Santalo (1976) p. 282. Kellerer (1985, 1986) used simulation to consider corresponding estimation variances.

Empirical contact distribution functions can be used as follows. From formula (3.3.1) it is seen that $\ln(1 - H_B(r))$ is quadratic in r with coefficients related to $\bar{U}$ and $\bar{A}$. Thus for the linear contact distribution function

$$a(B) = \lambda \bar{U}/\pi$$
$$b(B) = 0$$

and for the spherical contact distribution function

$$a(B) = \lambda \bar{U}$$
$$b(B) = \lambda \pi.$$

In a further step, one can attempt to transfer this information into a size distribution for the primary grains, on the basis of some shape assumption, and then use the least-squares method for fitting a theoretical function to observed values. Often it makes sense to assume circular primary grains; this assumption is made in most of the papers cited above and such an assumption seems a good starting point in Example 3.1.

Knowledge of $\bar{U}$ and $\bar{A}$ then enables determination of the mean and variance of the radius distributions for the primary grains. If a family of theoretical radius distributions has been assumed (for example, Gamma or Weibull) then estimates of the mean and variance help to determine a suitable member of the family. DUPAČ (1980) presented approximations for estimator variances in the case of Gaussian radius distribution.

Another possible shape assumption supposes the primary grains to be Poisson polyhedra as defined in 10.5.2; see also 84. In such a case estimation of just one of $\overline{U}$ or $\overline{A}$ suffices in order to estimate the parameter ϱ.

The final method uses the covariance in order to estimate λ and to make inference concerning the primary grains; see SERRA (1980). Sometimes no other information on the Boolean model is available and this method is the only possibility. Example 3.3 supplies a demonstration. Model parameters can as before be determined by the use of least-squares.

MASE (1982a, b) studied asymptotic properties and in particular asymptotic normality of estimators for volume fractions. The results can be used to obtain properties of estimators for p, C, et cetera, for the Boolean model.

Example 3.1 *(continued)*

Known data for the sample are

$$\lambda_1^{(2)} = 0.008$$

$$c_0 = 0.0109$$

$$c_1 = 0.00015$$

$$A_A = 0.352$$

$$L_A = 0.011$$

with units of mm^{-1} or mm^{-2} as appropriate.

Approximate values of λ, $\overline{U}$, and $\overline{A}$ fitting these data are

$$\lambda = 0.00015\ mm^{-2}$$

$$\overline{U} = 150\ mm$$

$$\overline{A} = 3000\ mm^2.$$

For the case of circular primary grains the first two moments of the radius distribution will then be

$$m_R = 23.9\ mm$$

$$m_R^{(2)} = 955\ mm^2$$

Example 3.2 (LIESENBERG and OHSER, 1983). *Lamellar graphite in cast iron.*

Figure 3.6 shows a pattern of lamellar graphite in a polished section of cast iron. The graphite particles are distributed so irregularly that it seems natural to assume a Boolean model. Figure 3.7 shows the corresponding empirical linear and rectangular contact distribution functions in logarithmic form. The rectangle W here has dimensions $[0, 1] \times [0, 0.75]$ (in mm). The empirical contact distribution functions were obtained by using a QTM image analyser.

As an consequence of this analysis Ohser concluded that the planar intersection figure, and also the original spatial arrangement of graphite particles, could both be described by Boolean models. Figure 3.8 shows a simulation of a Boolean model with segments as primary grains and with parameters fitted to the data provided. Broadly speaking there is a similarity between Figures 3.6 and 3.8.

However a closer examination reveals a distinction. Neighbouring lamellae in Figure 3.6 are apparently correlated both in orientation and in length. In particular, lamellae tend not to intersect each other. Thus it is clear that there are interactions between the lamellae. Nevertheless, the relatively good fit of the Boolean model as an approximation leads to the tentative conclusion that the lamellae to some extent form independently from each other, originating from randomly placed locations. Presumably the interactions occur in some growth phase of the lamellae.

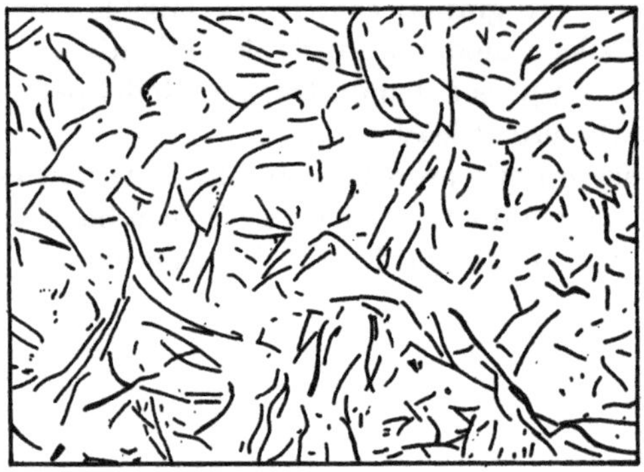

Fig. 3.6. A portion of a probe of cast iron (using a polished section). The black strips are sections of lamellar graphite bodies. The enlargement of the picture is about 60 times

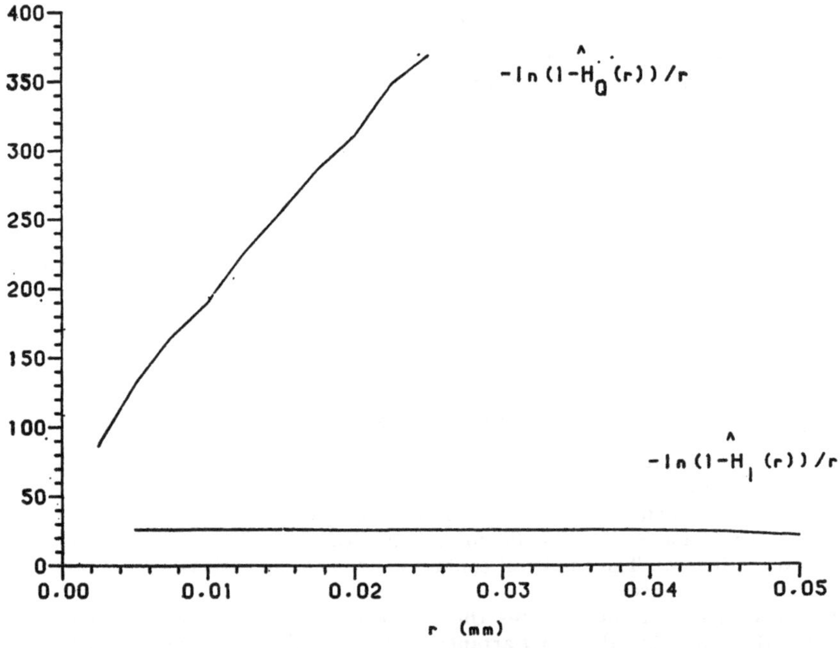

Fig. 3.7. Graphs of the logarithms of the empirical linear and quadrat contact functions for the lamellar graphite pattern of Figure 3.6. Corrections for edge effects have been applied

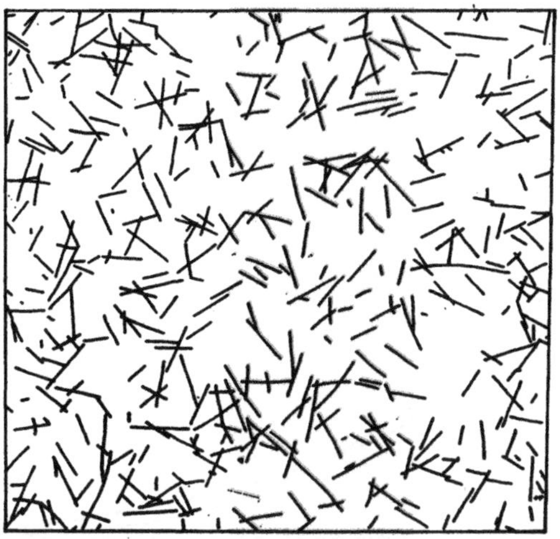

Fig. 3.8. A simulated Boolean model, using segments as primary grains with length distribution a Weibull distribution. The parameters of the Weibull distribution were fitted to the pattern of Figure 3.6

Example 3.3 (SONNTAG, STOYAN, and HERMANN, 1981). *X-ray analysis of inhomogeneities in an amorphous body.*

Small-angle scattering yields information on random three-dimensional structures at the scale of 1 to 100 nm. The small-angle intensity $I(s)$ is connected with the correlation function k by

$$k(r) = (2\pi^2 r)^{-1} \int_0^\infty sI(s) \sin(sr)\, ds/(\varrho_e[p - \varrho_e]^2)$$

Here $s = 4\pi \sin(\theta/\lambda)$, 2θ is the scattering angle, λ is the wavelength, and ϱ_e is the electron-density in the system Ξ of scattering regions; see HOSEMANN and BAGCHI (1962) and GLATTER and KRATKY (1982).

Figure 3.9 shows the experimental correlation function $\hat{k}$ obtained from small-angle scattering from a $Co_{75}Fe_5B_{20}$ alloy. One possible model for the set Ξ of Co precipitate is a Boolean model with Poisson polyhedra with parameter ϱ as primary grains. The model correlation function is k, where

$$k(r) = \{\exp\{6\lambda[\exp(-\pi\varrho r) - 1]/(\pi\varrho^3)\} - 1 + p\}/p \quad \text{for } r \geqq 0$$

Fitting this to the experimental correlation function $\hat{k}$, via minimisation of

$$\sigma^2 = \sum_{i=1}^{N} [\hat{k}(r_i) - k(r_i)]^2$$

using a Marquadt procedure yields estimates

$$\hat{p} = 0.1$$

$$\hat{\varrho} = 0.027 \text{ nm}^{-1}.$$

Here the r_i are equidistant values of r for which the experimental correlation function has been calculated from the small-angle scattering data. The second curve on Figure 3.9 is the model correlation function k for these parameters.

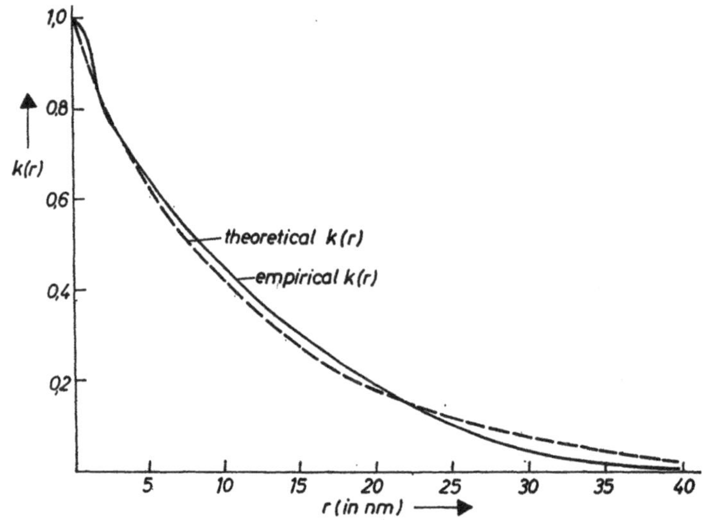

Fig. 3.9. The empirical correlation function for Co-precipitate in a $Co_{75}Fe_5B_{20}$ alloy. The broken line indicates the correlation function of a Boolean model with Poisson polyhedra as primary grains, in which the parameters of the Poisson polyhedra have been fitted by a least-squares method

3.4 Generalizations and variations of the Boolean model

Some generalizations of the Boolean model may be of use for description of structures of greater complexity. A straightforward generalization involves the weakening of model assumptions. It seems especially useful to suspend the assumption that the germs constitute a Poisson process. Random structures that are essentially Boolean models, but that use general point processes to provide the germs, are called germ-grain models. They are studied in Chapter 6. Such models may display a greater or a lesser irregularity of structure than the Boolean model, according to the particular properties of the germ process.

Two independent Boolean models $\Xi^{(1)}$ and $\Xi^{(2)}$ considered together can be used to produce patterns of greater irregularity. Three variations give the general idea. Proofs of the formulae are left to the reader as an exercise.

Variation 1. Suppose that $\Xi^{(1)}$ and $\Xi^{(2)}$ have volume fractions $p^{(1)}$ and $p^{(2)}$ and covariance functions $C^{(1)}$ and $C^{(2)}$. Then the intersection $\Xi^{(i)} = \Xi^{(1)} \cap \Xi^{(2)}$ is a closed random set (see Figure 3.10 (a) for an illustration). Its basic characteristics are: volume fraction $p = p^{(1)} p^{(2)}$, covariance function $C(r) = C^{(1)}(r) \times C^{(2)}(r)$ for $r \geqq 0$, and capacity functional $T(K) = T^{(1)}(K)\, T^{(2)}(K)$ for K compact.

Variation 2. Another closed set which can be built from $\Xi^{(1)}$ and $\Xi^{(2)}$ is

$$\Xi^{(c)} = \Xi^{(1)} \cap \left((\Xi^{(2)})^c\right)^{cl}.$$

Here $\left((\Xi^{(2)})^c\right)^{cl}$ is the closure of the complement of $\Xi^{(2)}$.

The basic characteristics of $\Xi^{(c)}$ are: volume fraction $p = p^{(1)}(1 - p^{(2)})$, and covariance function $C(r) = C^{(1)}(r)\,[1 - 2p^{(2)} + C^{(2)}(r)]$ for $r \geqq 0$, and capacity functional $T(K) = T^{(1)}(K)\,[1 - T^{(2)}(K)]$ for compact K. This is illustrated in Figure 3.10(b).

In a similar way, but departing from the province of random closed sets pure and simple, models can be constructed for multicomponent structures. A simple example is the three-component model below, studied by Greco, Jeulin, and Serra (1979) and by Serra (1982).

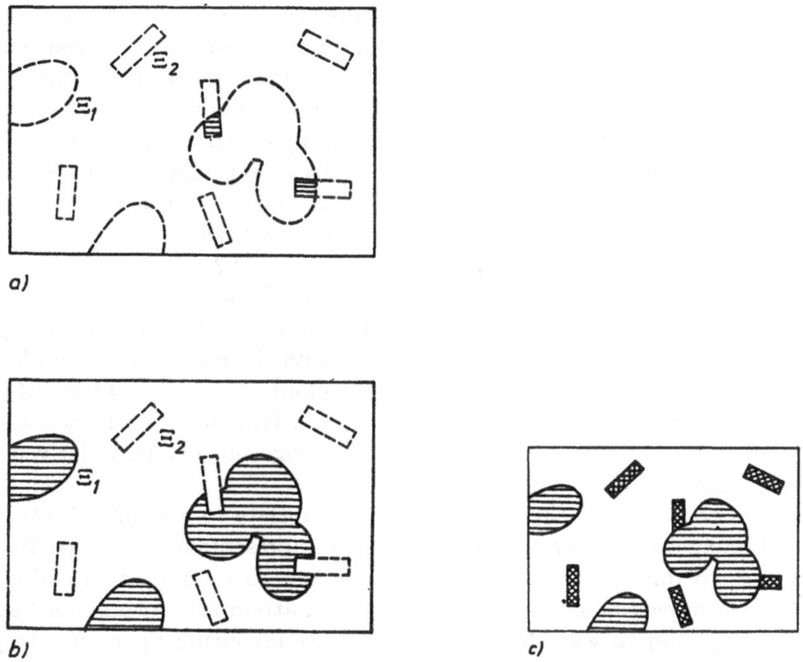

Fig. 3.10. Three variations on the basic Boolean model: (a) Intersection of two Boolean models; (b) Intersection of a Boolean model with the complement of another; (c) A simple three-phase model

Variation 3. The sets $\Xi^{(1)}$ and $\Xi^{(2)}$ may be combined to produce a pattern as in Figure 3.10(c). The complete Boolean model $\Xi^{(1)}$ forms component 1. Component 2 is made up of the set-difference $\Xi^{(2)} \setminus \Xi^{(1)}$; it can be interpreted as a pattern destroyed in part by $\Xi^{(1)}$. The remainder, $(\Xi^{(1)} \cup \Xi^{(2)})^c$, is component 3. This model has been used for the description of sinter metals. In such a case component 1 models the haematites, component 2 the slag, and component 3 the calcium ferrites.

A further model closely related to the Boolean model is Matheron's 'dead-leaves' model; see Serra (1982). Finally, there are models of rough surfaces that are very similar to the Boolean model; see some papers in Kališnik (1981).

3.5 A critical assessment of Boolean models

To close this chapter we make a few remarks about difficulties and intrinsic limitations of the Boolean model.

Interpretation

In many cases, the suggestive nature of the germ process ('germs' as origins of growth, or locations of physical entities) will not coincide with physical reality. This is not strictly speaking a disadvantage of the Boolean model, but more a tendency to be resisted in its interpretation. The Boolean model may fit a dataset to a greater or lesser degree, but that of itself does not justify a physical interpretation of the germ/grain mechanism.

Two kinds of interpretation can be distinguished. In the first, which might be called the *genetic* interpretation, the individual primary grains of the Boolean model correspond to physically identifiable entities. For example there are the early applications concerning particle counting cited in 3.1, and also the application to lamellar graphite discussed as Example 3.2. The second interpretation, the *representation* interpretation, scrupulously avoids drawing out such a correspondence but rather uses the Boolean model as a moderately flexible means of producing patterns to compare with the image. An example is Example 3.1 on lichen: it is tempting here to attempt to identify grains of the Boolean model with individual lichen entities but in fact such an identification is not made, nor can it be substantiated without rigorous evidence derived from the botanical context.

The point here is that conclusions appropriate to a genetical interpretation will be misleading, even nonsensical, in a representation context. In an application to dust-particle counting clearly one is interested in estimating the number of particles. Conceivably one might attempt to estimate lichen 'growth sites' in a similar way: but without evidence favouring a genetic interpretation such an attempt is ill-judged. However, the more restrictive representation interpretation provides a possible description of the dataset as an image.

Interaction between grains

The previous section has already introduced the generalization of replacing the Poisson germ process by a more general point process. In many cases, the assumption of independence between the grains is implausible. Consider for example various packing and covering problems, where intersection of grains is prohibited. In such a case the Boolean model is clearly inadequate. In fact, there is a general absence of constructive models for such a situation (however, germ-grain models can be used in a descriptive manner; see Chapter 6.).

When some interaction between the grains is merely a possibility to be considered the Boolean model can serve as a convenient benchmark. With it as a comparison the strength of the interaction can be assessed. This is illustrated in Example 3.2, in which the general absence of intersections between the lamellae (relative to the Boolean model) is evidence for an interaction between the lamellae.

Non-convex grains

Many of the useful formulae for the Boolean model arise from the case of convex grains. However in many data-sets arising in practise the grain-boundaries are clearly non-convex. Such an instance arises in Example 3.1. It is indeed the case that larger-scale features of the dataset can still be modelled by a Boolean model with convex grains; however care must be taken over the interpretation of quantities calculated from the model. See the comments on mean-chord length in Example 3.1. As shown above, it is possible to derive formulae for the case of non-convex grains.

4. Point processes II – General theory

4.1 Basic properties

Point processes provide models for irregular patterns of points. The theory was developed in response to various problems of physics, biology, and queueing theory; important pioneers were NEYMAN (1939) (as reported by NEYMAN and SCOTT, 1972), BARTLETT (1954), KHINTCHINE (1955), COX (1955), MATERN (1960) and PALM (1943). The modern phase of the theory was developed from 1960 to 1975 by Kallenberg, Kerstan, Krickeberg, Matthes, and Ryll-Nardzewski. Books and surveys on the subject include: BRILLINGER (1978), COX and ISHAM (1980), DALEY and VERE-JONES (1972, 1986), DALEY and MILNE (1973), KALLENBERG (1976, 1983b), KRICKEBERG (1982), MATTHES, KERSTAN and MECKE (1978), and SNYDER (1975). This chapter presents the general theory, which is a development of the special case of the Poisson process that was discussed in Chapter 2. Chapter 5 presents some special point process models.

The mathematical definition of a point process on $\mathbb{R}^d$ is as a random variable taking values in a measurable space $[\mathsf{N}, \mathcal{N}]$, where N is the family of all subsets φ of $\mathbb{R}^d$ satisfying the following two regularity conditions. These conditions are (i) that the set φ is *locally finite* (each bounded subset of $\mathbb{R}^d$ must contain only a finite number of the x_n), and (ii) that the set is *simple* (so $x_i \neq x_j$ if $i \neq j$). The condition (ii) of simplicity is relaxed in some extensions of the theory. For convenience we somtimes write $\varphi = \{x_n\}$.

If both conditions above are enforced then each φ in N can be regarded as a closed subset of $\mathbb{R}^d$. An element φ of N can also be regarded as a measure on $\mathbb{R}^d$ so that $\varphi(B)$ is the number of points of φ in B. The σ-algebra $\mathcal{N}$ is defined as the smallest σ-algebra on N to make measurable all mappings $\varphi \to \varphi(B)$ (for B running through the bounded Borel sets).

Thus formally a point process Φ is a measurable mapping of the probability space $[\Omega, \mathcal{A}, \mathsf{P}]$ into $[\mathsf{N}, \mathcal{N}]$. More intuitively it is a random choice of one of the φ in N. It generates a distribution on $[\mathsf{N}, \mathcal{N}]$, the *distribution* P of Φ.

Remarks. (1) The term 'random process' does not here imply a dynamic evolution over time. Historically, the first researchers into the theory had in mind random sequences of temporal events, such as instants of arrivals of calls at telephone exchanges, arrivals of customers at a queue, etc. However a notion of time is often absent in applications of point processes in $\mathbb{R}^2$ or $\mathbb{R}^3$. The phrase 'random point field' would be a more exact term. *Spatio-temporal point processes* (explicitly involving temporal as well as spatial dispersion of points) have their own theory; see CLIFF and ORD (1981).

A similar point can be made about the adjective 'stationary', as in 'stationary point processes', occurring later in the chapter. In this context 'stationary' and 'homogeneous' are equivalent terms.

(2) Point processes can be considered either as random sets of discrete points or as random measures counting the numbers of points lying in spatial regions. Corresponding to this are two different notations:

$x \in \Phi$ asserts the point x belongs to the random sequence Φ,

$\Phi(B) = n$ asserts the set B contains n points of Φ.

Since Φ is a random closed set, results from the theory of random closed sets can also be applied to point processes; see KALLENBERG (1973), RIPLEY (1976d), and BAUDIN (1984).

The distribution P of a point process

The *distribution* P of a point process Φ is determined by the probabilities

$$\begin{aligned} P(Y) &= \mathsf{P}(\Phi \in Y) \\ &= \mathsf{P}(\{\omega \in \Omega : \Phi(\omega) \in Y\}) \quad \text{for } Y \text{ in } \mathcal{N}. \end{aligned}$$

The *finite-dimensional distributions* are of particular importance. These are probabilities of the form

$$\mathsf{P}\big(\Phi(B_1) = n_1, \ldots, \Phi(B_k) = n_k\big)$$

where $B_1, \ldots, B_k$ are bounded Borel sets and $n_1, \ldots, n_k \geqq 0$. The distribution of Φ on $[\mathsf{N}, \mathcal{N}]$ is uniquely determined by the system of all these values for $k = 1, 2, \ldots$. In fact, it is determined by the subsystem for which the constituent B_i are pairwise disjoint.

A still smaller subsystem is that of the *void probabilities*

$$v_B = P\big(\{\varphi \in \mathsf{N} : \varphi(B) = 0\}\big) = \mathsf{P}\big(\Phi(B) = 0\big) = \mathsf{P}(\Phi \cap B \text{ is empty})$$

for Borel sets B. If the point process is simple, as described above, then P is determined by the system of values of v_K as K ranges through the compact sets.

This last and strongest characterization follows by interpreting Φ as a random closed set, and then applying the Choquet theorem. For details see RIPLEY (1976d), KALLENBERG (1973, 1983, Section 3.3). The system of void probabilities specifies the probabilities that points do *not* occur in a given region. This makes the characterization plausible.

An example of the use of void probabilities is the short proof of (4.4.6) below.

Notation

It has already been noted that Φ can be interpreted either as random sequence or as random measure. Correspondingly there is a variety of notation for point process formulae. The process Φ can be written as

$$\Phi = \{x_1, x_2, \ldots\} = \{x_n\}$$

which emphasizes the interpretation of Φ as a random closed set. Let f be a measurable function on $\mathbb{R}^d$. The sum of $f(x)$ over x in Φ can be written variously as

$f(x_1) + f(x_2) + \cdots$, $\sum_{x \in \Phi} f(x)$, or, as physicists would write,

$\int f(x)\, p(x)\, dx$ where $p(x) = \sum_{y \in \Phi} \delta(x - y)$ and δ is the Dirac delta function,

or $\int f(x)\, \Phi(dx)$.

The mean value of the sum is written variously as

$$\mathsf{E} \sum_{x \in \Phi} f(x), \quad \int \sum_{x \in \varphi} f(x)\, P(d\varphi) \quad \text{or} \quad \iint f(x)\, \varphi(dx)\, P(d\varphi)$$

The expression for the number of points in a set B is written as

$$\Phi(B) = \sum_{x \in \Phi} \mathbf{1}_B(x).$$

Its mean value is written as

$$\begin{aligned} \mathsf{E}\Phi(B) &= \int \varphi(B)\, P(d\varphi) = \mathsf{E} \sum_{x \in \Phi} \mathbf{1}_B(x) \\ &= \int \sum_{x \in \varphi} \mathbf{1}_B(x)\, P(d\varphi) \\ &= \iint \mathbf{1}_B(x)\, \varphi(dx)\, P(d\varphi). \end{aligned}$$

Such a plethora of notation may be confusing for the beginner but it reflects the variety of approaches to the theory.

Stationarity and Isotropy.

A point process Φ, or its distribution P, is said to be *stationary* if its statistics are invariant under translation; the processes $\Phi = \{x_n\}$ and $\Phi_x = \{x_n + x\}$ have the same distribution for all x in $\mathbb{R}^d$. So

$$\mathsf{P}(\Phi \in Y) = \mathsf{P}(\Phi_x \in Y) \tag{4.1.1}$$

for all Y in $\mathcal{N}$, all x in $\mathbb{R}^d$. If we put $Y_x = \{\varphi_x \in \mathsf{N} : \varphi \in Y\}$ for Y in $\mathcal{N}$ then equation (4.1.1) becomes

$$P(Y) = P(Y_x) \quad \text{for all } Y \text{ in } \mathcal{N}.$$

The notation of *isotropy* is entirely analogous; Φ is isotropic if its statistics are invariant under rotation; Φ and $\mathfrak{r}\Phi$ have the same distribution for every rotation $\mathfrak{r}$ around the origin. So

$$P(Y) = P(\mathfrak{r}Y)$$

for $\mathfrak{r}Y = \{\mathfrak{r}\varphi \in \mathsf{N} : \varphi \in Y\}$, Y in $\mathcal{N}$. Stationarity and isotropy together yield *motion-invariance*. A motion-invariant point process Φ has the same distribution as $\mathfrak{m}\Phi$ for all Euclidean motions $\mathfrak{m}$ of $\mathbb{R}^d$.

Figures 2.3, 2.4, 4.1 show realizations of non-stationary point processes. Figure 4.1 illustrates a subtlety; non-stationarity does not only imply variable point density in space. In Figure 4.1 the kind of distribution of the points depends on spatial location; lower in the figure there is greater clustering.

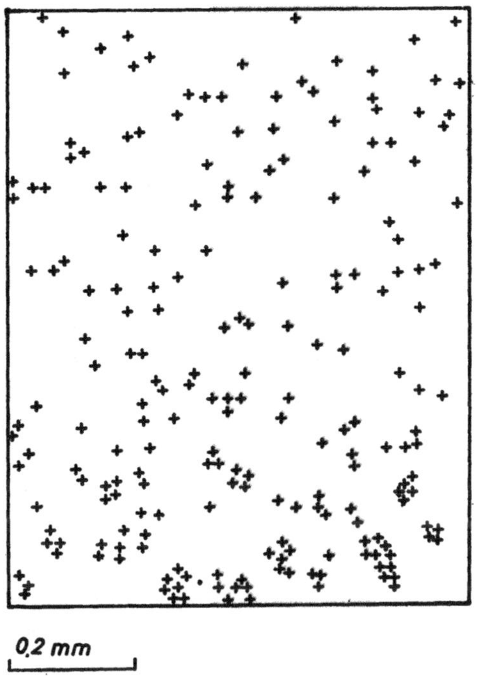

Fig. 4.1. A point pattern which it is appropriate to model by a non-stationary point process. The crosses are the midpoints of cell nuclei in a specimen of joint cartilage. The foot of the figure is the side of the bone tissue, while the upper part of the figure is nearer the joint.

The intensity measure

The *intensity measure* Λ of Φ is analogous to the mean of a real-valued random variable. Its definition is

$$\Lambda(B) = \mathsf{E}\Phi(B) = \int \varphi(B)\, P(\mathrm{d}\varphi) \quad \text{for Borel } B.$$

So $\Lambda(B)$ is the mean number of points in B. If Φ is stationary then the intensity measure simplifies; it must be translation-invariant since

$$\Lambda(B) = \mathsf{E}\Phi(B) = \mathsf{E}\Phi_x(B) = \mathsf{E}\Phi(B_{-x}) = \Lambda(B_{-x})$$

for all x. By (1.8.1) this translation-invariance implies that

$$\Lambda(B) = \lambda \nu_d(B)$$

for some (possibly infinite) non-negative constant λ, which is called the *intensity* of Φ. Choosing B to have measure 1 shows that λ may be interpreted as the mean number of points of Φ per unit volume. We shall always assume that $0 < \lambda < \infty$.

Calculations may be simplified by use of the *Campbell theorem*; for any non-negative measurable function f

$$\mathsf{E}\sum_{x\in\Phi} f(x) = \int \sum_{x\in\varphi} f(x)\, P(\mathrm{d}\varphi)$$
$$= \iint f(x)\, \varphi(\mathrm{d}x)\, P(\mathrm{d}\varphi) = \int f(x)\, \Lambda(\mathrm{d}x). \tag{4.1.2}$$

This is an application of Fubini's theorem. In the stationary case the Λ-integral in (4.1.2) becomes a volume integral so

$$\mathsf{E}\sum_{x\in\Phi} f(x) = \lambda \int f(x)\, \mathrm{d}x. \tag{4.1.3}$$

The intensity λ can be further interpreted in a local sense. Suppose $\{K_n\}$ is a bounded sequence of compact sets with $\nu_d(K_n) \to 0$ and the property $d(K_n) \to 0$, where, for a set A, $d(A)$ denotes the diameter of the smallest ball containing A. Then for a motion-invariant point process (but *not* a merely stationary one) when $d \geqq 2$

$$\mathsf{P}\big(\Phi(K_n) = 1\big) = \lambda\nu_d(K_n) + o\big(\nu_d(K_n)\big) \tag{4.1.4}$$

$$\mathsf{P}\big(\Phi(K_n) \geqq 2\big) = o\big(\nu_d(K_n)\big) \tag{4.1.5}$$

as shown in AMBARTZUMIAN (1976). These formulae generalize the wellknown formulae of Dobrushin and Korolyuk for the case of point processes on $\mathbb{R}^1$.

Ergodicity

A stationary distribution P on $[\mathsf{N}, \mathcal{N}]$ is said to be *ergodic* if

$P(Y)$ is zero or one for all invariant sets Y in $\mathcal{N}$.

Here an *invariant set* Y in $\mathcal{N}$ is one for which

$$P(Y \setminus Y_x \cup Y_x \setminus Y) = 0, \text{ for all } x \text{ in } \mathbb{R}^d.$$

The stationary Poisson process is ergodic whereas the mixed Poisson process is not.

Section 6.1 gives an ergodic theorem for random closed sets which also applies to point processes. If Φ is an ergodic point process

$$\lim_{K_n\uparrow\mathbb{R}^d} \Phi(K_n)/\nu_d(K_n) = \lambda$$

almost everywhere where '$K_n \uparrow \mathbb{R}^d$' is interpreted under the conditions given in Section 6.1.

Contact distributions

Let B be a closed Borel set in $\mathbb{R}^d$ with $o \in B$ and $\nu_d(B) > 0$. The *contact distribution function* H_B *(with respect to the test set* B*)* is defined by

$$H_B(r) = 1 - \mathsf{P}\big(\Phi(rB) = 0\big) \quad \text{for } r \geqq 0.$$

It is indeed a probability distribution function. In the special case of $rB = b(o, r)$ the contact distribution function is written as $H_s(r)$ and called the *spherical contact distribution function* or *first contact distribution function*.

4.2 Marked point processes

A point process is made into a *marked point process* by attaching a characteristic (the *mark*) to each point of the process. Thus a marked point process on $\mathbb{R}^d$ is a random sequence $\Psi = \{[x_n; m(x_n)]\}$ from which the points x_n together constitute a point process (not marked) in $\mathbb{R}^d$ and the $m(x_n)$ are the marks corresponding to each x_n. The marks $m(x_n)$ may have complicated structure. They belong to a given *space of marks* M^* which we assume to be a Polish space. The Borel σ-algebra of M^* is denoted by $\mathcal{M}^*$.

MATTHES (1963) and his co-workers established the theory of marked point processes. Its initial application was in queueing theory (KÖNIG, MATTHES, and NAWROTZKI, 1967); such applications up to the early 1980s are discussed in the book of FRANKEN, KÖNIG, ARNDT, and SCHMIDT (1981).

Specific examples are:

for x the centre of a particle, $m(x)$ the volume of the particle;
for x the position of a tree, $m(x)$ the stem diameter of the tree;
for x the centre of an atom, $m(x)$ the type of the atom;
for x the location (suitably defined) of a random convex compact set, $m(x)$ the centred random set itself.

The marks can be essentially continuous variables, as in the first two examples, indicators of types as in the third example (in which case the terms 'multivariate point process' or 'multitype point process' are often used; see DIGGLE (1983), or LOTWICK (1984), and DIGGLE and COX (1983) and UPTON and FINGLETON (1985) for the important case of 'bivariate point processes') or actually very complicated indeed, as in the last example which occurs in the marked point process interpretation of a germ-grain model (see 6.4). Figures 4.2 and 4.3 show two point patterns that can be interpreted as samples of marked point processes.

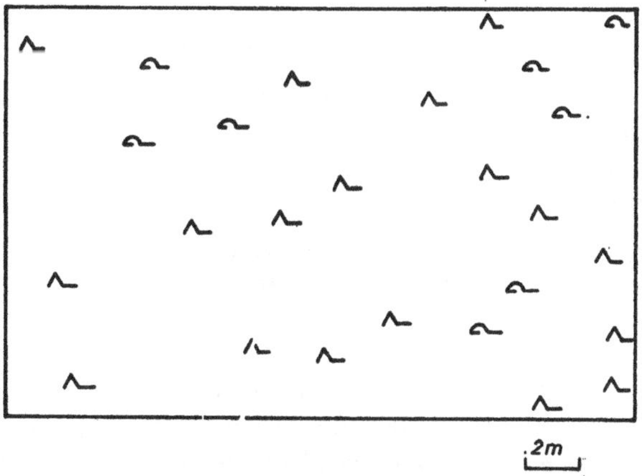

Fig. 4.2. Positions of spruces and birches in a forest. The point pattern can be interpreted as a sample of a bivariate point process; that is to say a marked point process with marks 0 and 1.

It is always possible to interpret Ψ as an unmarked point process in $\mathbb{R}^d \times \mathsf{M}^*$. The notation of a random counting measure can be used; for Borel B and L in $\mathcal{M}^*$ the number of points of Ψ in B with marks in L is denoted by $\Psi(B \times L)$.

Euclidean motions of marked point processes are defined as moving the points but leaving the marks unchanged. So Ψ_x, the translate of Ψ by x, is given by

$$\Psi_x = \{[x_1 + x; m(x_1)], [x_2 + x; m(x_2)], \ldots\}.$$

Consequently a marked point process is said to be *stationary* if for all x the translated process Ψ_x has the same distribution as Ψ. It is *motion-invariant* if for all Euclidean motions $\mathfrak{m}$ the process $\mathfrak{m}\Psi$ has the distribution of Ψ.

The definition of the *intensity measure* Λ of a general Ψ is analogous to that of the intensity measure of Ψ when Ψ is interpreted as an ordinary point process:

$$\Lambda(B \times L) = \mathsf{E}\Psi(B \times L).$$

Thus there is a *Campbell theorem*:

$$\mathsf{E} \sum_{[x, m(x)] \in \Psi} f(x, m(x)) = \int f(x, m)\, \Lambda(d(x, m)) \tag{4.2.1}$$

for any non-negative measurable function f on $\mathbb{R}^d \times \mathsf{M}^*$.

The measure $B \to \Lambda(B \times L)$ for fixed L in $\mathcal{M}^*$ is absolutely continuous with respect to Λ_p where Λ_p is the intensity measure of the point process $\{x_n\}$ (that is Ψ stripped of marks). Consequently, it can be shown that

$$\Lambda(d(x, m)) = M_x(\mathrm{d}m)\, \Lambda_p(\mathrm{d}x)$$

where M_x is a probability measure on $[\mathsf{M}^*, \mathcal{M}^*]$, which can be interpreted as the *distribution of the mark* of a point at x. Section 4.4 explains the close relationship of M_x to Palm distribution theory. The Polish space property of M^* is needed here to establish that M_x is indeed a measure.

When Ψ is stationary

$$\Lambda(B \times L) = \mathsf{E}\Psi(B \times L) = \mathsf{E}\Psi(B_x \times L) = \Lambda(B_x \times L) \quad \text{for all } x.$$

In that case if L is a fixed member of $\mathcal{M}^*$ then $B \to \Lambda(B \times L)$ is a translation-invariant measure. Hence, by (1.7.1)

$$\Lambda(B \times L) = \lambda_L \nu_d(B).$$

The quantity λ_L is the *intensity of Ψ with respect to L*; it is the mean number of points of Ψ per unit volume with marks in L. Of course

$$\lambda_{\mathsf{M}^*} = \lambda$$

where λ is the intensity of the unmarked point process.

As a function of L the quantity λ_L is a measure on $[\mathsf{M}^*, \mathcal{M}^*]$; the quotients λ_L/λ provide the so-called *mark distribution* M of the stationary marked point process Ψ:

$$M(L) = \lambda_L/\lambda \tag{4.2.2}$$

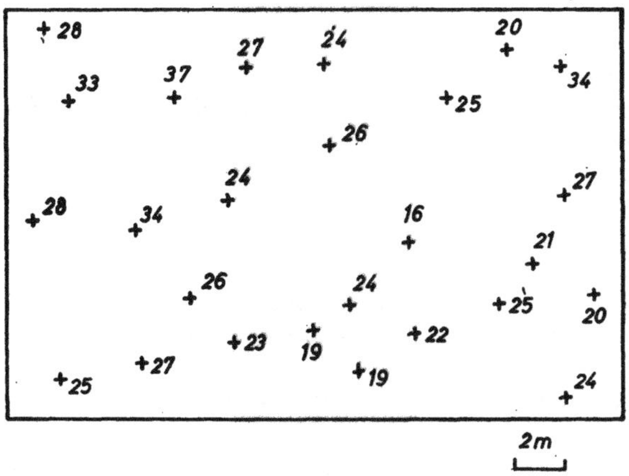

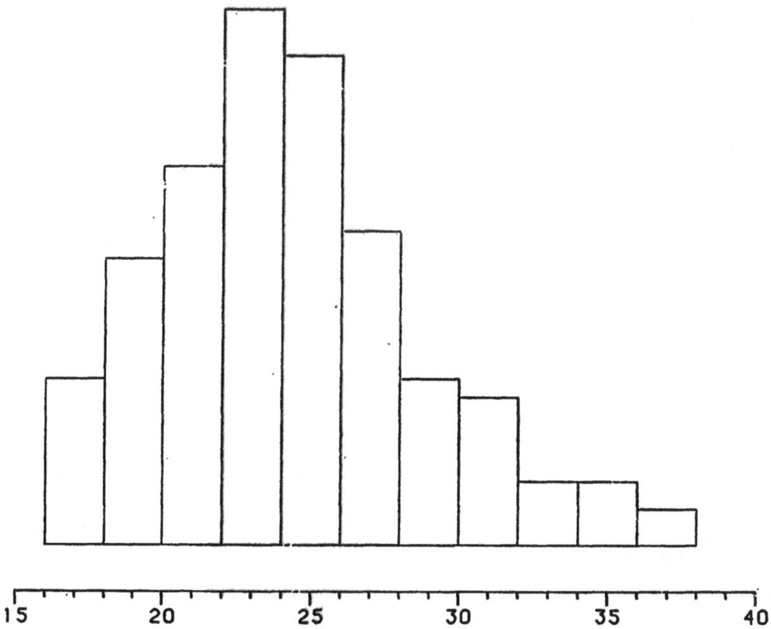

Fig. 4.3(a). Locations of spruces (*Pices abies*) in a forest. The numbers refer to the stem diameters; they can be interpreted as marks for the point process of locations; (b) Statistically determined mark (stem diameter) distribution for the spruce tree point pattern (taken from a larger area than that shown in Figure 4.3a)

and in this case of stationarity

$$\Lambda = \lambda \cdot M \times \nu_d. \tag{4.2.3}$$

In the stationary case $M = M_x$ for all x.

In the ergodic case M can be interpreted as the distribution of a mark of a 'randomly chosen point' of Ψ:

$$M(L) = \lim_{K_n \uparrow \mathbb{R}^d} \frac{\Psi(K_n \times L)}{\Psi(K_n \times \mathsf{M}^*)}$$

for sequences $\{K_n\}$ as in the ergodic theorem of Section 6.1.

In the stationary case the *Campbell theorem* takes the form

$$\mathsf{E} \sum_{[x,m(x)]\in\Psi} f(x, m(x)) = \lambda \iint f(x, m)\, M(\mathrm{d}m)\, \mathrm{d}x. \tag{4.2.4}$$

A typical application of this theorem is in the proof of formula (6.4.5). An important particular case is

$$\mathsf{E} \sum_{[x,m]\in\Psi} \mathbf{1}_B(x)\, h(m) = \lambda \int h(m)\, M(\mathrm{d}m) \tag{4.2.5}$$

where $\nu_d(B) = 1$ and h is a non-negative measurable function on M^*.

Example 4.1. *Spruces in a forest* (Stoyan, 1984b).

Figure 4.3a shows the position of spruces (Picea abies) in a rectangular region of a forest. The marks (the numbers attached to the points) are the stem diameters. Figure 4.3b shows the mark distribution (the stem diameter distribution) as estimated by means of a histogram (it was determined from a sample much larger than that of Figure 4.3a). The mean mark is

$$\overline{m} = 23.9 \text{ cm}$$

the intensity of the (unmarked) point process is

$$\lambda = 0.062 \text{ m}^{-2}.$$

In the remainder of this chapter most formulae are given for unmarked point processes. Corresponding formulae for a marked point process Ψ can be deduced either by interpreting Ψ as a point process in $\mathbb{R}^d \times \mathsf{M}^*$ or by noting that for each fixed L in $\mathcal{M}^*$ the process $\{x_n : [x_n; m(x_n)] \in \Psi, m(x_n) \in L\}$ is an unmarked point process.

4.3 Moment measures and related quantities

In the classical theory of random variables the moments (particularly mean and variance), generating functions, and Laplace-Stieltjes transforms are all important tools and provide useful mean of describing various distributions. In the theory of point processes their analogues are more complicated than mere scalars. For example numerical means and variances must be replaced by *moment measures*, as described below.

The following text is based mainly on unpublished work by MECKE (1976; 1977). Throughout the section (except when noted otherwise), $B, B_2, \ldots, B_n$ stand for general Borel sets in $\mathbb{R}^d$.

Moment measures

The nth-order moment measure of the point process Φ is the measure $\mu^{(n)}$ defined on $\mathcal{B}^{nd}$ by

$$\int f(x_1, \ldots, x_n)\, \mu^{(n)}\big(d(x_1, \ldots, x_n)\big) = \int \sum_{x_1,\ldots,x_n \in \varphi} f(x_1, \ldots, x_n)\, P(d\varphi)$$
$$= \mathsf{E} \sum_{x_1,\ldots,x_n \in \Phi} f(x_1, \ldots, x_n) \qquad (4.3.1)$$

where f is any non-negative measurable function on $\mathbb{R}^{nd}$. In particular

$$\mu^{(n)}(B_1 \times \cdots \times B_n) = \mathsf{E}\big(\Phi(B_1) \cdots \Phi(B_n)\big)$$

and if $B_1 = \cdots = B_n = B$ then

$$\mu^{(n)}(B^n) = \mathsf{E}\Phi(B)^n .$$

Thus $\mu^{(n)}$ yields the nth moment of the real-valued random variable $\Phi(B)$.

Particular cases

$$n = 1 : \mu^{(1)}(B) = \mathsf{E}\Phi(B) = \Lambda(B)$$
$$n = 2 : \mu^{(2)}(B_1 \times B_2) = \mathsf{E}\Phi(B_1)\, \Phi(B_2)$$
$$\operatorname{var}\big(\Phi(B)\big) = \mu^{(2)}(B \times B) - \big(\Lambda(B)\big)^2 .$$

The covariance of the random variables $\Phi(B_1)$ and $\Phi(B_2)$ satisfies

$$\operatorname{cov}\big(\Phi(B_1), \Phi(B_2)\big) = \mathsf{E}\Phi(B_1)\, \Phi(B_2) - \mathsf{E}\Phi(B_1)\, E\Phi(B_2)$$
$$= \mu^{(2)}(B_1 \times B_2) - \Lambda(B_1)\, \Lambda(B_2) .$$

If Φ is stationary then the $\mu^{(n)}$ are translation invariant in an extended sense

$$\mu^{(n)}(B_1 \times \cdots \times B_n) = \mu^{(n)}\big((B_1 + x) \times \cdots \times (B_n + x)\big)$$

for all x in $\mathbb{R}^d$.

Section 4.5 gives a simplified description of the second-order moment measure

Factorial moment measures

The nth-order factorial moment measure $\alpha^{(n)}$ of the point process Φ is defined on $\mathcal{B}^{nd}$ by

$$\int f(x_1, \ldots, x_n)\, \alpha^{(n)}\big(d(x_1, \ldots, x_n)\big) = \int \sum_{x_1,\ldots,x_n \in \varphi}^{\neq} f(x_1, \ldots, x_n)\, P(\mathrm{d}\varphi) . \quad (4.3.2)$$

Here again, f is any non-negative measurable function on $\mathbb{R}^{nd}$. The summation $\sum^{+}$ is over all n-tuples of *distinct* points in φ. This is the difference between $\alpha^{(n)}$ and $\mu^{(n)}$; in (4.3.2) the sum omits all n-tuples with two or more equal members. So if $B_1, \ldots, B_n$ are pairwise disjoint then

$$\mu^{(n)}(B_1 \times \cdots \times B_n) = \alpha^{(n)}(B_1 \times \cdots \times B_n). \tag{4.3.3}$$

For $n = 2$,

$$\mu^{(2)}(B_1 \times B_2) = \Lambda(B_1 \cap B_2) + \alpha^{(2)}(B_1 \times B_2) \tag{4.3.4}$$

as can be seen from the definitions of $\mu^{(2)}$, Λ, $\alpha^{(2)}$, and the relation

$$\sum_{x_1, x_2 \in \varphi} \mathbf{1}_{B_1}(x_1)\, \mathbf{1}_{B_2}(x_2) = \sum_{x_1 \in \varphi} \mathbf{1}_{B_1 \cap B_2}(x_1) + \sum_{x_1, x_2 \in \varphi}^{+} \mathbf{1}_{B_1}(x_1)\, \mathbf{1}_{B_2}(x_2).$$

Since

$$\alpha^{n}(B^{n}) = \mathsf{E}\big[\Phi(B)\,\big(\Phi(B) - 1\big) \cdots \big(\Phi(B) - n + 1\big)\big]$$

the quantity $\alpha^{n}(B^{n})$ is the nth-order factorial moment of the random variable $\Phi(B)$: thus the term 'factorial moment measure' is appropriate.

The factorial moment measures for a Poisson process are given in Section 2.4.2.

Product densities

Suppose that $\alpha^{(n)}$ is locally finite and absolutely continuous with respect to Lebesgue measure ν_{nd}. Then $\alpha^{(n)}$ has a density $\varrho^{(n)}$, the *product density*:

$$\alpha^{(n)}(B_1 \times \cdots \times B_n) = \int_{B_1} \cdots \int_{B_n} \varrho^{(n)}(x_1, \ldots, x_n)\, \mathrm{d}x_1 \cdots \mathrm{d}x_n \tag{4.3.5}$$

moreover, for any non-negative bounded measurable function f

$$\mathsf{E} \sum_{x_1, \ldots, x_n \in \Phi}^{+} f(x_1, \ldots, x_n) = \int f(x_1, \ldots, x_n)\, \varrho^{(n)}(x_1, \ldots, x_n) \cdot \mathrm{d}x_1 \cdots \mathrm{d}x_n.$$

The product densities have an intuitive interpretation, which probably accounts for their historical precedence over product measures. Suppose that $C_1, \ldots, C_n$ are pairwise disjoint spheres with centres $x_1, \ldots, x_n$ and infinitesimal volumes $\mathrm{d}V_1, \ldots, \mathrm{d}V_n$. Then $\varrho^{(n)}(x_1, \ldots, x_n)\, \mathrm{d}V_1 \cdots \mathrm{d}V_n$ is the probability that there is a point of Φ in each of $C_1, \ldots, C_n$.

For pairwise disjoint $B_1, \ldots, B_n$ formulae (4.3.3) and (4.3.5) yield

$$\mu^{(n)}(B_1 \times \cdots \times B_n) = \int_{B_1 \ldots B_n} \cdots \int \varrho^{(n)}(x_1, \ldots, x_n)\, \mathrm{d}x_1 \cdots \mathrm{d}x_n.$$

If Φ is stationary then $\varrho^{(1)} = \lambda$ and $\varrho^{(2)}$ depends only on the difference of its arguments:

$$\varrho^{(2)}(x_1, x_2) = f_s(x_1 - x_2) \quad \text{for } x_1, x_2 \text{ in } \mathbb{R}^d.$$

If furthermore Φ is motion-invariant then this f_s depends only on the distance r between x_1 and x_2:

$$\varrho^{(2)}(x_1, x_2) = f_i(\|x_1 - x_2\|) = f_i(r).$$

With the usual abuse of notation in such circumstances, f_i is often denoted by $\varrho^{(2)}$ and called the product density.

Product densities for the Poisson process can be found in 2.4.2 and 2.5.

The Campbell measure

The Campbell measure $\mathscr{C}$ is defined as a measure on $[\mathbb{R}^d \times \mathsf{N}, \mathscr{B}^d \times \mathscr{N}]$ by

$$\int \sum_{x \in \varphi} f(x, \varphi)\, P(\mathrm{d}\varphi) = \int f(x, \varphi)\, \mathscr{C}\big(\mathrm{d}(x, \varphi)\big) \tag{4.3.6}$$

where f is any non-negative measurable function on $\mathbb{R}^d \times \mathsf{N}$. Since

$$\mathscr{C}(B \times Y) = \int \varphi(B)\, \mathbf{1}_Y(\varphi)\, P(\mathrm{d}\varphi) \tag{4.3.7}$$

for B Borel and Y in $\mathscr{N}$, one can also write

$$\mathscr{C}(B \times Y) = \mathsf{E}\big(\Phi(B); \Phi \in Y\big).$$

It is sometimes useful to consider the *reduced Campbell measure* $\mathscr{C}^!$ defined by

$$\int \sum_{x \in \varphi} f(x, \varphi \setminus \{x\})\, P(\mathrm{d}\varphi) = \int \sum_{x \in \varphi} f(x, \varphi - \delta_x)\, P(\mathrm{d}\varphi)$$
$$= \int f(x, \varphi)\, \mathscr{C}^!\big(d(x, \varphi)\big).$$

Here $\varphi \setminus \{x\}$, $\varphi - \delta_x$ are alternative notations, in set-theoretic and measure theoretic language respectively, for the point-pattern φ with the point $x \in \varphi$ deleted.

Campbell measures and reduced Campbell measures of higher order can also be defined, as in Hanisch (1982), Jagers (1973), Kallenberg (1983 b), and Mecke (1979 a). For example, the second-order Campbell measure is

$$\mathscr{C}^{(2)}(B_1 \times B_2 \times Y) = \mathsf{E}\big(\Phi(B_1)\, \Phi(B_2); \Phi \in Y\big).$$

The Campbell measure for a Poisson process is calculated for special Y on page 115.

A covariance function for marked point processes

Moment measures for marked point processes can be defined in a fashion similar to the unmarked case. Details are in Stoyan (1984 a, b). Diggle (1983) and Hanisch and Stoyan (1979) consider moment measures for the case of discrete marks. Here we consider only one example, a special case in which the marks are real-valued. We define the *mark covariance function* k_{mm} for a motion-invariant marked point process $\Psi = \{[x_n; m(x_n)]\}$. Heuristically $k_{mm}(r)$ is the mean of the product of marks at two positions separated by a distance r, under the condition that there are indeed points of Ψ in these two positions. By motion-invariance it suffices to consider the positions o and $\boldsymbol{r}$, with $r = \|\boldsymbol{r}\|$. So

$$k_{mm}(r) = E_{o\boldsymbol{r}} m(o)\, m(\boldsymbol{r})$$

where $E_{o\boldsymbol{r}}$ is expectation subject to the conditioning that Ψ has points at positions o and $\boldsymbol{r}$.

Example 4.1 (continued). *Spruces in a forest.*

The mark covariance function k_{mm} was calculated using as marks the stem diameters of the spruce trees as indicated in Figure 4.3. The results were surprising: the mark covariance function is practically constant for all values of r, being approximately equal to the value $(\overline{m})^2$, where

$$\overline{m} = \int_{[0,\infty]} x M(\mathrm{d}x)$$

is the mean mark. This means that for this particular forest there is no observable dependence of stem diameter on inter-tree distance.

For illustrative purposes new marks were taken to be the areas of Dirichlet cells from the Dirichlet tessellation of the tree pattern (see Section 10.2). The corresponding normalized mark covariance function

$$k_{mm}(r)/(\overline{m})^2$$

is displayed in Figure 4.4. Here of course the mark does depend on the inter-tree distance.

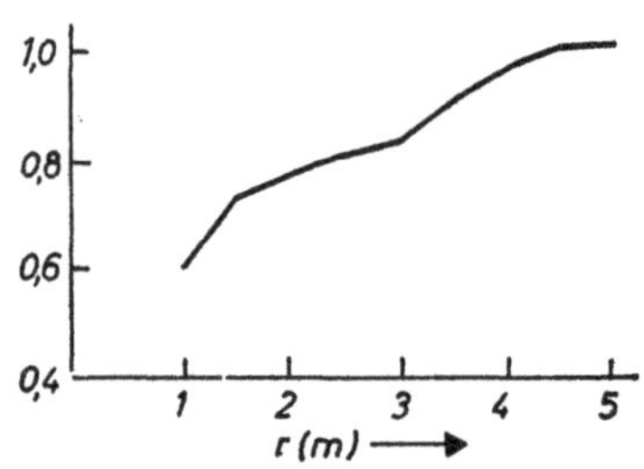

Fig. 4.4. Mark covariance function $k_{mm}(r)$ for the spruce tree point pattern, when the marks are chosen to be the areas of the corresponding Dirichlet cells. For the estimation of k_{mm} a kernel estimator was used (see page 126)

Stochastic process covariance functions typically do *not* behave in the manner of k_{mm}: instead they generally *increase* with decreasing r.

Stoyan (1984a) gave a mathematical description of $k_{mm}(r)$. Consider the second-order factorial moment measure of the marked point process in question

$$\alpha_m^{(2)}(B_1 \times B_2 \times L_1 \times L_2) = \int \sum\nolimits^{+} \mathbf{1}_{B_1}(x_1)\, \mathbf{1}_{B_2}(x_2)\, \mathbf{1}_{L_1}(m(x_1))\, \mathbf{1}_{L_2}(m(x_2))\, P\,(\mathrm{d}\psi)$$

where the sum is taken over all pairs $[x_1; m(x_1)]$ and $[x_2; m(x_2)]$ in Ψ (but excluding the cases when $x_1 = x_2$ since the summation is $\sum^{+}$). Let $\alpha^{(2)}$ be the second-order factorial moment measure of the unmarked point process $\Phi = \{x_n\}$. Then for fixed Borel subsets L_1, L_2 of $[0, \infty)$ the measure $B_1 \times B_2 \to \alpha^{(2)}(B_1 \times B_2 \times L_1 \times L_2)$ is absolutely continuous with respect to $\alpha^{(2)}$. If $\alpha^{(2)}$ is σ-finite then by the Radon-Nikodym theorem there is a density M_{x_1,x_2} such that

$$\alpha_m^{(2)}(B_1 \times B_2 \times L_1 \times L_2) = \int_{B_1 \times B_2} M_{x_1 x_2}(L_1 \times L_2)\, \alpha^{(2)}(\mathrm{d}(x_1, x_2)).$$

For fixed x_1, x_2 the density $M_{x_1 x_2}$ can be interpreted as a measure on $\mathsf{M}^* \times \mathsf{M}^*$, the 'two-point mark distribution'. Its intuitive interpretation follows the lines given above.

Under the assumption of motion-invariance $M_{x_1x_2}$ depends only on $r = \|x_1 - x_2\|$ and the simpler notation M_r is used. In that case

$$k_{mm}(r) = \int m_1 m_2 M_r(\mathrm{d}(m_1, m_2)).$$

The generating functional

The generating function G_X of a non-negative integer-valued random variable X is given by

$$G_X(t) = \mathsf{E}t^X = \sum_{n=0}^{\infty} t^n \mathsf{P}(X = n) \quad \text{for } t \text{ in } [0, 1].$$

The mean and variance of X can be derived from G_X:

$$\mathsf{E}X = G'(1), \qquad \operatorname{var}(X) = G''(1) + G'(1) \cdot (1 - G'(1))$$

under suitable regularity conditions if these moments exist.

The generating functional G of a point process $\Phi = \{x_n\}$ is defined by analogy with G_X. Let U be the family of all non-negative bounded measurable functions u on $\mathbb{R}^d$ whose support $\{x \in \mathbb{R}^d : u(x) > 0\}$ is bounded. Furthermore let V be the family of all functions $v = 1 - u$ for u in U, $1 \geqq u \geqq 0$. Then G is defined by

$$G(v) = \mathsf{E}\{v(x_1)\, v(x_2) \cdots\} = \mathsf{E} \prod_{x \in \Phi} v(x)$$
$$= \int \prod_{x \in \Phi} v(x)\, P(\mathrm{d}\varphi) \quad \text{for } v \text{ in } V.$$

The distribution P of Φ is determined uniquely by G. The joint generating functional $G_{B_1,\ldots,B_n}$ of the random vector $(\Phi(B_1), \ldots, \Phi(B_n))$ is given (for bounded Borel sets $B_1, \ldots, B_n$) by

$$G_{B_1,\ldots,B_n}(t_1, \ldots, t_n) = \mathsf{E}t_1^{\Phi(B_1)} \ldots t_n^{\Phi(B_n)} = G(v)$$

where $v(x) = v_1(x) \cdot \cdots \cdot v_n(x)$ and $v_i(x) = 1 + (t_i - 1)\, \mathbf{1}_{B_i}(x)$ and $t_1, \ldots, t_n$ all belong to $[0, 1]$.
Thus in particular for $n = 1$

$$G_B(t) = G(v) \quad \text{with } v(x) = 1 + (t - 1)\, \mathbf{1}_B(x).$$

The factorial moment measures can be derived from G

$$\alpha^{(n)}(B_1 \times \cdots \times B_n) = (-1)^n \left[\left(\frac{\partial}{\partial s_1} \cdots \frac{\partial}{\partial s_n}\right) G(1 - s_1 \mathbf{1}_{B_1} - \cdots - s_n \mathbf{1}_{B_n})\right]$$

with the derivatives taken at $s_1 = \cdots = s_n = 0$. Conversely, under suitable regularity conditions the generating functional is determined by the factorial moment measures; for u in U with $0 \leqq u \leqq 1$

$$G(1 - u) = 1 + \sum_{n=1}^{\infty} \frac{(-1)^n}{n!} \int u(x_1) \cdots u(x_n)\, \alpha^{(n)}(d(x_1 \cdots x_n))$$

if the $\alpha^{(n)}$ are Radon measures for all n and if the right-hand side converges.

Example 4.2 *The generating functional of the Poisson process.*
Let Φ be a Poisson process of intensity measure Λ. Then

$$G(v) = \exp\left(-\int [1 - v(x)]\, \Lambda(\mathrm{d}x)\right) \quad \text{where } v \in \boldsymbol{V}. \tag{4.3.7}$$

Proof of (4.3.7). Consider $v(x) = 1 - \sum_{i=1}^{n} (1 - z_i)\, \mathbf{1}_{C_i}(x)$ for z_i all in [0, 1] and $C_1, \ldots, C_n$ pairwise disjoint compact sets. Then

$$\begin{aligned} G(v) &= \int z_1^{\varphi(C_1)} \cdots z_n^{\varphi(C_n)} P(\mathrm{d}\varphi) \\ &= \mathsf{E} z_1^{\Phi(C_1)} \cdots z_n^{\Phi(C_n)} \\ &= \mathsf{E} z_1^{\Phi(C_1)} \cdots \mathsf{E} z_n^{\Phi(C_n)} \end{aligned}$$

the last step using the independence of $\Phi(C_1), \ldots, \Phi(C_n)$ when $C_1, \ldots, C_n$ are disjoint. Then each $\mathsf{E} z_i^{\Phi(C_i)}$ can be determined, since $\Phi(C_i)$ is Poisson of mean $\Lambda(C_i)$. Therefore

$$\begin{aligned} G(v) &= \exp\left(-(1 - z_1)\, \Lambda(C_1)\right) \cdots \exp\left(-(1 - z_n)\, \Lambda(C_n)\right) \\ &= \exp\left(-\sum_{i=1}^{n} \int_{C_i} (1 - z_i)\, \Lambda(\mathrm{d}x)\right) \\ &= \exp\left(-\int [1 - v(x)]\, \Lambda(\mathrm{d}x)\right). \end{aligned}$$

Thus (4.3.7) holds for piecewise constant v. The general result follows by standard arguments from measure theory.

A point process can be considered as a random counting measure. Hence the definition of a Laplace functional of a point process follows from Section 7.2.

4.4 Palm distributions

Heuristic introduction

In the study of point processes it is often natural to condition on there being a point of the process at the fixed location x. Section 2.4 gives an elementary example. Suppose $\mathsf{P}(\Phi(b(x, r)) = 1 \parallel x)$ represents the conditional probability that given that Φ has a point at x there is no further point within the sphere of radius r centred at x. This conditional probability is ambiguous without further explanation since it is conditioned on an event of probability zero. It must be defined by requiring it to satisfy some extra properties. What properties might one reasonably require of such conditional distributions $\mathsf{P}(\cdot \parallel x)$? To begin with one would require in the stationary case that the relation (I) hold:

$$\mathsf{P}(\Phi \in Y \parallel x) = \mathsf{P}(\Phi \in Y_{-x} \parallel o) \tag{I}$$

for x in $\mathbb{R}^d$ and Y in $\mathcal{N}$ and Y_{-x} defined in the same manner as Y_x after equation (4.1.1). In such a case one could define a distribution function

$$D(r) = 1 - \mathsf{P}(\Phi(b(o, r)) = 1 \parallel o)$$

and this would be the distribution function of the distance to its nearest neighbour for a 'typical' point.

In order to introduce further properties that one might require of such conditional distributions we consider a specific example. Suppose that Φ is a point process, not necessarily stationary, of distribution P and locally finite intensity measure Λ. For the set Y in $\mathcal{N}$ given by

$$Y = \{\varphi \in \mathsf{N} : \varphi(b(o, r)) = 1\}$$

consider the function h on $\mathbb{R}^d \times \mathsf{N}$ given by

$$h(x, \varphi) = \mathbf{1}_B(x)\, \mathbf{1}_Y(\varphi - x)$$

for some bounded Borel set B; as usual if $\varphi = \{x_n\}$ then

$$\varphi_{-x} = \varphi - x = T_x\varphi = \{x_n - x\}) .$$

With the distribution function given above in mind one might wish to evaluate the mean number of points of Φ in B whose neighbours are all at distance at least r:

$$\mathsf{E} \sum_{x \in \Phi} h(x, \Phi) = \int \sum_{x \in \varphi} h(x, \varphi)\, P(\mathrm{d}\varphi) .$$

More generally one might wish to evaluate this for non-negative measurable functions h on $\mathbb{R}^d \times \mathsf{N}$.

If $\mathbb{R}^d$ is partitioned into domains $D_1, D_2, \ldots$ each of positive volume, then the expectation above can be written as

$$\mathsf{E} \sum_{x \in \Phi} h(x, \Phi) = \sum_k \mathsf{E} \Big[\sum_{x \in \Phi \cap D_k} h(x, \Phi) \mid \Phi(D_k) > 0 \Big] \cdot \mathsf{P}\big(\Phi(D_k) > 0\big) .$$

Suppose that each D_k tends to an infinitesimal volume element $\mathrm{d}x$. Then $\mathsf{P}\big(\Phi(D_k) > 0\big)$ should converge to $\Lambda(\mathrm{d}x)$, while the conditional mean should converge to the mean $\mathsf{E}\big(h(x, \Phi) \parallel x\big)$ of $h(x, \varphi)$ with respect to $\mathsf{P}(\cdot \parallel x)$. Hence one is led to the relation

$$\mathsf{E} \sum_{x \in \Phi} h(x, \Phi) = \int \mathsf{E}\big(h(x, \Phi) \parallel x\big)\, \Lambda(\mathrm{d}x) . \tag{II*}$$

Under the assumption that Φ is stationary, $\Lambda = \lambda \nu_d$ with $\lambda \in (0, \infty)$. The relations (I) and (II*) then yield

$$\mathsf{E} \sum_{x \in \varphi} h(x, \Phi) - \lambda \int \mathsf{E}\big(h(x, \Phi_x) \parallel o\big)\, \mathrm{d}x . \tag{II}$$

If the notation $P_o(Y) = \mathsf{P}(\Phi \in Y \parallel o)$ is used (for $Y \in \mathcal{N}$) then (II) takes the form

$$\int \sum_{x \in \varphi} h(x, \varphi)\, P(\mathrm{d}\varphi) = \lambda \iint h(x, \varphi_x)\, P_o(\mathrm{d}\varphi)\, \mathrm{d}x \tag{III}$$

and so for $h(x, \varphi) = \mathbf{1}_B(x) \cdot \mathbf{1}_Y(\varphi - x)$ as above

$$\int \sum_{x \in \varphi \cap B} \mathbf{1}_Y(\varphi_{-x})\, P(\mathrm{d}\varphi) = \lambda \nu_d(B)\, P_o(Y) \tag{IV}$$

for Borel B and Y in $\mathcal{N}$.

This depends on the details of the convergence of the arbitrary partition $D_1, \ldots, D_n$. In the following, a distribution on $[\mathsf{N}, \mathcal{N}]$ will be constructed which will have the behaviour of the hypothetical P_o above: it will satisfy (III) and its definition will be inspired by (IV).

The Palm distribution: first definition

Suppose that Φ is a stationary point process, with finite and nonzero intensity λ. The *Palm distribution* (at o) of P is a distribution defined on $[\mathsf{N}, \mathcal{N}]$ by

$$P_o(Y) = \int \sum_{x \in \varphi \cap B} \mathbf{1}_Y(\varphi - x)\, P(\mathrm{d}\varphi)/(\lambda \nu_d(B)) \quad \text{for } Y \text{ in } \mathcal{N}. \tag{4.4.1}$$

Here B is an arbitrary Borel set of positive volume: the definition does not depend on the choice of B. The definition can be understood intuitively using marked point processes as follows. To each point x of Φ a mark of 1 or 0 is given according as to whether the shifted process Φ_{-x} belongs to Y or not. For example, consider $Y = \{\varphi \in \mathsf{N} : \varphi \cap b(o, r) \text{ is a singleton}\}$. Then x has mark 1 precisely when its nearest neighbour is further away than r. The result of this is a stationary marked point process of mark distribution M. Recalling (4.2.2), $P_o(Y)$ is given by a mark distribution probability:

$$P_o(Y) = M(\{1\}) = \lambda_{\{1\}}/\lambda \tag{4.4.2}$$

By the definition of M for fixed B the number of points of Φ in B of mark 1 has mean $\lambda_{\{1\}} \cdot \nu_d(B)$. So if $P_o(Y)$ is defined by (4.4.1) then it does not depend on the choice of the test set B. This approach is closely linked to methods of statistical estimation of Palm probabilities.

If Φ is ergodic then the Palm probability $P_o(Y)$ may be interpreted as the probability that a 'typical point' x of Φ has the property that Φ_{-x} belongs to Y. This is exactly comparable with the ergodic interpretation of a mark distribution.

Because of (4.4.1) the Palm distribution can be shown to satisfy the so-called *refined Campbell theorem* (see also Section 7.1).

Theorem (MECKE, 1967). *For any non-negative measurable function h on* $\mathbb{R}^d \times \mathsf{N}$,

$$\begin{aligned} \mathsf{E} \sum_{x \in \Phi} h(x, \Phi) &= \int \sum_{x \in \varphi} h(x, \varphi)\, P(\mathrm{d}\varphi) \\ &= \lambda \iint h(x, \varphi_x)\, P_o(\mathrm{d}\varphi)\, \mathrm{d}x. \end{aligned} \tag{4.4.3}$$

Taking $h(x, \varphi) = f(x)$, (4.4.3) coincides with (4.1.3).

Many problems in point process theory can be resolved by calculations of quantities in the form of the left-hand side of (4.4.3). The refined Campbell theorem reduces these calculations to relatively simple integrations.

The Palm distribution: Second definition

This approach uses the Radon-Nikodym theorem and the Campbell measure $\mathcal{C}$. Suppose that the intensity measure Λ is σ-finite: this implies that $\mathcal{C}$ is likewise σ-finite. Furthermore, for $Y \in \mathcal{N}$ the measure

$$B \to \mathcal{C}(B \times Y)$$

is absolutely continuous with respect to Λ. So there is a density $P_x(Y)$ with

$$\mathcal{C}(B \times Y) = \int_B P_x(Y)\, \Lambda(\mathrm{d}x). \tag{4.4.4}$$

For fixed x, $P_x(\cdot)$ can be taken to be a distribution on $[\mathsf{N}, \mathcal{N}]$: the *Palm distribution of P with respect to x*. In the stationary case, with $0 < \lambda < \infty$, it can be taken to satisfy (I) and so

$$P_o(Y) = P_x(Y_x) \quad \text{for } Y \text{ in } \mathcal{N} \tag{4.4.5}$$

as

$$\lambda \int_B P_z(Y)\,\mathrm{d}z = \mathcal{C}(B \times Y) = \mathcal{C}(B_x \times Y_x)$$
$$= \lambda \int_{B_x} P_y(Y_x)\,\mathrm{d}y = \lambda \int_B P_{x+z}(Y_x)\,\mathrm{d}z$$

for all B, x, and Y. This implies $P_z(Y) = P_{x+z}(Y)$. Putting $z = 0$ yields (4.4.5).

The elliptical language here ('can be taken to satisfy ...') is used because relations such as (4.4.4) define $P_x(Y)$ for all x off a null-set that may *a priori* depend on Y. The measure-theoretic machinery of *regular conditional probabilities* can be used to resolve this problem. It can be ignored for the practical purposes of Palm distribution theory.

The refined Campbell theorem for stationary point processes is a consequence of this second definition:

$$\mathsf{E} \sum_{x \in \Phi} h(x, \Phi) = \int h(x, \varphi)\,\mathcal{C}(d(x, \varphi))$$
$$= \lambda \iint h(x, \varphi)\,P_x(\mathrm{d}\varphi)\,\mathrm{d}x = \lambda \iint h(x, \varphi_x)\,P_o(\mathrm{d}\varphi)\,\mathrm{d}x\,.$$

The two definitions agree in the stationary case (the first definition is not otherwise applicable). A further possible way to define the Palm distribution is sketched in Section 2.4; it is to consider the conditional probabilities of point patterns given $\Phi(b(o, \varepsilon)) = 1$, and their limits as $\varepsilon \to 0$. Mathematical details can be found in MATTHES, KERSTAN, and MECKE (1978, Section 9.3), and KALLENBERG (1983b, Section 12.4).

Reduced Palm distributions

As the ordinary Palm distributions are related to $\mathcal{C}$, so the reduced Palm distributions are related to the reduced Campbell measure $\mathcal{C}^!$. If the appropriate replacement is made in (4.4.4) then the *reduced Palm distribution* $P_x^!$ is defined. So

$$P_o^!(Y) = P(\Phi \setminus \{o\} \in Y \,\|\, o) \quad \text{for } Y \in \mathcal{N}.$$

In the stationary case (4.4.1) can be employed to define $P_o^!(Y)$ if $\mathbf{1}_Y(\varphi - x)$ is replaced by $\mathbf{1}_Y((\varphi - x) \setminus \{o\})$.

The *nearest neighbour distribution function* D can be expressed via P_o, or $P_o^!$, by

$$D(r) = 1 - P_o(\varphi \in \mathsf{N} : \varphi(b(o, r) = 1))$$
$$= 1 - P_o^!(\varphi \in \mathsf{N} : \varphi(b(o, r) = 0))\,.$$

This distribution function is a valuable characteristic of point processes, and an aid to the description of empirical point patterns. Figure 4.5 shows D for various point process models.

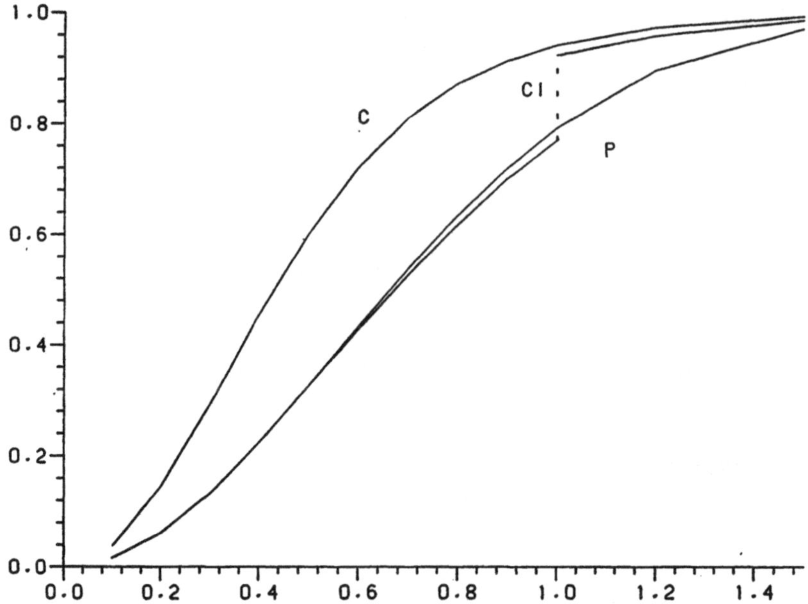

Fig. 4.5. Distribution function D of nearest neighbour distance for various point process models. The point process models are all normalised to have intensity 1. The Gauss-Poisson process is discussed in Example 5.4. The jump at $r = 1$ results from the probability of 1/3 of a cluster being two points separated by a distance one. The Cox process is discussed in Example 5.2; it is a mixture of two stationary Poisson processes of intensities $\lambda_1 = 0.5$ and $\lambda_2 = 1.5$. Poisson process, P; Cluster process (Gauss-Poisson), Cl; Cox process, C

n-fold Palm distributions

These are derived from higher order Campbell measures: see Hanisch (1982), Jagers (1973), Kallenberg (1983b), and Mecke (1979a). An n-fold Palm distribution is a conditional distribution for a point process given that n fixed positions hold points of the process. Stoyan (1984a, b, e) and Heinrich (1984b) gave examples of its use in the description of point processes. The mark covariance function k_{mm} described above is closely related to a two-fold Palm distribution.

Example 4.3 *The Palm distribution of a Poisson process.*

The *theorem of Slivnyak* gives the Palm distribution P_x of a Poisson process of intensity measure Λ and distribution P:

$$P_x = P * \delta_{\delta_x} \quad \text{for all } x. \tag{4.4.6}$$

Here '$*$' denotes convolution of distributions, which corresponds to the superposition of point processes (see 5.1). The symbol δ_{δ_x} denotes the distribution of the point process that consists solely of the (non-random) point x. Equation (4.4.6) can be interpreted as

$$P_x(Y) = \mathsf{P}(\Phi \in Y \parallel x) = \mathsf{P}(\Phi \cup \{x\} \in Y) \quad \text{for } Y \text{ in } \mathcal{N},$$

or

$$\int f(\varphi)\, P_x(\mathrm{d}\varphi) = \int f(\varphi \cup \{x\})\, P(\mathrm{d}\varphi)$$

for all measurable non-negative functions f.

If the reduced Palm distribution is used then (4.4.6) takes on a more elegant form:

$$P_x^{!} = P \quad \text{for all } x. \tag{4.4.7}$$

The Poisson process is characterized by (4.4.6) or (4.4.7): see JAGERS (1973), MATTHES KERSTAN, and MECKE (1978), MECKE (1968).

Proof that a Poisson process satisfies (4.4.6). Both $P * \delta_{\delta_x}$ and P_x are distributions of (simple) point processes. So by the discussion in 4.1 their equality is established if

$$P * \delta_{\delta_x}(V_K) = P_x(V_K)$$

for all compact K in $\mathbb{R}^d$, where $V_K = \{\varphi \in \mathsf{N} : \varphi(K) = 0\}$.

Suppose A is a bounded Borel set. Then

$$\int_A P * \delta_{\delta_x}(V_K)\, \Lambda(\mathrm{d}x) = \int_{A \setminus K} P(V_K)\, \Lambda(\mathrm{d}x) = P(V_K) \cdot \Lambda(A \setminus K)$$

$$= \mathsf{E}(\Phi(A \setminus K); \Phi(K) = 0)$$

(using the independence property of the Poisson process)

$$= \mathscr{C}((A \setminus K) \times V_K).$$

But $\mathscr{C}((A \cap K) \times V_K) = \mathsf{E}(\Phi(A \cap K); \Phi(K) = 0) = 0$.

Hence

$$\int_A P * \delta_{\delta_x}(V_K)\, \Lambda(\mathrm{d}x) = \mathscr{C}(A \times V_K) = \int_A P_x(V_K)\, \Lambda(\mathrm{d}x)$$

using (4.4.4). Since A was arbitrary, this establishes (4.4.6).

Clearly P_o is never a stationary distribution since under P_o a point process must always contain o (but Example 4.3 shows that $P_o^{!}$ can be stationary). However if the point process Φ is motion-invariant then its Palm distribution is isotropic (invariant with respect to rotations about the origin o).

Proof. Rotation invariance means

$$P_o(Y) = P_o(\mathfrak{r}Y) \quad \text{for } Y \text{ in } \mathcal{N} \tag{4.4.8}$$

for every rotation $\mathfrak{r}$ about o. To establish this consider

$$\nu_d(b(o, r)) \cdot P_o(Y) = \iint_{b(o,r)} \mathbf{1}_Y(\varphi - x)\, \varphi(\mathrm{d}x)\, P(\mathrm{d}\varphi)$$

$$= \iint_{b(o,r)} \mathbf{1}_Y(\mathfrak{r}^{-1}\varphi - x)\, \mathfrak{r}^{-1}\varphi(\mathrm{d}x)\, P(\mathrm{d}\varphi) \quad \text{(by rotational invariance of } \Phi)$$

$$= \iint_{b(o,r)} \mathbf{1}_Y(\mathfrak{r}^{-1}\varphi - \mathfrak{r}^{-1}y)\, \varphi(\mathrm{d}y)\, P(\mathrm{d}\varphi) \quad \text{(changing variables to } y = \mathfrak{r}^{-1}x)$$

$$= \iint_{b(o,r)} \mathbf{1}_{\mathfrak{r}Y}(\varphi - y)\, \varphi(\mathrm{d}y)\, P(\mathrm{d}\varphi) = P_o(\mathfrak{r}Y) \cdot \nu_d(b(o, r)).$$

Suppose a system of polar coordinates is based on a 'typical' point of a motion-invariant point process. Then as a direct consequence of (4.4.8) the random angle α of the co-ordinates of the nearest neighbour is uniformly distributed over $(0, 2\pi]$ (see Figure 4.6).

Palm distributions for marked point processes

The definitions of these follow the case of an un-marked point process, whether using the Campbell measure or (in the stationary case) the formula (4.4.1). For a marked point process Ψ the Campbell measure is defined by

$$\mathscr{C}(B \times L \times Y) = \int \psi(B \times L)\, \mathbf{1}_Y(\psi)\, P_\Psi(\mathrm{d}\psi)$$

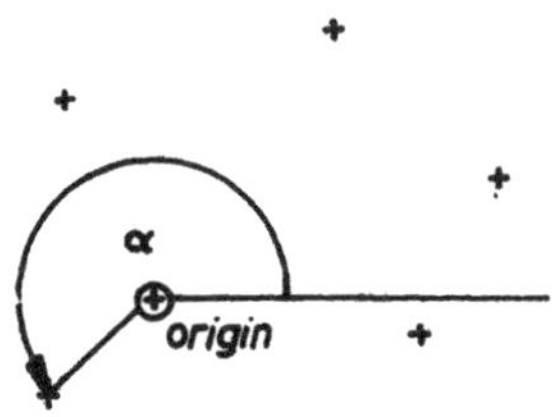

Fig. 4.6. The angle α between a point x and its nearest neighbour $n(x)$

for Borel B, L in $\mathscr{M}^*$, Y in $\mathscr{N}_M$. Here $\mathscr{N}_M$ is the smallest σ-algebra on N_M (the family of marked point sequences) that makes every mapping

$$\psi \to \psi(B \times L)$$

measurable for all Borel B and L in $\mathscr{M}^*$.

Thus for all Y the measure

$$B \times L \to \mathscr{C}(B \times L \times Y)$$

is absolutely continuous with respect to Λ, and so $P_x{}^m$ can be found such that

$$\mathscr{C}(B \times L \times Y) = \int_{B\times L} P_x{}^m(Y)\, \Lambda\big(d(x, m)\big). \tag{4.4.9}$$

If x and m are fixed then $P_x{}^m$ is a distribution on $\mathscr{N}_M$; it can be interpreted as the probability of Y given that Ψ has a point x of mark m.

From (4.2.3) in the stationary case (4.4.9) takes the form

$$\mathscr{C}(B \times L \times Y) = \lambda \int_B \int_L P_o{}^m\,(Y_{-x})\, M(\mathrm{d}m)\, \mathrm{d}x$$

where M is the mark distribution of Ψ.

The *refined Campbell theorem* for stationary marked point processes is

$$\int \sum_{[x,m]\in\psi} h(x, m, \psi)\, P(\mathrm{d}\psi) = \lambda \iiint h(x, m, \psi_x)\, P_o{}^m(\mathrm{d}\psi)\, M(\mathrm{d}m)\, \mathrm{d}x \tag{4.4.10}$$

for any non-negative measurable function h on $\mathbb{R}^d \times \mathsf{M}^* \times \mathsf{N}$.

Again in the stationary case, suppose that $L \in \mathscr{M}^*$. Then we can define the *Palm distribution of Ψ with respect to the mark set L* as P_L where

$$P_L{}^m(Y) = \int_L P_o(Y)\, M(\mathrm{d}m)/M(L)$$

(see Franken, König, Arndt, and Schmidt, 1981, and Matthes, 1963). It has an interpretation as the conditional distribution of Ψ given that Ψ has a point at o with mark in L.

The mark distribution itself can be thought of as a Palm distribution. Set $Y_L = \{\psi \in \mathsf{N}_M : [o; m] \in \psi, m \in L\}$. Then

$$P_{\mathsf{M}^*}(Y_L) = \int_{\mathsf{M}^*} P_o^m(Y_L)\, M(\mathrm{d}m) = M(L)$$

since $P_o^m(Y_L)$ is 1 if $m \in L$, 0 otherwise.

For brevity and convenience throughout this book and in the literature the phrase 'typical point' is used. This is to be interpreted using a Palm distribution. For example, if under the Palm distribution the distance from o to the nearest point has an exponential distribution, then it is said that the distance of the 'typical point' to its nearest neighbour is exponentially distributed. Similarly the statement that the mark of the 'typical point' has probability p of lying in L implies that $M(L) = p$.

It is theoretically possible to determine the distribution of Φ from λ and P_o if the point process Φ is stationary. Hanisch (1982) and Thiele (1979) give such 'inversion formulae' for point processes in $\mathbb{R}^d$. Here for example is a simple relationship between Φ and P_o:

$$k\mathsf{P}\big(\Phi(B) = k\big) = \lambda \int_B P_o\big(\varphi \in \mathsf{N} : \varphi(B - x) = k\big)\, \mathrm{d}x \tag{4.4.11}$$

for $k = 1, 2, \ldots$, and bounded Borel B.

4.5 The second moment measure

Just as the variance is a fundamental parameter of the distribution of a random variable so the second moment measure is an important characteristic of a point process. A simple description is provided in the stationary case by the Palm distribution. The second factorial moment measure $\alpha^{(2)}$ can be written

$$\begin{aligned}\alpha^{(2)}(B_1 \times B_2) &= \lambda^2 \int_{B_1} \mathcal{K}(B_2 - x)\, \mathrm{d}x \\ &= \lambda^2 \iint \mathbf{1}_{B_1}(x)\, \mathbf{1}_{B_2}(x + h)\, \mathcal{K}(\mathrm{d}h)\, \mathrm{d}x \end{aligned} \tag{4.5.1}$$

where the measure $\mathcal{K}$ is defined for Borel B by

$$\lambda\mathcal{K}(B) = \int \varphi(B \setminus \{o\})\, P_o(\mathrm{d}\varphi) = \int \varphi(B)\, P_o^!(\mathrm{d}\varphi).$$

So $\lambda\mathcal{K}(B)$ is the mean number of points in $B \setminus \{o\}$ under the Palm distribution; $\mathcal{K}$ is called the *second reduced moment measure*. There is a symmetry property under reflection:

$$\mathcal{K}(B) = \mathcal{K}(\check{B}) \quad \text{for Borel } B. \tag{4.5.2}$$

Formula (4.5.1) follows from the refined Campbell theorem:

$$\begin{aligned}\alpha^{(2)}(B_1 \times B_2) &= \mathsf{E}\left[\sum_{x_1, x_2 \in \Phi}^{\neq} \mathbf{1}_{B_1}(x_1)\, \mathbf{1}_{B_2}(x_2)\right] \\ &= \int \sum_{x \in \varphi} \mathbf{1}_{B_1}(x)\, \varphi(B_2 \setminus \{x\})\, P(\mathrm{d}\varphi) \\ &= \lambda \iint \mathbf{1}_{B_1}(x)\, \varphi\big((B_2 - x) \setminus \{o\}\big)\, P_o(\mathrm{d}\varphi)\, \mathrm{d}x. \end{aligned}$$

From (4.3.4) a formula follows for the second moment measure

$$\mu^{(2)}(B_1 \times B_2) = \lambda \nu_d(B_1 \cap B_2) + \lambda^2 \iint \mathbf{1}_{B_1}(x)\, \mathbf{1}_{B_2}(x+h)\, \mathcal{K}(\mathrm{d}h)\, \mathrm{d}x \tag{4.5.3}$$

By these formulae the second moment measure $\mu^{(2)}$ (a measure on $\mathcal{B}^d \times \mathcal{B}^d$) can be described in terms of λ and $\mathcal{K}$ (a measure on $\mathcal{B}^d$). By the work of Section 4.3 the second-order product density $\varrho^{(2)}$ will also describe $\mu^{(2)}$ if it exists. So there is a relationship between $\varrho^{(2)}$ and $\mathcal{K}$

$$\lambda^2 \mathcal{K}(B) = \int_B \varrho^{(2)}(x)\, \mathrm{d}x \quad \text{for Borel } B. \tag{4.5.4}$$

There are similar reductions for higher moment measures. Suppose the nth-reduced moment measure $\mathcal{K}^{(n)}$ has a density $g^{(n)}$ with respect to the Lebesgue density $\nu_{(n-1)d}$. Then

$$\lambda^n \cdot g^{(n)} = \varrho^{(n)}.$$

Brillinger (1975), Krickeberg (1982), and Jolivet (1983, 1984) considered estimation of $g^{(n)}$, Hanisch (1983a) studied estimation of $\mathcal{K}^{(n)}$ and furthermore discussed a directional analysis using the third reduced moment measure.

The measure $\mathcal{K}$ can be used to investigate anisotropy as in Ohser and Stoyan (1981). In the planar case ($d = 2$) this is achieved by defining a distribution function $\Theta_{r_1 r_2}$

$$\Theta_{r_1 r_2}(\alpha) = \frac{\mathcal{K}\big(S(\alpha; r_1, r_2)\big)}{\mathcal{K}\big(S(\pi; r_1, r_2)\big)} \quad \text{for } \pi > \alpha \geqq 0$$

where $S(\alpha; r_1, r_2)$ is the intersection of a closed wedge and an annulus: both wedge and annulus are based on o; the annulus radii are r_1, and r_2; the wedge is defined by the positive x-axis and a ray making the angle α with the axis. The construction is illustrated in Figure 4.7. The function $\Theta_{r_1 r_2}$ is well defined if the denominator is positive; by the symmetry property (4.5.2) it suffices to consider α in $[0, \pi]$.

The function $\Theta_{r_1 r_2}$ is a *directional distribution* for the point process Φ. If Φ is isotropic then $\Theta_{r_1 r_2}$ is uniform on $[0, \pi]$. Conversely, departure of $\Theta_{r_1 r_2}$ from uniformity indicates that there are favoured directions for points of Φ in the annulus of radii r_1, r_2 centred on a 'typical point' of Φ. Of course such anisotropy can easily depend on the parameters r_1, r_2, as is shown in the following example.

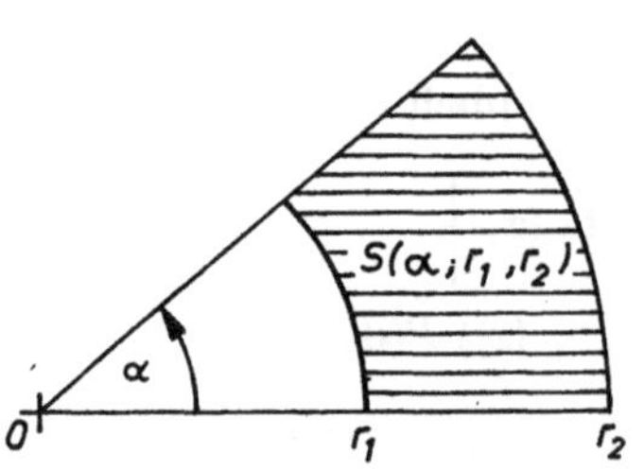

Fig. 4.7. Definition of the set $S(\alpha; r_1, r_2)$

Example 4.4. *Directional analysis of an anisotropic point pattern; centres of microtubuli.*

Figure 4.8a shows the centres of microtubuli in a cross section through a system of microtubuli. At the scale of display there are various anisotropies. Figure 4.8b shows two directional distributions $\Theta_{r_1r_2}$ for the point pattern. For $r_1 = 1$, $r_2 = 2$ dominant directions appear to be approximately 70° and 160° while for $r_1 = 2$, $r_2 = 3$ they appear to be 20° and 110°. Thus the apparent isotropy exhibits a scale dependence.

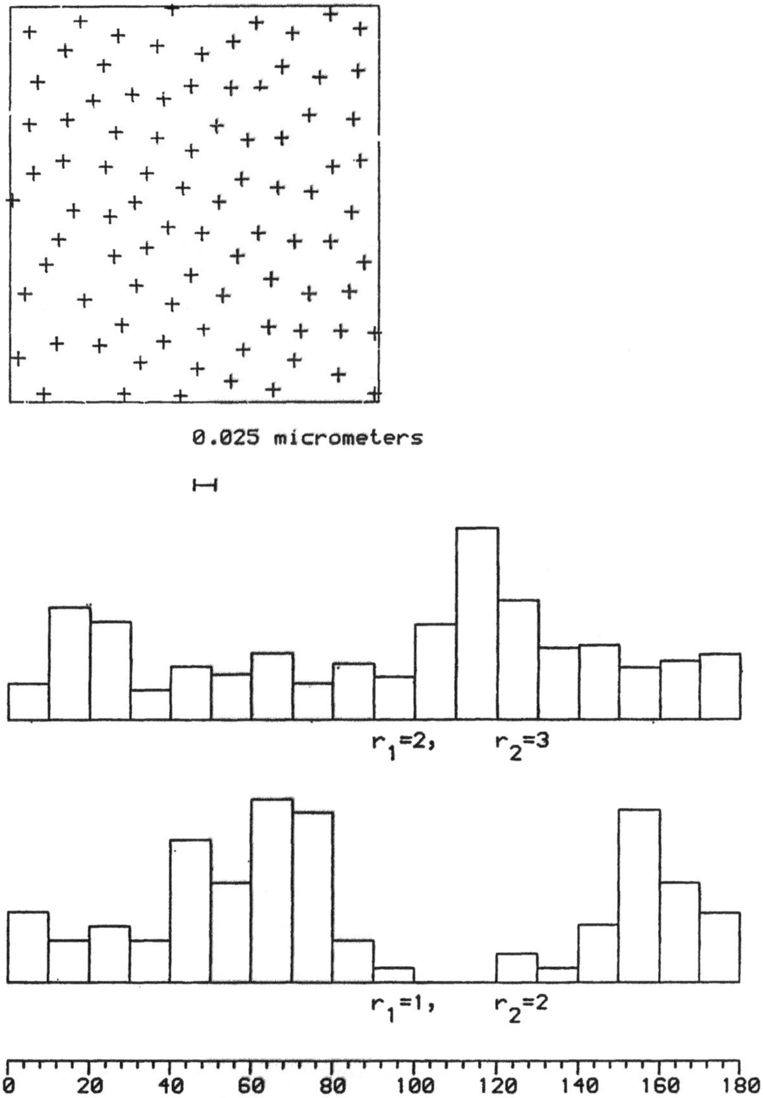

Fig. 4.8(a). Centres of cross sections through a system of microtubuli from an electromicrogram; (b) Directional distributions $\Theta_{r_1r_2}$ for $r_1 = 1$ and $r_2 = 2$ and for $r_1 = 2$ and $r_2 = 3$ for the microtubuli pattern. The distributions show quite different anisotropies at the two ranges of distances (distances in units as in Figure 4.8a)

The description of the second moment measure simplifies still further in the motion-invariant case (when isotropy is added to stationarity). It suffices to consider the *second reduced moment function* K;

$$K(r) = \mathcal{K}(b(o, r)) \quad \text{for } r \geqq 0.$$

The quantity $\lambda K(r)$ is the mean number of points of Φ within a sphere of radius r centred on a 'typical point' which is not itself counted. In the case of a stationary Poisson process

$$K(r) = \omega_d r^d \quad \text{for } r \geqq 0 \tag{4.5.5}$$

which follows from Slivnyak's theorem

$$\begin{aligned}\lambda K(r) &= \int \varphi(b(o, r))\, P_o^!(\mathrm{d}\varphi) = \int \varphi(B)\, P(\mathrm{d}\varphi) \\ &= \mathsf{E}\Phi(b(o, r)) = \lambda \nu_d(b(o, r)).\end{aligned}$$

Other functions than K are often used to describe the second-order behaviour of a point process. Which function is to be preferred depends mainly on convenience but also on statistical considerations. Some functions originate from the physical literature in which they have been used for a long time. Examples are in Ornstein and Zernicke (1914) and Girling (1982).

Here is a list of such functions.

The *product density* $\varrho^{(2)}$:

$$\varrho^2(r) = \lambda^2 \frac{\mathrm{d}K(r)}{\mathrm{d}r} \Big/ (d \cdot \omega_d r^{d-1}).$$

The *pair correlation function* g:

$$g(r) = W(r) = \varrho^{(2)}(r)/\lambda^2.$$

The *radial distribution function* RDF:

$$\begin{aligned}RDF(r) &= \lambda \frac{\mathrm{d}K(r)}{\mathrm{d}r} = d \cdot \omega_d r^{d-1} \varrho^{(2)}(r)/\lambda \\ &= 4\pi r^2 \varrho^{(2)}(r)/\lambda \quad \text{in the case } d = 3.\end{aligned}$$

The *L-function*:

$$L(r) = \sqrt[d]{(K(r)/\omega_d)}.$$

In the Poisson process case all these functions satisfy simple formulae

$$\varrho^{(2)} = \lambda^2;\ g(r) = 1;\ RDF(r) = 4\pi r^2 \lambda \quad (d = 2)$$
$$L(r) = r.$$

The forms of these functions correspond to various properties of the underlying point process. Maxima of g, or values of $K(r)$ larger than $\omega_d r^d$ for r in specific intervals, indicate frequent occurrences of interpoint distances at such values; equally minima of g or low values of $K(r)$ indicate inhibition at these distances. Either way, some form of inner order of clustering in the point pattern may be responsible. Model identification may be suggested by comparison of empirical pair correlation functions, or reduced second moment functions, with their theoretical versions. Figure 4.9 exhibits pair correlation functions for four different spatial point process models.

The K-function does *not* represent all the important information about a stationary isotropic point process. BADDELEY and SILVERMAN (1984) gave an

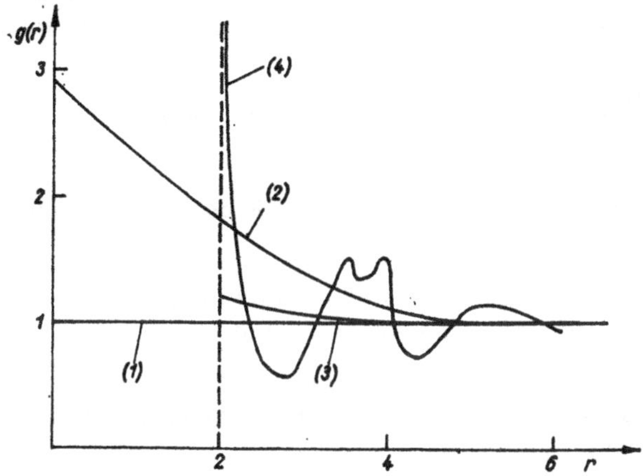

Fig. 4.9. Pair correlation functions g for various point process models: (1) Poisson process; (2) Matern's cluster process with parameters $R = 2.5$ and $\mu = 5$ (the values of $g(r)$ greater than 1 for small r result from clustering); (3) Matern hard-core process with parameters $R = 1$ and $p_h \simeq 0.03$ (the hard-core distance is 2); (4) Dense random packing of spheres; an approximation for $R = 1$ (again the hard-core distance is 2. The peaks occur in a typical pattern at $2R$, at less than $4R$, etc.)

example of a point process which is quite different from a Poisson process and yet has the same K-function

$$K(r) = \pi \cdot r^2 \quad (d = 2).$$

However, if $\mathcal{K}$, K, or g is given then (at least in principle) all variances and covariances of point counts can be calculated. In particular if B is a bounded Borel set then in isotropic case

$$\operatorname{var}\big(\Phi(B)\big) = \lambda^2 \int \gamma_B(h)\, \mathcal{K}(\mathrm{d}h) + \lambda \nu_d(B) - \big(\lambda \nu_d(B)\big)^2 \tag{4.5.6}$$

$$= \lambda^2 \int \bar{\gamma}_B(r)\, \mathrm{d}K(r) + \lambda \nu_d(B) - \big(\lambda \nu_d(B)\big)^2 \tag{4.5.7}$$

where

$$\gamma_B(x) = \nu_d(B \cap B_x) \quad \text{for } x \text{ in } \mathbb{R}^d$$

and

$$\bar{\gamma}_B(r) = \int \nu_d(B \cap B_{(r,u)})\, \mathrm{d}u/(d \cdot \omega_d).$$

Here the point (r, u) is expressed in spherical polar coordinates so that u is a point of the unit sphere. Thus if $l_B(x, r)$ is the proportion in B of the surface of a sphere of radius r and centred at x then

$$\bar{\gamma}_B(r) = \int_B l_B(x, r)\, dx$$

(STOYAN, 1983).

The determination of $\bar{\gamma}_B$ for a given B is no simple matter in general; see page 73. If $B = b(o, s)$ then the formula for $\gamma_h(r)$ on 147 can be used. If B is large ('spherically infinite' in the terminology of GIBLING, 1982) then

$$\frac{\operatorname{var}[\Phi(B)]}{\nu_d(B)} \simeq \lambda + 4\pi^2 \int_0^\infty [g(r) - 1]\, r^2\, dr \quad (d = 3).$$

In the case of multivariate point processes it is useful to consider the function K_{ij} defined by

$\lambda_j \cdot K_{ij}(r) =$ the mean number of j-points in $b(o, r)$ given that there is an i-point at o.

Appropriate statistical methods for estimating K_{ij} are discussed in BRILLINGER (1976), HANISCH and STOYAN (1979), LOTWICK and SILVERMAN 1982), DIGGLE (1983), and LOTWICK (1984).

(4.6 Introduction to statistics for planar point processes

General remarks

The theory of statistics for point processes comprises part of spatial statistics in the sense of RIPLEY (1981). The ideas presented in this section are aimed at finding suitable models for given point patterns, and then at expressing their statistical properties by means of a small number of convenient parameters or functions. Hence, the objective is to estimate basic characteristics of stationary point processes (such as λ, $\mathcal{K}$, K, D, and H_s). Methods for specific models and for other characteristics are further discussed in Section 2.7 and in Chapter 5.

Here we confine ourselves to the planar case, which has been considered in detail in the books of RIPLEY (1981), DIGGLE (1983) and UPTON and FINGLETON (1985). The statistics of point processes on the real line (often described as series of events) has a relatively large literature; see BRILLINGER (1975, 1978), COX and LEWIS (1966), SNYDER (1975), and KARR (1986). The special nature of the real line makes this case more amenable. By contrast, three-dimensional point patterns provide difficult statistical problems, especially because in many practical applications only two-dimensional projections of the spatial co-ordinates of points are available. This leads to stereological problems, which are treated in Chapter 11.

Generally statistical analysis of a point process will depend on observation of *one sample* only, and that via a compact sampling window W. Patterns arising in geology, geography, and biology are often truly unique samples of a stochastic phenomenon; in other cases data collection is so complicated that only one sample is collected. Often, therefore, it is assumed that the observed patterns are samples of stationary ergodic point processes; an assumption not susceptible to statistical analysis if there is only one sample, but one that is necessary if any statistical analysis is to be possible. We will assume ergodicity throughout this section. In practise, either it is plausible from the very nature of the data, or else one must proceed on an *ad hoc* basis as if the assumption were true, and subject one's conclusions to the proviso that while they may be of possible value in the non-stationary case they will not then have the same interpretation. For example, an empirical point density (a mean number of points per unit area) can still be calculated even if the observed point pattern is non-stationary. It has value as a description of the (spatial) average behaviour of the pattern but does not possess all the properties of an estimator of a stationary point process intensity.

An ever present problem of spatial statistics is that of *edge-effects*; RIPLEY (1982) provides an instructive discussion. OHSER, RIPLEY, and STOYAN (1985) calculated the bias due to edge-effects of a well-known estimator for K. The problem intervenes in point process statistics in the estimation of $\mathcal{K}, K, H_s$, and D. As an illustration, consider the problem of estimating the nearest-neighbour distribution function D. Two naive methods suggest themselves: plus-sampling and minus-sampling (named according as to whether they require more, or less, information than is given through the window W).

Plus-sampling

In this method D is estimated by the empirical distribution function of the sample of all distances $\|x - h(x)\|$, where x runs through the points of the pattern in the window W, and $h(x)$ is the nearest-neighbour of x *whether in the sampling window or not*. The objection to this method is that if x is near the boundary of W then its true nearest neighbour may lie outside W. So plus-sampling may require more information than is contained in the window W.

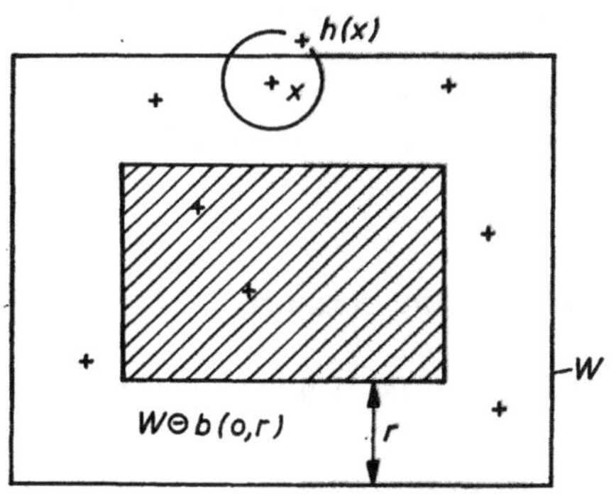

Fig. 4.10. The window W and the eroded window $W \ominus b(o, r)$ for minus-sampling of D. The shaded area is the eroded window. The nearest neighbour $h(x)$ is not in the window and so cannot be used for estimation of $D(r)$ by minus-sampling

When the point pattern can only be observed within W this means that for some x it may not be possible to determine $h(x)$. Figure 4.10 gives an illustration. If one simply ignores this difficulty, and uses the nearest-neighbour to x *lying within* W then bias is introduced; plus-sampling itself is unbiased.

Minus-sampling

In estimating $D(r)$, one might use only the points x in $W \ominus b(o, r)$ in creating the sample $\|x - h(x)\|$. This will avoid bias but loses much information for large r.

In the remainder of the section more economical edge-corrections will be suggested. The methods above are studied further in the section on estimation of Palm quantities and mark distributions, and also on p. 192.

Naturally the statistical methods employed depend on the methods of measurement. For the statistician's purposes the study of a map, or image of a point pattern, is ideal; however such maps or images may be expensive to construct. In geological or ecological analyses an exhaustive *in situ* study may be hindered by the inaccessibility of some points of the pattern. In the case of a map or an image the method of quadrat counts is typically convenient. In an *in situ* study 'distance methods' may be more economical than an exhaustive study; see p. 56.

Estimation of the intensity λ

The intensity λ can be estimated by $\hat{\lambda}$, where

$$\hat{\lambda} = \Phi(W)/\nu_2(W) \tag{4.6.1}$$

the number of points in W per area of W. Straightforward calculation shows it to be unbiased. If Φ is ergodic then $\hat{\lambda}$ is strongly consistent, in the sense that $\hat{\lambda} \to \lambda$ almost surely as the window size is increased. The sense in which the window size must be increased is that of the $\{K_n\}$ on p. 171.

The other estimators of this section are likewise consistent if the point process is ergodic.

Distance methods can also be used to estimate $\hat{\lambda}$; see Section 2.7, the references therein, and also Ripley (1981), Diggle (1983) and Upton and Fingleton (1985).

Estimation of the reduced second moment measure

Estimation of $\mathcal{K}$ (anisotropic case) In the anisotropic case $\lambda^2\mathcal{K}(B)$ can be estimated by $\hat{\kappa}_s(B)$, where

$$\hat{\kappa}_s(B) = \sum_{\substack{x,y \in \Phi \cap W \\ y-x \in B}}^{\neq} \frac{1}{\nu_2(W_x \cap W_y)} \tag{4.6.2}$$

valid for bounded Borel B such that $\nu_2(W \cap W_z)$ is positive for all z in B.

A proof that $\hat{\kappa}_s(B)$ is unbiased is as follows; OHSER and STOYAN (1981). Set $f(x_1, x_2) = \mathbf{1}_B(x_2 - x_1)\, \mathbf{1}_W(x_1)\, \mathbf{1}_W(x_2)/\nu_2(W_{x_1} \cap W_{x_2})$ and apply (4.3.2) and (4.5.1)

$$\begin{aligned} \mathsf{E}\hat{\kappa}_s(B) &= \int f(x_1, x_2)\, \alpha^{(2)}\big(d(x_1, x_2)\big) \\ &= \lambda^2 \iint f(x, x + h)\, \mathrm{d}x\, \mathcal{K}(\mathrm{d}h) \\ &= \lambda^2 \int \mathbf{1}_B(h)\, \mathcal{K}(\mathrm{d}h) = \lambda^2 \mathcal{K}(B). \end{aligned}$$

Estimation of $\mathcal{K}$ (isotropic case) Here $\hat{\kappa}_R(r)$ is an unbiased estimator of $\lambda^2 K(r)$

$$\hat{\kappa}_R(r) = \sum_{\substack{x, y \in \Phi \cap W \\ 0 < \|x - y\| < r}} \frac{k(x, y)}{\nu_2(W^{\|x-y\|})} \quad \text{for } 0 \leqq r < r^* \tag{4.6.3}$$

where $r^* = \sup\{r : \nu_2(W^{(r)}) > 0\}$, and $W^{(r)} = \{x \in W : \partial\big(b(x, r)\big) \cap W$ is not empty$\}$, and $k(x, y) = k(x; r) = 2\pi/\alpha_{xy}$ where α_{xy} is the sum of all angles of the arcs in W of a circle with centre x and radius $r = \|x - y\|$. Figure 4.11 illustrates the definition of $k(x, y)$. If $\alpha_{xy} = 0$ then $k(x, y) = 0$.

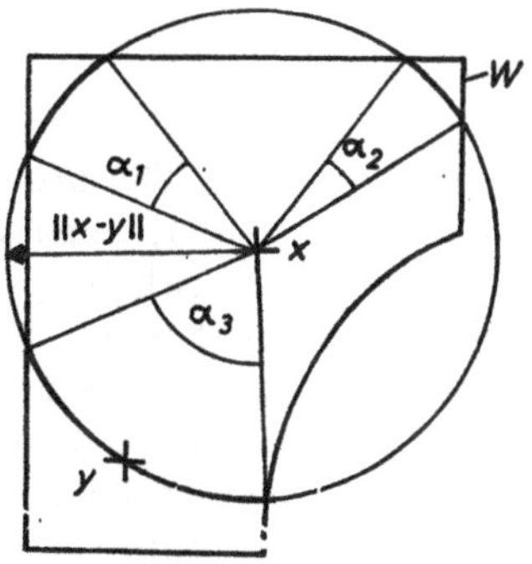

Fig. 4.11. Definition of α_{xy} for estimation of $\hat{\kappa}_R$. In the example $\alpha_{xy} = \alpha_1 + \alpha_2 + \alpha_3$

The estimator $\hat{\kappa}_R$ was originally suggested by RIPLEY (1976a); the formula (4.6.3) is as modified by OHSER (1983). The modification extends estimation to the case of large r; Ripley's original form was unbiased only for

$$r \leqq \min\{r : \text{there is } x \text{ in } W \text{ with } \partial\big(b(x, r)\big) \cap W \text{ not empty}\}.$$

OHSER (1983) contains a proof that $\hat{\kappa}_R(r)$ is unbiased, using moment measures.

OHSER and STOYAN (1981) and OHSER (1983) proposed a further unbiased estimator for $\lambda^2 \cdot K(r)$

$$\hat{\kappa}_i(r) = \sum_{\substack{x, y \in \Phi \cap W \\ 0 < \|x - y\| < r}} \frac{2\pi\|x - y\|}{\nu_2(W_x \cap W_y)\, v(\|x - y\|)} \quad \text{for } 0 \leqq r \leqq r^* \tag{4.6.4}$$

where

$$v(r) = r \int_{W_r^*} \overline{\omega}(\mathrm{d}u)$$

with $W_r^* = \{u \in \partial b(o, 1) : W \cap W_{(r,u)}$ is not empty$\}$ and (r, u) a point expressed in spherical/polar coordinates.

Which of $\hat{\kappa}_R(r)$, $\hat{\kappa}_i(r)$ is a better estimator is not clear; see RIPLEY (1982, 1984) for a comparative discussion. The differences are unlikely to be significant in practice for reasonable windows W. Numerical case studies for the Poisson process indicate that when Φ has many points in W then $\hat{\kappa}_s(b(o,r))$ and $\hat{\kappa}_i(r)$ are nearly equal. Indeed $\hat{\kappa}_i(r)$ can be interpreted as a rotation average of $\hat{\kappa}_s$.

Ohser (1983) considers further estimators of $\lambda^2 K(r)$. References to the direct estimation of the density of K are given below and on page 118.

Generally λ will be unknown, so that in application of (4.6.2), (4.6.3) and (4.6.4) λ must be replaced by $\hat{\lambda}$. Of course this introduces bias in the resulting estimators of $\mathcal{K}(B)$ and $K(r)$.

All three estimators have edge-correction built in. For example the naive estimator corresponding to (4.6.3) is

$$\hat{\kappa}(r) = \left[\frac{\begin{array}{l}\text{(Number of point pairs in } W\\ \text{separated by distance less than } r)\end{array}}{\nu_2(W)}\right] \tag{4.6.3a}$$

and for small r (such that $W = W^{(r)}$) the edge-bias is corrected by a weighting factor $k(x, y)$

$$\hat{\kappa}_R(r) = \sum_{\substack{x,y\in\Phi\cap W\\ 0<\|x-y\|<r}} \frac{k(x,y)}{\nu_2(W)} \quad (r \text{ small}).$$

This is the form originally suggested by Ripley (1976a). The principle is to correct the bias, by setting the weight $k(x, y)$ greater than 1, whenever one or both of the x, y are near the boundary of W. Deep in W (when $b(x, \|x-y\|)$, $b(y, \|x-y\|)$ lie completely in W) the weight is set equal to 1. The isotropy assumption is used to estimate the mean number of pairs missed out by the window at a particular location, and thus to calculate the weight. The full form of (4.6.3) makes a further correction which is necessary for large r.

In practical statistics it is useful to stabilise variances; this can be done by using the L-function, estimating $L(r)$ by

$$\hat{L}(r) = \sqrt{(\hat{K}(r)/\pi)}$$

where $\hat{K}$ is either $\hat{\kappa}_R/\hat{\lambda}^2$ or $\hat{\kappa}_i/\hat{\lambda}^2$. Example 2.2 illustrates the use of L.

The product density $\varrho^{(2)}$ and the pair-correlation function g can be estimated by applying numerical differentiation to an estimate of $\lambda^2 \cdot K(r)$ and then appealing to the formulae on page 120. Alternatively $\varrho^{(2)}$ can be estimated directly by using an edge-corrected density estimator such as

$$\hat{\varrho}^{(2)}(r) = \sum_{x,y\in\Phi\cap W} \frac{k\,(\|x-y\|-r)}{2\pi\,\|x-y\|\,\nu_2(W_x\cap W_y)}$$

where k is a kernel function; see Fiksel (1987a).

Estimation of the contact distribution H_s and of the generating functional G

The methods of estimating H_s for random closed sets (see Section 6.3) can be used in the particular case of point processes. Since

$$H_s(r) = \mathsf{P}(o \in \Phi \oplus b(o,r)) \quad \text{for } r \geqq 0$$

it is the areal fraction of the random closed set

$$\Xi = \Phi \oplus b(o, r) = \bigcup_{x \in \Phi} b(x, r).$$

An unbiased estimator is given by

$$\hat{H}_s(r) = \frac{\nu_2\Big(\big(W \ominus b(o, r)\big) \cap \bigcup_{x \in \Phi} b(x, r)\Big)}{\nu_2\big(W \ominus b(o, r)\big)} \quad \text{for } r \geqq 0. \tag{4.6.5}$$

This is edge-correction by minus-sampling; see Figure 4.12. Modifications of the ideas of the next section could be used to improve on this estimator. Ripley (1977) and Lotwick (1984) compared the use of H_s and K in point process statistics.

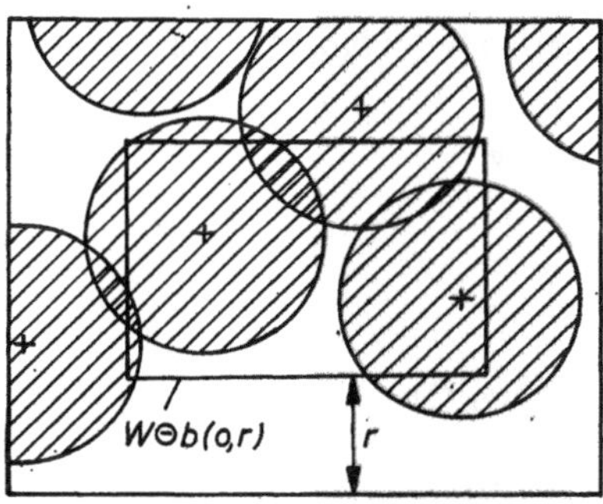

Fig. 4.12. Edge-effects and their effect on the estimation of H_s. The points outside of W contribute to $\nu_2(W \cap \Phi \oplus b(o, r))$. If their positions are not known then only points falling in the eroded window $W \ominus b(o, r)$ may be used.

Estimation of the generating functional G follows a similar route. Take v in V with $v(x) = 1$ if x is not in $b(o, r)$. Then

$$\hat{G}(v) = \sum_{i=1}^{n} \prod_{x \in \Phi \cap (W \ominus b(o,r))} v(x + z_i)/n$$

is an unbiased estimator of $G(r)$ if the $z_1, \ldots, z_n$ form a grid in $W \ominus b(o, r)$. Again this estimator is of the minus-sampling type.

Estimation of the nearest-neighbour-distance distribution function

Recall that the nearest-neighbour distribution function is defined (for a stationary point process) as

$$D(r) = 1 - P_o^{!}\big(\{\varphi \in \mathsf{N} : \varphi(b(o, r) = 0\}\big) \quad \text{for } r \geqq 0.$$

An equivalent formulation uses the marked point process obtained by marking each point x of the process with the distance $d(x)$ from x to its nearest neighbour $h(x)$. The resulting marked point process Ψ inherits a stationarity property from $\Phi = \{x_n\}$. As can be seen from the mark distribution approach to Palm distributions on page 112 the mark distribution function for Ψ is precisely D. This formulation via Ψ clarifies the logic behind the expression for estimating D.

The first estimator to be described is similar to the minus-sampling estimator of (4.6.5). It is $\hat{D}_1$, where

$$\hat{D}_1(r) = \sum_{[x,d] \in \Psi \cap (W \ominus b(o,r))} \mathbf{1}_{[0,r]}(d) / \Phi(W \ominus b(o,r)) \quad \text{for } r \geqq 0. \tag{4.6.6}$$

This is probably the most natural estimator. It is used in Ripley (1981). Only the points in the eroded window $W \ominus b(o, r)$ are used. If Φ is ergodic then $\hat{D}_1$ is asymptotically unbiased.

The quantity

$$\hat{\mathbf{D}}_2(r) = \sum_{[x,d] \in \Psi \cap (W \ominus b(o,r))} \mathbf{1}_{[0,r]}(d) / \nu_2(W \ominus b(o,r)) \quad \text{for } r \geqq 0 \tag{4.6.7}$$

is an unbiased estimator of $\lambda D(r)$. Clearly

$$\hat{\mathbf{D}}_2(r) = \lambda_r^* \hat{D}_1(r)$$

where

$$\lambda_r^* = \Phi(W \ominus b(o,r)) / \nu_2(W \ominus b(o,r))$$

and this suggests a further estimator of $D(r)$

$$\hat{D}_2(r) = \hat{\mathbf{D}}_2(r) / \hat{\lambda}.$$

These estimators appear natural, but have unfortunate properties. They need not be monotonic increasing in r, and they can exceed 1 in value; see Figure 4.13. Hanisch (1984a) has suggested other estimators which do not suffer from these disadvantages

$$\hat{D}_3(r) = \frac{\sum_{[x,d] \in \Psi} \mathbf{1}_{[0,r]}(d)\, \mathbf{1}_{W \ominus b(o,d)}(x)}{\sum_{[x,d] \in \Psi} \mathbf{1}_{W \ominus b(o,d)}(x)}. \tag{4.6.8}$$

$$\hat{D}_4(r) = \hat{\mathbf{D}}_4(r) / \hat{\mathbf{D}}_4(R) \quad \text{for } 0 \leqq r \leqq R \tag{4.6.9}$$

where

$$\hat{\mathbf{D}}_4(r) = \sum_{[x,d] \in \Psi} \mathbf{1}_{W \ominus b(o,d)}(x)\, \mathbf{1}_{[0,r]}(d) / \nu_2(W \ominus b(o,d))$$

and

$$R = \sup \{r > 0 : \nu_2(W \ominus b(o,r)) > 0\}.$$

The principle underlying both $\hat{D}_3(r)$ and $\hat{D}_4(r)$ is simple; to use precisely those points x for which it is known that the nearest neighbour $h(x)$ is both within W and closer than r to x.

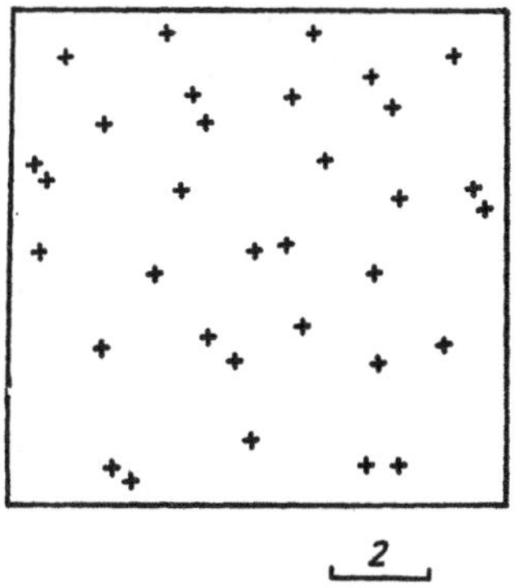

(a)

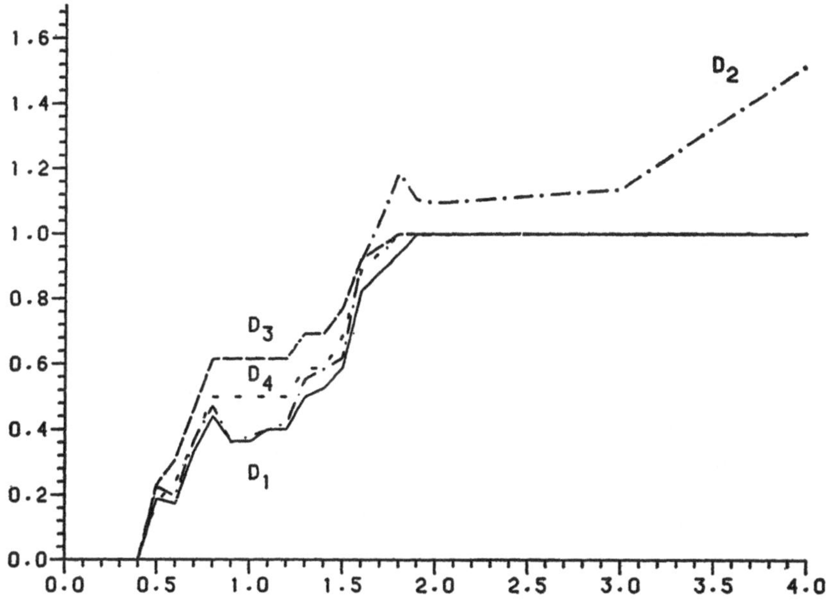

(b)

Fig. 4.13(a) A point pattern which is assumed to be a part of a sample of a stationary point process; (b) Estimation of D for the point pattern in Figure 4.13a. Note that $\hat{D}_1$ is not monotonic and that $\hat{D}_2$ exceeds 1. This example is taken from HANISCH (1984c).

Both estimators $\hat{D}_3$ and $\hat{D}_4$ are asymptotically unbiased. Numerical trials show that there is no simple relation between the precisions of any of the estimators; for example the relative sizes of estimation variances depend both on r and on the underlying model.

Estimation of Palm and mark distributions

The estimation procedures used for D above are prototypes for estimation of quantities related to Palm or mark distributions; see HANISCH (1984c) and HEINRICH (1984b). If information is accessible beyond that contained in the window then possibly the *plus-sampling* method will be useful. Consider for example estimation of λz_f, where f is a non-negative measurable function on N,

$$z_f = \int f(\varphi)\, P_o^!(\mathrm{d}\varphi).$$

Then

$$(\widehat{\lambda z_f}) = \sum_{x \in \Phi \cap W} f(\Phi_{-x} \setminus \{o\})/\nu_2(W) \tag{4.6.10}$$

is unbiased for λz_f.

For a marked point process Ψ if the marks are given for all points that fall in W then it is possible to carry out unbiased estimation of the mark distribution M

$$M(L) = \sum_{[x,m] \in \Psi \cap W} \mathbf{1}_L(m)/\nu_2(W) \quad \text{for } L \text{ in } \mathcal{M}^*. \tag{4.6.11}$$

If on the other hand, as in the two cases above, the information in W does not suffice for the calculation of $f(\varphi)$, or for the determination of all marks, then *minus-sampling methods* must be used. For example, suppose f in (4.6.10) is such that

$$f(\varphi) = f(\varphi \cap B)$$

for some set B, so that one only needs the information local to B. Then an unbiased estimator of λz_f is

$$(\widehat{\lambda z_f}) = \sum_{[x,m] \in \Phi \cap (W \ominus B)} f(\Phi_{-x} \setminus \{o\})/\nu_2(W \oplus \check{B}) \tag{4.6.12}$$

where $\check{B} = \{x : -x \in B\}$ as in Section 1.3. Calculation of $\nu_2(W \ominus \check{B})$ is described in MATHERON (1978) and WEIL (1982). The estimator $\hat{D}_2(r)$ above is a particular case of (4.6.12) using $f(\varphi) = 0$ if $\varphi(b(o, r)) = 0$, 1 otherwise, and here $B = b(o, r)$.

An analogous procedure for estimation of $M(L)$ can be carried out if the mark $m(x)$ is determined only when all points of Φ in B_x are known.

Simulation methods in point process statistics

In order to construct statistical tests or confidence intervals, and in order to compare differing estimators of the same parameter, it is necessary to know something about the distributions of the statistical quantities observed. Unfortunately such knowledge is hard to come by for the point process estimators considered above. A series of authors have considered asymptotic distributions for estimators of D and K, and other characteristics, in the case of binomial and stationary Poisson processes: see RIPLEY (1981) and references therein (pp. 161—163), DIGGLE (1983), and HEINRICH (1984a; 1986). Their results can be summarized as Central Limit Theorems and theorems ensuring weak convergence to Poisson processes. More general point process models have been studied, but only rarely: see JOLIVET (1980, 1983, 1984), KARR (1986), and KRYSCIO and SAUNDERS (1983).

In this situation it is expedient to consider simulation methods. *Simulation tests*, first suggested by DWASS (1957) and BARNARD (1963), and advocated in the context of point processes by RIPLEY (1977), BESAG and DIGGLE (1977), and DIGGLE (1979, 1983), provide a flexible means of statistical investigation when combined with use of a computer. A general discussion of simulation techniques is given by RIPLEY (1986).

A cursory description of a typical test runs as follows. Suppose the hypothesis to be considered is that a given point pattern φ, observed through a window W, is a sample of a point process Φ. One chooses a test characteristic χ, such as

$$\chi = \hat{\lambda}$$

$$\chi = \sup_{r_1 \leq r \leq r_2} |\hat{L}(r) - r|$$

$$\chi = \sup_{r_1 \leq r \leq r_2} |D(r) - \hat{H}_s(r)|$$

and calculates its value χ_0 for the given sample $\varphi|_W$. This value is compared with the ordered sample

$$\chi_{(1)} \leqq \chi_{(2)} \leqq \cdots \leqq \chi_{(m)}$$

obtained by simulating Φ m times, observing it through the window W, and then calculating the statistic χ for each simulation.

It is then possible to perform a test of the hypothesis at an *exact* significance level: for example a two-sided test of significance level $\alpha = 2k/(m + 1)$ is obtained by the rejection of the hypothesis when

$$\chi_0 \leqq \chi_{(k)} \quad \text{or} \quad \chi_0 \geqq \chi_{(m-k+1)}.$$

MARRIOTT (1979) suggested the choice of $m = 99$ for a 5% level of significance and $m = 999$ for a 1% level; see also DIGGLE (1983).

Any knowledge of the probabilistic behaviour of Φ can generally be incorporated into the test — for example, weak convergence results of the form mentioned above can be exploited to justify the testing of a derived process being a Poisson process. Such extra knowledge leads to increased economy of simulation, which otherwise can involve much effort.

Parameter estimation is also possible using simulation tests: see RIPLEY (1977) or DIGGLE (1979, 1983). A trial-and-error method chooses a set of parameters for a fixed point process model, and then conducts a simulation test to see if the parameters are compatible with the empirical data. If they are, then the parameter set can be used as an estimate of the model parameters. If not, then the procedure is repeated for a new set of parameters, and so forth until a satisfactory set is found. This is a rather *ad hoc* procedure, but it can be refined using techniques involving iterative maximization of likelihood: cf. DIGGLE and GRATTON (1984).

5. Point processes III – Construction of models

5.1 Operations on point processes

Readers will need to construct suitable point processes to study practical problems. This section describes three fundamental operations which produce new point processes from old ones. Several important models are derived from simpler ones (such as the Poisson process) by means of these operations. The resulting models, as well as some others, are described in this chapter.

The three fundamental operations to be described in this section are:

thinning;
clustering;
superposition.

Kallenberg (1976, 1983b), and Kerstan, Matthes and Mecke (1978, 1982), and others report mathematical investigations into limit theorems for the results of repeated use of these operations both separately and in combination. We confine ourselves to describing the operations.

Thinnings

A thinning operation uses some definite rule to delete points of a point process Φ, thus yielding the *thinned point process* Φ_{th}. As a random closed set Φ_{th} is a subset of Φ so that

$$\Phi_{th} \subset \Phi.$$

The simplest thinning is *p-thinning*. In this each point of Φ has probability $1 - p$ of suffering deletion, and its deletion is independent both of the locations and of the possible deletions of any other points of Φ. A natural generalization allows the thinning probability p to depend on the location x of the point. A (deterministic) function $p(x)$ is given on $\mathbb{R}^d$, with $0 \leqq p(x) \leqq 1$. A point x in Φ is deleted with probability $1 - p(x)$ and again its deletion is independent of locations and possible deletions of any other points. To emphasize the spatial dependence the generalized operation is called *$p(x)$-thinning*.

In a further generalization the function p is chosen randomly. Formally, a random field $\boldsymbol{\pi} = \{\pi(x) : x \in \mathbb{R}^d\}$ is given independent of Φ. A sample φ_{th} of the thinned process Φ_{th} is constructed by taking a sample φ of Φ and applying $p(x)$-thinning to φ, using for $p(x)$ a sample $\{p(x) : x \in \mathbb{R}^d\}$ of the random field $\boldsymbol{\pi}$. Given $\boldsymbol{\pi} = p$ and given $\Phi = \varphi$, the probability of x in Φ also belonging to Φ_{th} is $p(x)$.

All these thinnings can be called *independent thinnings*. This means there is no interaction between the points, so that the thinning functions (which are independent of Φ) determine the operation completely. Yet another generalization allows dependence on the configuration of Φ, giving the class of *dependent thinnings*. A practical context for a dependent thinning would be a point pattern representing plant locations. Mutual inhibition between plants would suggest a point process constructed by a thinning with deletion probability higher in regions of higher intensity. Figure 5.1 gives an example of the outcome of independent and dependent thinnings applied to the same initial point pattern. The Matern hard-core process of Section 5.4 is a point process resulting from dependent thinning.

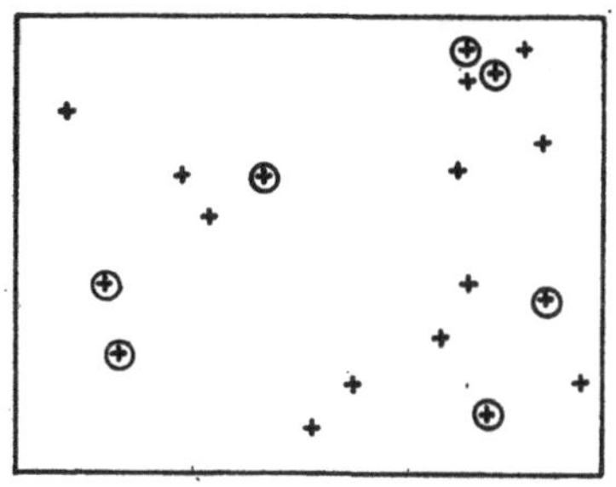

a)

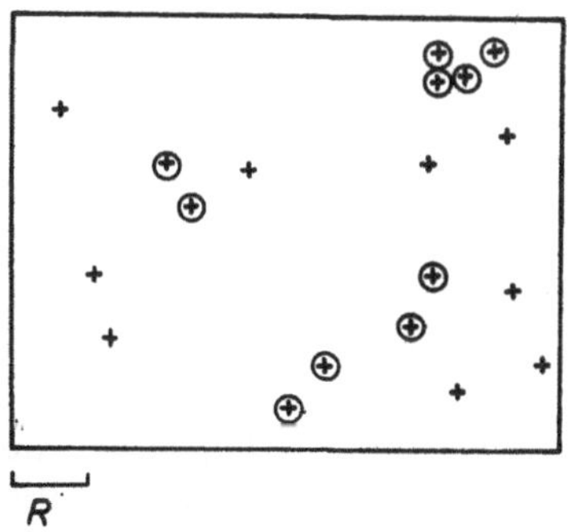

b)

Fig. 5.1. Thinnings of a point pattern. Points of the original pattern, +; Deleted points, ○ (a) Independent thinning; (b) Dependent thinning deleting all points with a nearest neighbour with in distance R

If the characteristics of the original process are known then it is relatively easy to calculate the characteristics of point processes produced by independent thinning. Thus if Φ_{th} is the result of $p(x)$-thinning of Φ, the intensity measure Λ_{th} is given by

$$\Lambda_{th}(B) = \int_B p(x)\, \Lambda(\mathrm{d}x), \tag{5.1.1}$$

where Λ is the intensity measure of Φ. If G_{th} is the generating functional of Φ_{th} then

$$G_{th}(v) = G(v_p) \quad \text{for } v \text{ in } \boldsymbol{V} \tag{5.1.2}$$

where

$$v_p(x) = v(x)\,p(x) + 1 - p(x)$$

and G is the generating functional of Φ. Analogous formulae for $\pi(x)$-thinning follow by averaging with respect to the distribution of the random field $\boldsymbol{\pi}$.

If Φ is stationary then the p-thinned process Φ_{th} is also stationary (in general this is false for $p(x)$-thinning). From (5.1.1) the intensity is given by

$$\lambda_{th} = p\lambda. \tag{5.1.3}$$

The $\pi(x)$-thinned process is stationary if $\boldsymbol{\pi}$ is a stationary random field; (5.1.3) still applies with $p = \mathsf{E}\boldsymbol{\pi}(x)$.

When both Φ and $\boldsymbol{\pi}$ are motion-invariant so is Φ_{th}. The existence of a second-order product density $\varrho^{(2)}(r)$ for φ implies the same for Φ_{th} if $\boldsymbol{\pi}$ is a second-order random field. We consider formulae for the thinned second-order product density $\varrho^{(2)}_{th}(r)$.

In the case of a p-thinning

$$\varrho^{(2)}_{th}(r) = p^2\varrho^{(2)}(r) \tag{5.1.4}$$

and the reduced second-order moment functions coincide. In the case of a $\pi(x)$-thinning let $k(r)$ denote the covariance function of the random field $\boldsymbol{\pi}$;

$$k(r) = \mathsf{E}\boldsymbol{\pi}(o)\,\boldsymbol{\pi}(\boldsymbol{r}) - p^2 \quad \text{where } \|\boldsymbol{r}\| = r.$$

Then

$$\varrho^{(2)}_{th}(r) = \big(k(r) + p^2\big)\,\varrho^{(2)}(r) \tag{5.1.5}$$

and the reduced second-order moment function K_{th} of Φ_{th} satisfies

$$K_{th}(r) = \int_0^r \big(k(x) + p^2\big)\,\mathrm{d}K(x)/p^2.$$

If $\varrho^{(2)}(r)$ is given the intuitive interpretation of 4.3 then formulae (5.1.4) and (5.1.5) are plausible; if a pair of points are both to contribute to the product density then both of them must survive.

It is easy to describe Φ_{th} if Φ is a Poisson process. Let Φ be Poisson with intensity measure Λ. Under a $p(x)$-thinning Φ_{th} is again Poisson with intensity measure given by (5.1.1). This follows easily from (5.1.2) and the formula for the generating functional of a Poisson process. Alternatively one can compute the void probabilities of a p-thinned Poisson process;

$$\begin{aligned}\mathsf{P}\big(\Phi_{th}(K) = 0\big) &= \sum_{k=0}^{\infty} \mathsf{P}\big(\Phi(K) = k\big)\cdot \mathsf{P}\,(\text{these } k \text{ points are deleted})\\ &= \sum_{k=0}^{\infty} \mathrm{e}^{-\Lambda(K)}\,\frac{\Lambda(K)^k}{k!}\,(1-p)^k = \mathrm{e}^{-p\Lambda(K)}.\end{aligned}$$

In particular if Φ is a stationary Poisson process of intensity λ then its p-thinning is stationary Poisson of intensity $p\lambda$. The $\pi(x)$-thinnings of Poisson processes are doubly stochastic Poisson processes (i.e. Cox processes) as described in Section 5.2.

Example 5.1. *An interrupted point process; buzzard nest locations.*

Figure 5.2 shows the positions of 37 buzzard nests in the New Forest in Southern England, together with the distribution of woodland. The data are taken from RIPLEY (1985c). Thinning can provide a simple model for the point pattern. The observed locations are taken to be the result of thinning a motion-invariant point process Φ (representing 'potential nests') by using the indicator of a motion-invariant random closed set

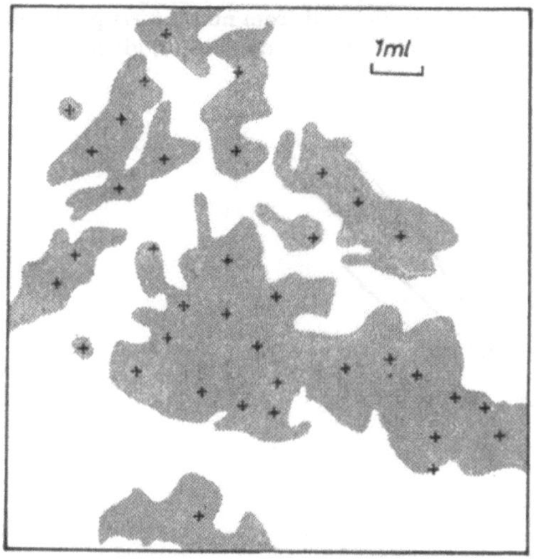

Fig. 5.2. Positions of Buzzards' nests and approximate boundaries of woodland. This figure is taken from RIPLEY (1985b) and is reproduced by kind permission of the author and Springer-Verlag.

Ξ (representing the woodland). In the notation of STOYAN (1979d) this is an *interrupted point process*; it is a $\pi(x)$-thinning with $\pi(x) = \mathbf{1}_\Xi(x)$. It can also be written as

$$\Phi_i = \Phi \cap \Xi .$$

Let Φ have intensity λ and reduced second-order moment function K and let Ξ have area fraction p and covariance C. Note that

$$k(r) = C(r) - p^2$$

if k is the covariance *function* of the random field $\mathbf{1}_\Xi$. From the formulae above the intensity of Φ_i is

$$\lambda_i = \lambda p$$

and its reduced second-order moment function is

$$K_i(r) = \int_0^r C(x) \, \mathrm{d}K(x)/p^2 . \tag{5.1.6}$$

Naturally this model is only an approximation, assuming in particular that buzzard nest locations depend only on the *presence* of woodland and on no other features of the woodland region. This entails implicit assumptions about the way in which birds choose their nest locations.

A simple statistical study is informative. Figure 5.3 displays estimates of $L_i(r) = \sqrt{(K_i(r)/\pi)}$ and $C(r)$. The estimated first-order statistics are

$$\hat{p} = 0.32$$
$$\hat{\lambda}_i = 0.29 \text{ miles}^{-2}$$
$$\hat{\lambda} = 0.91 \text{ miles}^{-2}.$$

Numerical solution of (5.1.6) yields the curve $L(r)$ shown in Figure 5.3a. It is similar to reduced second-order moment functions derived from non-interrupted point patterns of nest locations, as discussed in RIPLEY (1981). RIPLEY (1985b) gives a detailed discussion of these nest patterns.

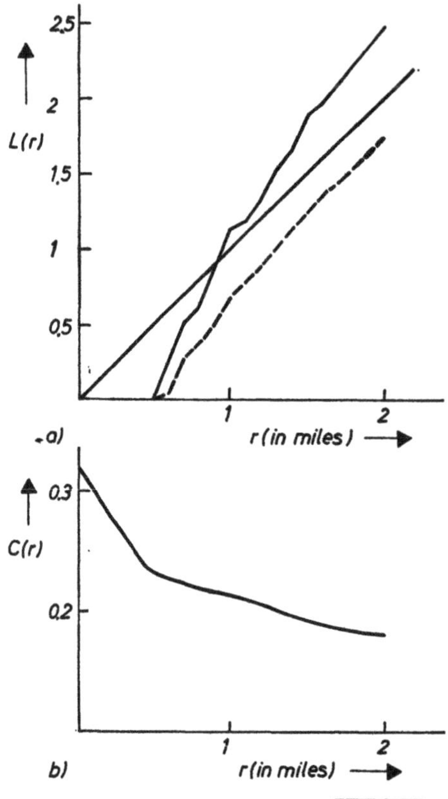

Fig. 5.3 (a). A graph of $L_i(r) = \sqrt{(\hat{K}_i(r)/\pi)}$ for the point pattern of Figure 5.2. The dashed curve is the L-function obtained by numerical solution of (5.1.6); (b) The covariance of the woodland area of Figure 5.2

Clustering

In clustering operations every point x of the point process Φ is replaced by a cluster N^x of points. The replacement clusters N^x are themselves point processes, but in most cases they have only a finite number of points each. The union of all these clusters is the *cluster point process* Φ_{cl};

$$\Phi_{cl} = \bigcup_{x \in \Phi} N^x.$$

It is tacitly assumed that with probability one none of the clusters intersect, so that $N^x \cap N^y$ is empty if $x \neq y$. Furthermore, it is assumed that Φ_{cl} is locally finite. Depending on the circumstances, x may or may not itself belong to N^x. Figure 5.4 shows a sample of a cluster point process.

Thinning operations are special cases of cluster operations. A thinning can be obtained by supposing N^x to be either the singleton set $\{x\}$ or the empty set, depending on the thinning criterion evaluated at x.

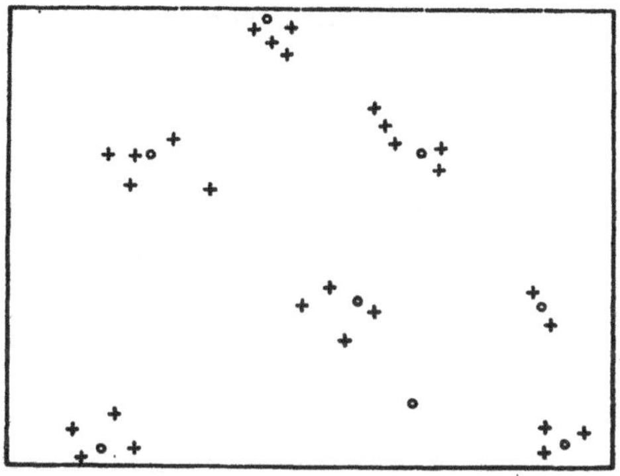

Fig. 5.4. A sample of a cluster process. Parent points, ○; Daughter points, +

Cluster processes have been used as models for many natural phenomena, as documented by the references in 5.3. A typical interpretation regards Φ as an assembly of 'parent points', perhaps locations of plants. Then the points of the clusters are the 'daughters' of the corresponding parent points, perhaps seed-fall locations. The cluster pattern is generated as the new plants grow.

An important special case is the Neyman–Scott process which is described in detail in Section 5.3. Here the formulae are given for a somewhat more general model. It is supposed that the parent point process $\Phi = \{x_1, x_2, \ldots\}$ is stationary and the clusters N^x are of the form

$$N^{x_i} = N_i + x_i$$

for each x_i in Φ. The N_i form a family of independent, identically distributed, finite point sets with distribution $\mathfrak{c}$, independent of the parent point process Φ. This is called *homogeneous independent clustering*. If Φ is a Poisson process the resulting process Φ_{cl} is called a *Poisson cluster process*.

Whatever the form of Φ in this model, the intensity λ_{cl} of Φ_{cl} is given by

$$\lambda_{cl} = \lambda \bar{c} \tag{5.1.7}$$

where $\bar{c}$ is the mean number of points in a finite point set of distribution $\mathfrak{c}$. (Here $\bar{c}$ and the intensity λ of Φ are assumed to be finite.)

If G, G_c, and G_{cl} denote the generating functionals of Φ, N_0, and Φ_{cl} then

$$G_{cl}(v) = G\big(G_{(\cdot)}(v)\big) = \mathsf{E} \prod_{x \in \Phi} G_{(x)}(v) \tag{5.1.8}$$

where

$$G_{(x)}(v) = G_c(v(x + \cdot)) = \mathsf{E} \prod_{y \in N_0} v(x + y).$$

Here $v \in \boldsymbol{V}$, and N_0 is a cluster with distribution $\mathfrak{c}$, a *representative cluster* of the cluster process.

Formula (5.1.8) expresses a close similarity to the theory of branching processes, and indeed the clustering operation is precisely what is used to create new generations of a *branching random walk* or *spatial branching process*; see BIGGINS (1977), MATTHES, KERSTAN, and MECKE (1978), or MOLLISON (1977).

Superposition

Let Φ_1, Φ_2 be two point processes with distributions P_1 and P_2. (Other characteristics of the two point processes will be suffixed in a similar fashion.) Consider the union

$$\Phi_s = \Phi_1 \cup \Phi_2.$$

It is assumed that points of Φ_1 and Φ_2 have probability zero of coinciding (equivalent in the case of independence to the corresponding intensity measures having no common atoms, by JAGERS, 1973). The set-theoretic union then coincides with the superposition operation of general point process theory (in the general theory it is legitimate for the result of superposition to have multiple points, a complication we choose to exclude).

It is clear that the intensity measure Λ_s and intensity λ_s (in case of stationarity) of Φ_s are given by

$$\Lambda_s = \Lambda_1 + \Lambda_2$$

$$\lambda_s = \lambda_1 + \lambda_2.$$

If Φ_1, Φ_2 are independent then the generating functional G_s of Φ_s satisfies

$$G_s(v) = G_1(v) \cdot G_2(v) \quad \text{for } v \text{ in } \boldsymbol{V}.$$

The distribution P_s of Φ_s is then written as

$$P_s = P_1 * P_2.$$

If Φ_1 and Φ_2 are independent Poisson processes then so is Φ_s. This is clear because the sum of two independent Poisson variables is itself Poisson.

5.2 Doubly stochastic Poisson processes (Cox processes)

An obvious generalization of a Poisson process is made by supposing that the intensity measure is itself random, with the point process being Poisson *conditional* on the realization of the intensity measure. Such a process is called a *doubly stochastic Poisson process* or *Cox process*.

Formally a Cox process is defined by choosing a distribution Q on $[\mathsf{M}, \mathcal{M}]$, the space of non-negative locally finite measures on $\mathbb{R}^d$. Suppose P_Λ is the distribution of a Poisson process of intensity measure Λ, and Ψ is a random measure with distribution Q. (Chapter 7 discusses the theory of random measures.) Then the *Cox process Φ_C with driving random measure Ψ* has distribution

$$\begin{aligned} P_{\Phi_C}(Y) &= \int P_\Lambda(Y)\, Q(\mathrm{d}\Lambda) \\ &= \mathsf{E}\big(P_\Psi(Y)\big) \quad \text{for } Y \in \mathcal{N}. \end{aligned}$$

To ensure that Φ_C is simple (that Φ_C has no multiple points) it is assumed that Q is concentrated on the subset of diffuse measures.

A Cox process can be thought of as arising from a two-step random mechanism (hence the term 'doubly stochastic'). The first step is to generate a measure Λ on $\mathbb{R}^d$ according to the driving random measure distribution Q. The second stage is to generate a Poisson process of intensity measure Λ.

Cox processes are both sufficiently general and amenable to calculation that they find important applications as stochastic models. They occur frequently in physics, for example, in optics. Systematic study began with Cox (1955), after earlier examples. Mecke (1968) and Krickeberg (1972a) made important contributions. Grandell (1976) is an excellent book on the subject; further material is to be found in Snyder (1975), Kallenberg (1983b), and Karr (1986).

Examples of Cox processes

(1) *The mixed Poisson process* The *mixed Poisson process* is a simple instance of a Cox process; it can be thought of as a stationary Poisson process with randomized intensity parameter. The driving random measure Ψ is the random measure

$$\Psi = X\nu_d$$

where X is a non-negative random variable (the randomized intensity). Every

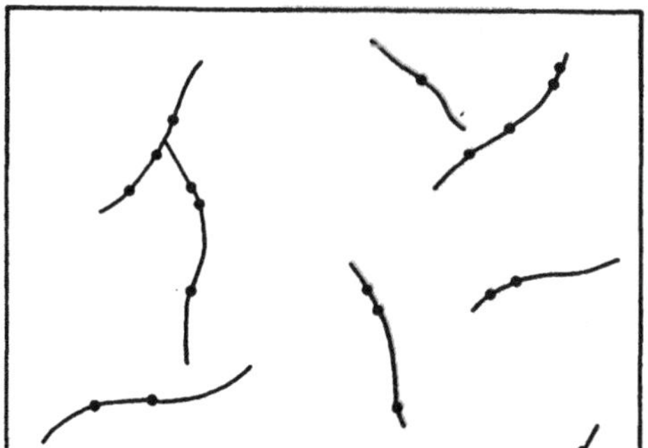

Fig. 5.5. A sample of a Cox process. The points are randomly scattered on a system of random fibres. (Compare with fig. 7.1 in Hodder and Orton (1976)! That figure shows the distribution of finds of late Iron Age inscribed coins and Roman roads in central and southern England. Many coin finds are on or very close to these roads.)

sample of such a process looks like a sample of some stationary Poisson process. This is a typical example of a non-ergodic Cox process.

(2) *A $\pi(x)$-thinning of a Poisson process.* In this case the driving random measure Ψ is given by

$$\Psi(B) = \int \pi(x)\, \Lambda(\mathrm{d}x) \quad \text{for Borel } B$$

where Λ is the intensity measure of the original Poisson process.

(3) *Points scattered on random fibres.* Let χ be a random fibre process (see Section 9.3). The random measure Ψ defined by

$$\Psi(B) = N_L \nu_1(\chi \cap B) \quad \text{for Borel } B$$

yields a driving measure for a Cox process. Here N_L is a positive constant and $\nu_1(\chi \cap B)$ is the total length of fibres intersected with B. The resulting Cox process is formed of points scattered at random on the fibres of χ, as illustrated in Figure 5.5. The parameter N_L is the number of points per unit length of fibre.

Formulae for characteristics of Cox processes

The generating functional G_C of the Cox process Φ_C is given by

$$G_C(v) = L_Q(1 - v) \quad \text{for } v \text{ in } \boldsymbol{V} \tag{5.2.1}$$

where L_Q is the Laplace functional of the driving random measure Ψ of Φ_C. This follows from

$$\begin{aligned} G_C(v) &= \int G_\Lambda(v)\, Q(\mathrm{d}\Lambda) = \int \exp\left(-\int (1 - v)\, \mathrm{d}\Lambda\right) Q(\mathrm{d}\Lambda) \\ &= L_Q(1 - v) \end{aligned}$$

where G_Λ is the generating functional of the Poisson process of intensity measure Λ.

The void probabilities of a Cox process are given by

$$\begin{aligned} v_K &= \int \exp\big(-\Lambda(K)\big)\, Q(\mathrm{d}\Lambda) \text{ for compact } K \\ &= L_Q(\mathbf{1}_K) \end{aligned} \tag{5.2.2}$$

since

$$\begin{aligned} v_K &= \mathsf{P}(\Phi_C \cap K \text{ is empty}) = \mathsf{P}\big(\Phi_C(K) = 0\big) \\ &= \mathsf{E} \exp\big(-\Psi(K)\big) = \int \exp\big(-\Lambda(K)\big)\, Q(\mathrm{d}\Lambda). \end{aligned}$$

The factorial moment measures $\alpha_C{}^{(n)}$ of Φ_C coincide with the moment measures $\mu^{(n)}$ of Ψ

$$\alpha_C{}^{(n)} = \mu^{(n)} \quad \text{for } n = 1, 2, \ldots \tag{5.2.3}$$

and in particular the intensity measures are equal

$$\Lambda_C = \Lambda. \tag{5.2.4}$$

If Ψ is stationary then so is Φ_C, and both have the same intensity parameter. In particular every mixed Poisson process is stationary.

The Palm distributions P_x for Φ_C are given by

$$P_x = \delta_{\delta_x} * P_{Q_x}, \tag{5.2.5}$$

where P_{Q_x} is the distribution of a Cox process with driving random measure Q_x, and Q_x is the Palm distribution of Ψ at x. Note that the Palm measures for a random measure are defined in Section 7.1.3.

Formula (5.2.5) is a good starting point for calculating the second-order reduced moment distribution function K, and the nearest-neighbour distance distribution function D.

For all Cox processes the numbers of counts in bounded sets is *super-Poissonian*. That is to say, their variances always exceed those for stationary Poisson processes of the same intensity. This is a consequence of the mixed Poisson distribution of the number of counts in a bounded set.

Example 5.2. *A mixed Poisson process using two intensities λ_1 and λ_2.*

Consider a mixed Poisson process Φ_C as above, with the randomised intensity X having a two-point distribution

$$\mathsf{P}(X = \lambda_1) = 1 - p, \qquad \mathsf{P}(X = \lambda_2) = p$$

where $\lambda_1 < \lambda_2$.

Its intensity is given by

$$\lambda_C = (1 - p)\,\lambda_1 + p\lambda_2.$$

Its Palm distribution P_o is derived from the Palm distributions $P_o^{(1)}$ and $P_o^{(2)}$ of Poisson processes with intensities λ_1 and λ_2 respectively

$$P_o(Y) = \frac{1}{\lambda_C}\left(\lambda_1(1 - p)\,P_o^{(1)}(Y) + \lambda_2 p P_o^{(2)}(Y)\right).$$

Consequently the nearest-neighbour distance distribution function is given by

$$D_C(r) = \frac{1}{\lambda_C}\left[\lambda_1(1 - p)\,(1 - \exp(-\lambda_1\omega_d r^d)) + \lambda_2 p(1 - \exp(-\lambda_2\omega_d r^d))\right];$$

see Figure 4.5. The reduced second-order moment function K_C of Φ_C is

$$K_C(r) = \frac{\lambda_1^2(1 - p) + \lambda_2^2 p}{\lambda_C^2}\,\omega_d r^d \quad \text{for } r \geqq 0.$$

Applications of Cox processes in spatial statistics are discussed by Bartlett (1975), Grandell (1981), and Matern (1971). Diggle and Milne (1983), Diggle (1983), and Lotwick (1984) discussed a marked Cox process as a model for a pattern of points of two types, such as locations of two different species of trees. Positive and negative dependence are combined in this model.

Stapper, McLaren, and Dreckmann (1980) used a mixed Poisson process to model the pattern of point defects on the surfaces of silicon wafers arising in semiconductor manufacture. The randomised intensity was given a Gamma distribution. See also Baudin (1984).

Statistical problems for Cox processes have been studied in many papers. Krickeberg and Karr have made major contributions; see Krickeberg (1982) and Karr (1986). The book of Synder (1975) considers statistical problems for Cox processes on the real line.

5.3 Neyman–Scott processes

Neyman–Scott processes are examples of cluster processes, and are often used in spatial statistics. They result from homogeneous independent clustering applied to a stationary Poisson process; the parent points form a stationary Poisson process of intensity λ while the daughter points of a representative cluster N_0 are random in number and are scattered independently and with identical distribution about o. The parent points do not occur in the observed point pattern; only daughter points are included.

Under these assumptions the resulting cluster process Φ_{cl} is stationary. If the scattering distribution is isotropic then so is Φ_{cl}.

The Neyman–Scott process can be interpreted as a particular kind of Boolean model. This is useful for the derivation of formulae for void probabilities.

The general formula (5.1.7) for the intensity λ_{cl} gives

$$\lambda_{cl} = \lambda \bar{c}$$

where $\bar{c}$ is the mean number of daughter points per parent. The generating functional formula (5.1.8) specializes to

$$G_{cl}(v) = \exp\left(-\lambda \int_{\mathbb{R}^d} \left[1 - g\left(\int_{\mathbb{R}^d} v(x + y)\, f(y)\, \mathrm{d}y\right)\right] \mathrm{d}x\right) \tag{5.3.1}$$

where g is the generating function for the random number of points of N_0 and f is the density for the position of a typical daughter point of N_0.

Further progress can be made by considering an important formula for the Palm distribution P_o of a Poisson cluster process. Let P_{cl} be the distribution of Φ_{cl} and $\mathfrak{c}$ the distribution of the cluster N_0. Let $\mathfrak{c}_o$ be the Palm distribution of N_0. Then

$$P_o = P_{cl} * \mathfrak{c}_o \tag{5.3.2}$$

as given by Ambartzumian (1966) and Mecke (1968). The Palm distribution $\mathfrak{c}_o$ is given by

$$\mathfrak{c}_o(Y) = \frac{1}{\bar{c}} \mathsf{E} \sum_{x \in N_0} \mathbf{1}_Y(N_0 - x) \quad \text{for } Y \in \mathcal{N} \tag{5.3.3}$$

which is the formula for the Palm distribution of a point process with finite mean total number of points. More general formulae than (5.3.2) can be found in Kerstan, Matthes and Mecke (1978, p. 430) and in Liemant (1981).

Formula (5.3.2) can be seen as plausible if it is considered as a generalization of Slivnyak's theorem for the Palm distribution of a stationary Poisson process. The resulting Palm distribution P_o for Φ_{cl} may be interpreted as the superposition of the Palm distribution $\mathfrak{c}_o$ for a 'typical cluster' and an independent point process with the distribution P_{cl} of the original Φ_{cl}. The $\mathfrak{c}_o$ derives from the cluster containing the 'typical point' of Φ_{cl} while P_{cl} describes the statistics of the Neyman–Scott process obtained by deleting the parent of the 'typical point'.

Let $\{p_n\}$ be the distribution of the number of points in the representative cluster N_0 so that $p_n = \mathsf{P}(N_0 = n)$. Then

$$\bar{c} = \sum_{n=0}^{\infty} n p_n .$$

Simple calculations with conditional probabilities can be applied to $Y_k = \{\varphi \in \mathsf{N} : \varphi(\mathbb{R}^d) = k\}$, the event that a point process has precisely k points, to show that

$$\begin{aligned}\pi_k &= \mathsf{c}_o(Y_k) = \mathsf{E} \sum_{x \in N_o} \mathbf{1}_{Y_k}(N_o - x)/\bar{c} \\ &= k p_k/\bar{c} \quad \text{for } k = 1, 2, \ldots \\ \pi_0 &= 0.\end{aligned}$$

Thus the Palm distribution c_o is *number-weighted* and different from the distribution of the representative cluster N_0.

By use of (5.3.2) calculations can be made of quantities that are related to Palm distributions of Neyman–Scott processes. In the isotropic case the reduced second-order moment function K_{cl} satisfies

$$\lambda_{cl} K_{cl}(r) = \lambda_{cl} \nu_d\big(b(o, r)\big) + \frac{1}{\bar{c}} \sum_{n=1}^{\infty} p_n n(n-1)\, \mathsf{F}(r) \tag{5.3.4}$$

where F is the distribution function of the distance between two independent random points whose common distribution is the scattering distribution used in the definition of N_0.

From (5.3.4) the second-order product density $\varrho_{cl}^{(2)}$ is given by

$$\varrho_{cl}^{(2)}(r) = \lambda_{cl}^2 + \lambda \sum_{n=1}^{\infty} p_n n(n-1) \frac{\mathsf{f}(r)}{d\omega_d r^{d-1}} \quad \text{for } r \geq 0$$

where f denotes the density function for the distribution F.

From (5.3.2) the nearest-neighbour distance distribution function D_{cl} is given by

$$1 - D_{cl}(r) = \mathsf{P}\big(\Phi_{cl}(b(o, r)) = 0\big) \cdot \mathsf{c}_o\big(\{\varphi \in \mathsf{N} : \varphi(b(o, r)) = 1\}\big).$$

Stoyan (1983) showed that the variance of the number of points in a compact convex set is super-Poissonian for a Neyman–Scott process. That is, it is greater than that for a Poisson process of the same intensity. A similar inequality holds for the corresponding K-functions.

Example 5.3. *Matérn's cluster process* (Matérn, 1960).

In this example, the representative cluster N_0 has distribution as follows. The number of points in N_0 is Poisson of parameter μ where $\mu > 0$. The points of N_0 are independently uniformly scattered in the ball $b(o, R)$ where R is a further model parameter. Hence

$$\lambda_{cl} = \lambda\mu$$

and the density f occurring in the formula for $\varrho_{cl}^{(2)}(r)$ is of the form

$$f(r) = \begin{cases} \left(1 - \frac{r}{2R}\right)\Big/R & \text{if } d = 1 \\ \frac{4r}{\pi R^2}\left\{\arccos\frac{r}{2R} - \frac{r}{2R}\sqrt{\left(1 - \frac{r^2}{4R^2}\right)}\right\} & \text{if } d = 2 \\ \frac{3}{2}\frac{r^2}{R^6}\left(R - \frac{r}{2}\right)^2\left(2R + \frac{r}{2}\right) & \text{if } d = 3 \end{cases}$$

for $0 < r < 2R$. (These results, due to various authors, can be found in SANTALÓ, 1976, p. 212, note 6.)

A sample of a simulated Matern cluster process is shown in Figure 5.6; Figure 4.9 shows the second-order product density for a particular example.

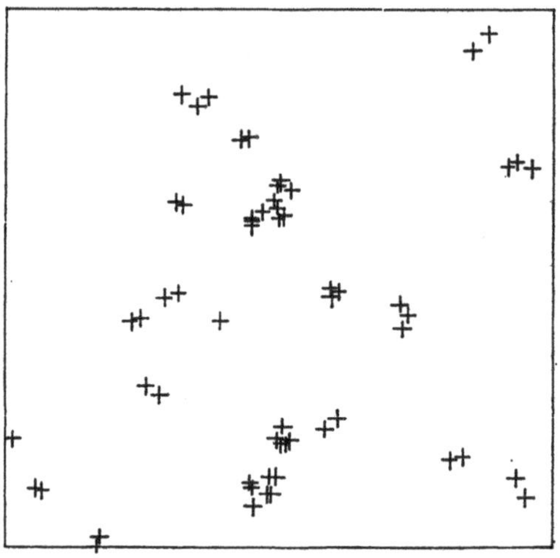

Fig. 5.6. A sample of a Matern cluster process generated by simulation. The parameters are: $\lambda = 25$; $R = 0.025$; $\mu = 2$. The sidelength of the square is one

Variations on the scatter distribution produce other models. If in the construction above the uniform distribution in the disc is replaced by the symmetric normal distribution then the result is the *modified Thomas process*. For this model in the case $d = 2$ the second-order product density is

$$\varrho_{cl}^{(2)}(r) = \lambda_{cl}^2 + \frac{\lambda\mu^2}{4\pi\sigma^2}\exp\left(-\frac{r^2}{4\sigma^2}\right) \quad \text{for } r \geqq 0$$

(COX and ISHAM, 1980). Here $\mu = \bar{c}$ is the mean of the Poisson distribution of cluster size and σ^2 is the variance of the symmetric normal distribution.

A subclass of Neyman–Scott processes is formed by the Gauss–Poisson processes. MILNE and WESTCOTT (1972) discuss this subclass in which each cluster consists of zero, one, or two points only.

Example 5.4. *Calculation of D_{cl} for a planar Gauss–Poisson process.*

In this example a typical cluster N_0 has an isotropic distribution, being composed of zero, one, or two points with probability p_0, p_1, or p_2 respectively. If N_0 is composed of one point then that point is the origin o. If N_0 is composed of two points then these are separated by a unit distance and have midpoint o.

The formulae developed above yield the following results

$$\mathsf{E}\nu_2[N_0 \oplus b(o, r)] = p_1\pi r^2 + p_2[2\pi r^2 - \gamma_r(1)]$$

where

$$\gamma_r(1) = 2r^2 \arccos\left(\frac{1}{2r}\right) - \frac{1}{2}\sqrt{(4r^2 - 1)} \quad \text{for } 2r \geq 1 .$$

Here $\gamma_r(1)$ is the area of intersection of two circles of radius r and centres separated by 1. Moreover,

$$\mathsf{C}_o\{\varphi \in \mathsf{N} : \varphi(b(o, r)) = 0\} = \frac{1}{\bar{c}}(p_1 + 2p_2\bar{\Theta}_1(r))$$

where $\bar{\Theta}_1(x) = 1$ if $x < 1$ and 0 otherwise. Therefore

$$1 - D_{cl}(r) = \frac{1}{\bar{c}}\exp\left[-\lambda\left(p_1\pi r^2 + p_2(2\pi r^2 - \gamma_r(1))\right)\right] \cdot (p_1 + 2p_2(\bar{\Theta}_1 r)) .$$

A graph of D_{cl} is given in Figure 4.5 for the particular case of

$$p_0 = p_1 = p_2 = 1/3 \quad \text{and} \quad \lambda = 0.5 .$$

Some processes are simultaneously Neyman–Scott *and* Cox processes. A Neyman–Scott process is Cox if the number of points per cluster is Poisson; this category includes the Matern cluster process, the modified Thomas process, and the Gauss–Poisson process. For example, the driving measure Ψ of the Matern cluster process is

$$\Psi(B) = \mu \sum_{x \in \Phi} \nu_d\big(B \cap b(x, R)\big)/\nu_d\big(b(o, R)\big)$$

where Φ denotes the Poisson process of parent points. See also Cliff and Ord (1981), and Bartlett (1963) who gives an example of a Poisson cluster process which is *not* Cox.

The Neyman–Scott process has found many applications as a model of spatial phenomena. Neyman and Scott (1958) introduced it as a model for the locations of galaxies in space. In this context the cluster centres are unobservable. The survey paper of Neyman and Scott (1972) contains further examples: the distribution of larvae in fields, and the geometry of bombing. In the case of larvae in fields the parent points are positions of egg masses and the daughter points are the positions of larvae. In the case of the geometry of bombing, the parent points are the aiming points of bomb release and the daughter points are the points of impact of individual bombs. Baudin (1979) studied the pattern of epicentres of Japanese micro-earthquakes. Warren (1962, 1971) considered the application of cluster processes in ecology.

Problems of statistical inference for Neyman–Scott processes have been considered by Baudin (1979, 1981, 1983), using the D and K functions. Kryscio and Saunders (1983) also used interpoint distances to give statistical tests using Poisson limit theorems, and they gave simulated examples.

5.4 Hard-core point processes

A hard-core point process is a point process in which the constituent points are forbidden to lie closer together than a certain minimum distance. In this section the minimum distance is denoted by h. These hard-core models serve to describe patterns produced by the locations of centres of non-overlapping circles or spheres of radius $R = h/2$. MATERN (1960) suggested two such models; here we describe the model yielding a higher eventual intensity of points (a similar hard-core process was proposed by POROD, 1952. The intensity that it yields is also not higher than the intensity yielded by the model described here).

The model, the *Matern hard-core process*, is essentially a dependent thinning applied to Φ a stationary Poisson process of intensity λ. The points of Φ are

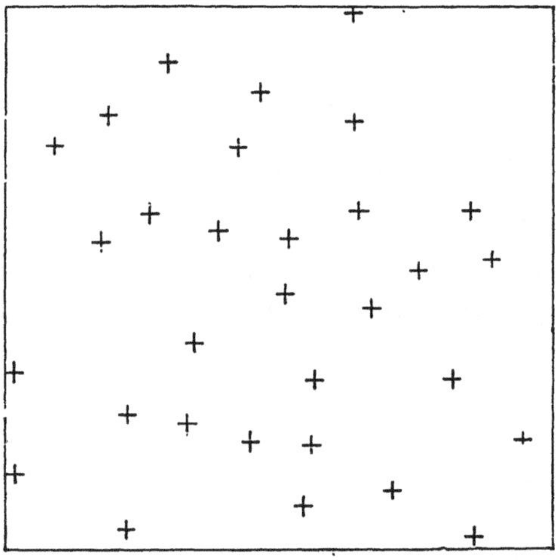

Fig. 5.7. A sample of a Matern hard-core process generated by simulation. The parameters are: $\lambda = 200$; $h = 0.1$. The sidelength of the square is one.

marked independently random numbers uniformly distributed over $(0, 1)$. The dependent thinning retains the point x of Φ (with mark $m(x)$) if the sphere $b(x, h)$ contains no points of Φ with marks smaller than $m(x)$. Formally the thinned process Φ_h is given by

$$\Phi_h = \{x \in \Phi : m(x) < m(y) \quad \text{for all } y \text{ in } \Phi \cap b(x, h)\}.$$

A situation which Φ_h might model is to be found in ecology; the points of Φ are the locations of seedlings while the marks are the instants at which the seedlings commence growth. So Φ_h consists of the positions x of those plants which are the eldest seedlings in their required growing area $b(x, h)$. A simulation is shown in Figure 5.7.

The intensity λ_h of Φ_h is given by

$$\lambda_h = p_h \lambda$$

by (4.4.2). Here p_h is the *Palm retaining probability* of a 'typical point' of Φ. It is given by

$$p_h = \int_0^1 r(t)\,\mathrm{d}t = \big(1 - \exp(-\lambda c)\big)/(\lambda c) \quad \text{for } c = \omega_d h^d$$

where $r(t) = \exp(-\lambda c t)$ is the retaining probability of a point of mark t. This formula follows from the observation that the point process

$$\{x \in \Phi : m(x) < t\}$$

is simply a t-thinning of a Poisson process, hence itself a Poisson process of intensity λt. So $r(t)$ is the probability that a sphere of radius h contains no points of the t-thinned process.

If $h = 1$ then the maximum intensity λ_h is approximately 0.32 if $d = 2$, and 0.24 if $d = 3$. These maxima result from letting λ tend to infinity.

The second-order product density $\varrho_h^{(2)}$ is given by

$$\varrho_h^{(2)}(r) = \begin{cases} 0 & \text{if } r \leqq h \\ \dfrac{2\Gamma_h(r)\,[1 - \exp(-\lambda c)] - 2c\big[1 - \exp\big(-\lambda\Gamma_h(r)\big)\big]}{c\Gamma_h(r)\,[\Gamma_h(r) - c]} \end{cases}$$

where

$$\Gamma_h(r) = \nu_d\big(b(o, h) \cup b(\boldsymbol{r}, h)\big) = 2\nu_d\big(b(o, h)\big) - \gamma_h(r) \quad \text{if } r > h$$

where $r = \|\boldsymbol{r}\|$.

In the important low-dimensional cases for $r \leqq 2h$ are

$$\gamma_h(r) = \begin{cases} 2h^2 \arccos \dfrac{r}{2h} - \dfrac{r}{2}\sqrt{(4h^2 - r^2)} & (d = 2) \\ \dfrac{4\pi}{3}h^3\left[1 - \dfrac{3r}{4h} + \dfrac{r^3}{16h^3}\right] & (d = 3). \end{cases}$$

The pair-correlation function of this process with particular values chosen for the parameters is graphed in Figure 4.9.

These formulae can be obtained by using $\varrho_h^{(2)}(r) = \lambda^2 k(r)$ where $k(r)$ is the (two-point Palm) probability that two points of Φ separated by distance r are both retained. Details of a calculation of the above using the second factorial moment measure and generalizations of the model are to be found in Stoyan and Stoyan (1985).

Hard-core models are important in branches of theoretical physics. The empirically-defined *dense random packing of hard spheres* is of great significance. This process is realized in the laboratory by compressing together an assembly of hard spheres while preventing the formation of crystalline regions. Cargill (1975, 1984) and Finney (1981) have discussed this; Finney emphasized the importance of 'statistical geometry' in his phrase. Such dense random packings have been proposed as models for the study for metallic glasses and for inert gas liquids.

Theoretical investigations of these dense random packings present enormous difficulties. Computer simulation studies have elicited some estimates, following the physical simulations by Bernal (1959); see for example, Jodrey and Tory (1979). Some estimated quantities are:

intensity of the point process of volume centres ($\simeq 0.15$ if $R = 1$)

volume fraction of space occupied by spheres ($\simeq 2/\pi = 0.637$)

pair correlation function or second-order product density (see Fig. 4.9)

and

angular correlation function.

Random dense packings are *not* the densest packings of hard spheres. According to Rogers (1964) and Bos (1979) regular packings produce higher volume fractions. Rogers expresses well the essentially empirical flavour of known results: 'many mathematicians believe, and all physicists know that the density' (volume fraction in our terminology) 'cannot exceed $\pi/\sqrt{18}$'.

A further example of a hard-core process is the So-called 'simple sequential inhibition' process (SSI process). This is defined in a bounded region and has a finite number of points. See Herczynski (1975), Diggle, Besag, and Gleaves (1976), and Diggle (1983).

Some other hard-core processes are considered in the next section, on Gibbs point processes.

5.5 Gibbs point processes

5.5.1 Introduction

Gibbs point processes (*Gibbs processes* for short) first arose in the theories of statistical physics. They are related to the so-called Gibbs distributions which describe the equilibrium states of subsystems of very large closed physical systems. In this book the physical background will be largely ignored; our treatment will concentrate on explaining the mathematical reasoning. Gibbs processes will be treated as a class of point processes, and described in terms of the characteristics used to describe the classes of point processes discussed above. Detailed discussion of the physical aspects may be found in Minlos (1968), Preston (1974, 1976), and Ruelle (1969, 1970); historical aspects of the mathematical theory are covered briefly in Kallenberg (1983 b, p. 177).

To commence the discussion in the spirit of this chapter note that apart from thinning, clustering, and superposition, another way of producing new point processes from old is to use a probability density to alter the distribution directly. The theory of Gibbs processes is a development of this idea.

Example. Altering the distribution of a Poisson process of finite total intensity measure.

Suppose that Ψ is a Poisson process with intensity measure Λ that is of total mass $\Lambda(\mathbb{R}^d) = 1$. Let the distribution of Ψ be Q. If a new point process distribution P_λ is defined by

$$P_\lambda(Y) = \int_Y f_\lambda(\varphi)\, Q(\mathrm{d}\varphi) \quad \text{for } Y \text{ in } \mathcal{N}$$

where

$$f_\lambda(\varphi) = \mathrm{e}^{1-\lambda} \cdot \lambda^{\varphi(\mathbb{R}^d)}$$

then under the new probability distribution P_λ the point process becomes Poisson of intensity measure $\lambda\Lambda$.

Proof. Let Φ be a Poisson process of intensity measure $\lambda\Lambda$. Then calculation shows that P_λ and P_Φ give the same measure to the set

$$\{\varphi \in \mathsf{N} : \varphi(\mathbb{R}^d) = n, \varphi(K) = 0\}$$

for each non-negative integer n and each compact set K. So their void probabilities must agree and they must be equal.

Underlying the theory of Gibbs processes is the idea of a general *base* or *weight process* Ψ of distribution Q. A new probability distribution P can be defined on $[\mathsf{N}, \mathcal{N}]$ by means of a probability density

$$P(Y) = \int_Y f(\varphi)\, Q(\mathrm{d}\varphi) \quad \text{for } Y \text{ in } \mathcal{N}. \tag{5.5.1}$$

If Φ is a point process of distribution P then $f(\varphi)$ is the *likelihood* that Φ takes the realization φ compared with Ψ taking on the same realization. Obviously this construction is very general. In particular whole classes of realizations of Φ can be prohibited by setting f to be zero on the corresponding subset of N.

The form of the density f is often suggested by the field of application. It is well adapted to modelling interactions between the points of the process. The Gibbs process approach is straightforward if the process to be modelled contains only a finite number of points confined to a bounded region B. Problems tarise when considering more general point processes.

Example. The stationary Poisson process $\Phi^{(1)}$ of intensity λ_1 cannot be obained in the simple manner described above from another stationary Poisson process $\Phi^{(2)}$ of intensity λ_2 if $\lambda_1 \neq \lambda_2$.

Proof. Consider a family $\{K_n\}$ of compact sets with $\nu_d(K_n) \to \infty$. Set

$$Y_\lambda = \{\varphi \in \mathsf{N} : \varphi(K_n)/\nu_d(K_n) \to \lambda\}$$

then clearly $Y_{\lambda_1} \cap Y_{\lambda_2}$ is empty but $\mathsf{P}(\Phi^{(i)} \in Y_{\lambda_i}) = 1$. So the distributions of the two processes are mutually singular and a representation of the form (5.5.1) is impossible.

To obtain Gibbs processes in the more general situation the density idea described above must be applied to conditional distributions confined to bounded regions. Section 5.5.3 describes the general case of infinite point processes; Section 5.5.2 discusses the simpler case of finite point processes.

5.5.2 Gibbs point processes in bounded regions

Processes with non-random total number of points

First we consider a point process Φ of distribution P and with *exactly n points in a specific bounded set* B (and no points outside B). It is assumed that the distribution of the point process is given by a probability density function $f\colon \mathbb{R}^{dn} \to [0, \infty)$ so that

$$\mathsf{P}(\Phi \in Y) = \int \cdots \int_{\{x_1, \ldots, x_n\} \in Y} f(x_1, \ldots, x_n) \, \mathrm{d}x_1 \ldots \mathrm{d}x_n$$

for Y in $\mathcal{N}$. Because point processes are unordered sets of points $f(x_1, \ldots, x_n)$ is taken not to depend on the order of the arguments. The form used to describe f is

$$f(\varphi) = \exp\big(-U(\varphi)\big)/Z = \exp\big(-U(x_1, \ldots, x_n)\big)/Z. \tag{5.5.2}$$

Here Z is a norming constant called the *configurational partition function*, and the function $U\colon \mathbb{R}^{dn} \to \mathbb{R} \cup \{\infty\}$ is called the *energy function* or *multiparticle potential* (the terms derive from statistical mechanics). The convention is adopted that

$$\exp(-\infty) = 0.$$

Frequently U is taken to be in a more specialized form as an *interaction potential*

$$U(\varphi) = \sum_{\psi \subset \varphi} V(\psi) \tag{5.5.3}$$

where V is zero on large subsets ψ of φ; or in the even more special form of a sum of *pair potentials*

$$U(x_1, \ldots, x_n) = \sum_{1 \leq i < j \leq n} \theta(\|x_i - x_j\|). \tag{5.5.4}$$

Again, in homage to the physical origins of the subject the function $\theta\colon \mathbb{R} \to (-\infty, \infty]$ is called the *pair potential*. A typical example of a pair potential is graphed in Figure 5.8. Its features correspond to features of the point process of density f constructed by using (5.5.1), (5.5.2), (5.5.4). For example, because $\theta(r)$ is infinite for $r < h$ the interpoint distance can never be less than h. So the point process is in fact a hard-core model. Because $\theta(r)$ is large when r is only a little larger than h such distances r must be unlikely to occur as interpoint distances. In contrast distances close to r_1, the minimum of θ, should occur relatively frequently.

From the theory of statistical mechanics distributions of the form (5.5.2) arise when the system of particles Φ is constrained to have fixed total energy

$$\int_{\mathsf{B}^n} U(x_1, \ldots, x_n) \, f(x_1, \ldots, x_n) \, \mathrm{d}x_1 \ldots \mathrm{d}x_n$$

and maximum entropy

$$H = -\int_{\mathsf{B}^n} f(x_1, \ldots, x_n) \log f(x_1, \ldots, x_n)\, \mathrm{d}x_1 \ldots \mathrm{d}x_n$$

among such fixed-energy systems.

If (5.5.2) is interpreted as an integral of a probability density, as in (5.5.1), then the underlying weight process Ψ has density proportional to the uniform density on B^n. Thus the form of B influences the distribution of the resulting

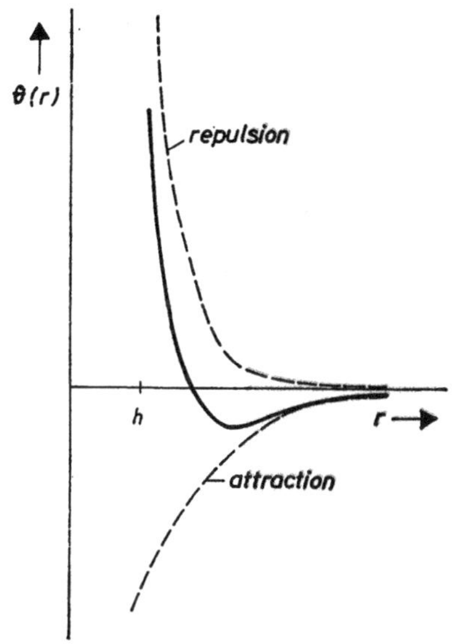

Fig. 5.8. A typical pair potential, the result of superposition of attractive and repulsive forces

Gibbs process Φ. (In the case of rectangular B, undesirable boundary effects are sometimes alleviated by supposing the point pattern to be continued periodically outside B).

The norming constant Z is given by

$$Z = \int_{\mathsf{B}^n} \exp\left(-U(x_1, \ldots, x_n)\right) \mathrm{d}x_1 \ldots \mathrm{d}x_n. \tag{5.5.5}$$

If the formulae above are applied to an infinite particle system in an unbounded B then it is possible to obtain an infinite value of U or zero value for Z. Computation of Z is rarely easy even in the bounded case for $d \geqq 2$. Ogata and Tanemura (1981) proposed an approximate method for the case of pair-potentials

$$Z \simeq \nu_d(\mathsf{B})^n \left[\frac{1-a}{\nu_d(\mathsf{B})}\right]^{n(n-1)/2} \tag{5.5.6}$$

where

$$a = \int_0^\infty \left[1 - \exp\left(-\theta(r)\right)\right] \nu_d\left(\partial b(o, r)\right) \mathrm{d}r.$$

Gates and Westcott (1986) and Ogata and Tanemura (1983) have discussed the quality of this approximation; it is poor if θ is negative for small r.

Example. *A simple pair potential.* Consider

$$\theta_{h,b,R}(r) = \begin{cases} \infty & \text{if } r \leqq h \\ -b & \text{if } h < r \leqq R \\ 0 & \text{if } r > R. \end{cases}$$

Then the density f has the form

$$f(\varphi) = \begin{cases} c \exp\left(b s_\varphi(R)\right) & \text{if } \varphi \subset \mathsf{B} \text{ and } s_\varphi(h) = 0 \\ 0 & \text{otherwise.} \end{cases}$$

The constant c is a norming constant. For $\varphi = \{x_1, \ldots, x_n\}$ we set

$$s_\varphi(r) = \sum_{1 \leqq i < j < n} \sum \alpha_r(x_i, x_j)$$

$$\alpha_r(x_i, x_j) = 1 \quad \text{if } \|x_i - x_j\| \leqq r, \text{ otherwise } 0.$$

So $s_\varphi(r)$ is the number of interpoint distances in φ that are less than r.

In this example h is a hard-core distance while R is a radius of influence. The case $b > 0$, $b < 0$ correspond respectively to attraction, and repulsion. Particular cases are

the uniform distribution of n points in B,

(when $b = 0$, $h = 0$)

and

the Strauss model with fixed number of points

(when $b > 0$, $h = 0$).

The Strauss model is discussed in Ripley (1981) and Gates and Westcott (1986).

In the case of 'sparse' point systems (n small, B large) Saunders, Kryscio, and Funk (1982) showed that when $b < 0$, or $b > 0$ and $h > 0$, the jump points of the step function $s_\varphi(r)$ $(0 \leqq r \leqq \varrho)$ behave approximately like a non-stationary Poisson process. Here ϱ is 'some radius beyond which influence is inconceivable'. The intensity measure Λ is given by

$$\Lambda([0, r]) = \mathsf{E}s_\Phi(r) = \begin{cases} 0 & \text{if } r \leqq h \\ \varkappa(r^2 - h^2)\,\mathrm{e}^b & \text{if } h < r \leqq R \\ \varkappa(R^2 - h^2)\,\mathrm{e}^b + \varkappa(r^2 - R^2) & \text{if } R < r \leqq \varrho \end{cases} \tag{5.5.7}$$

Here $\varkappa$ is an unknown factor.

This result allows statistical estimation of h, R, and b, and enables tests to be conducted of the hypothesis that the n points are uniformly distributed in B; see Saunders, Kryscio, and Funk (1983). Ripley (1981) used a trial-and-error method for the same estimation problem.

Example 5.5. *An example of sink-holes.* (Hanisch and Stoyan, 1983.)

Figure 5.9 shows a pattern of 31 sink-holes caused by sulphide Karst processes near the Harz region, G.D.R., Figure 5.10 shows an approximation to $s_\varphi(r)$ up to a factor of pro-

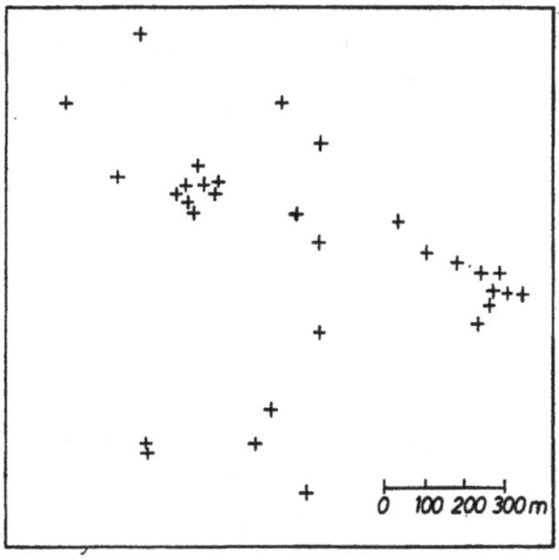

Fig. 5.9. Positions of 31 sinkholes near the Harz

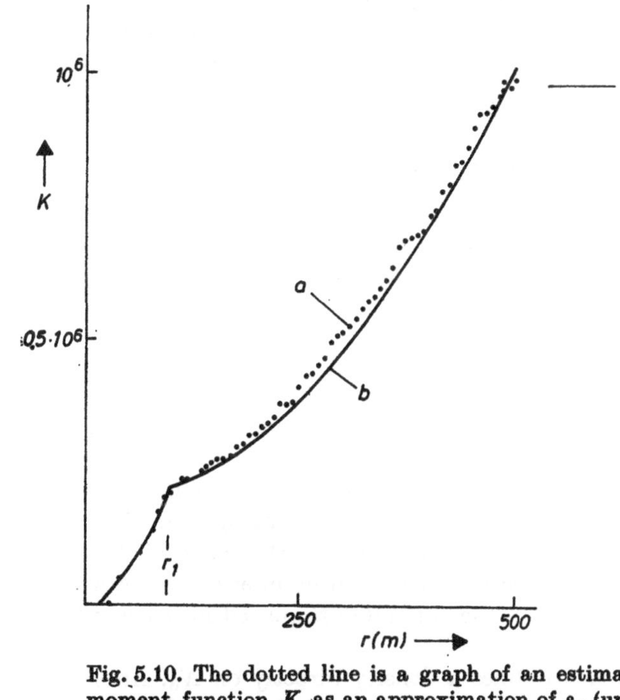

Fig. 5.10. The dotted line is a graph of an estimate of the reduced second moment function K as an approximation of s_φ (up to a factor of proportionality) for the pattern on Figure 5.9. The solid curves show the two corresponding approximating parabolic arcs. The cusp is at $r = 100$ m

portionality. This function is an edge-corrected estimate of the reduced second-order moment function. The minimum distance is at $r = 15$ m and there is a clear cusp at $r = 100$ m. Estimates for the $\theta_{h,R,b}$ model are therefore

$$\hat{h} = 15 \text{ m}$$
$$\hat{R} = 100 \text{ m}$$
$$\hat{b} = 2.0.$$

Possibly the model is too simple for the data; there are suggestions of further cusps in the empirical curve. However, these may be consequences of the hard-core property of the point-process model. This suggestion is supported by the presence of the peaks of the pair-correlation function (proportional to the product-density) for the hard-core model displayed in Figure 4.9.

OGATA and TANEMURA (1981) studied other parametric families of pair potentials θ, namely

$$\theta_1(r) = -\ln\{1 + (\alpha r - 1)\,\mathrm{e}^{-\beta r^2}\}, \quad \theta_2(r) = -\ln\{1 + (\alpha - 1)\,\mathrm{e}^{-\beta r^2}\}$$

$$\theta_3(r) = \beta(\sigma/r)^n - \alpha(\sigma/r)^m$$

for $\alpha \geqq 0$, $\beta, \sigma > 0$, and $n > m$.

The last potential θ_3 is a common model in statistical mechanics, called the Lennard–Jones potential. The case $n = 12$, $m = 6$ is of particular interest. It provides regions both of attraction and of repulsion.

Ogata and Tanemura describe maximum-likelihood estimation of parameters for these models in case of sparseness obtaining the approximations to the likelihood by means of (5.5.2) and (5.5.6).

PENTTINEN (1984) gives a similar approach. In the planar case, his method gives for the parameter b of $\theta_{h,b,R}$, if h and R are known, the estimator

$$\hat{b} = \ln\big(2\nu_2(\mathsf{B})\, s_\Phi(R)/(\pi n^2(R^2 - h^2))\big).$$

Processes with random total number of points

The definition of a Gibbs process Φ with a *random* total number of points N in B $\big(N = \Phi(\mathsf{B})\big)$ proceeds in a manner similar to the non-random case. Let $\boldsymbol{N}_{\mathsf{B}_n}$ denote the family of all sets of n distinct points in B, and

$$\boldsymbol{N}_{\mathsf{B}} = \bigcup_{n=0}^{\infty} \boldsymbol{N}_{\mathsf{B}_n}$$

with $\boldsymbol{N}_{\mathsf{B}_n} = \{\emptyset\}$ the family consisting only of the empty set. The Gibbs process Φ has a distribution P on $\boldsymbol{N}_{\mathsf{B}}$ which can be approached in one of two equivalent ways.

One approach defines a sequence of 'conditional densities' $f_n : \boldsymbol{N}_{\mathsf{B}_n} \to [0, \infty)$, and constructs Φ by the two-step process of first arranging for N to have an appropriate distribution and then (conditional on the value of N) using the construction of the preceding section with total number of points $N = n$ and with density f_n. So given $N = n$ the n points are placed in B using the joint density f_n.

A more elegant approach defines a density directly on $\boldsymbol{N}_{\mathsf{B}}$. An example is the model of STRAUSS (1975); see also KELLY and RIPLEY (1976). This model has density

$$f(\varphi) = c a^{\varphi(\mathsf{B})} \exp\big(b s_\varphi(R)\big) \quad \text{for } a \geqq 0. \tag{5.5.8}$$

As above $s_\varphi(R)$ is the number of interpoint distances no greater than R. For f to be a density it is necessary that $b \leqq 0$.

The density of (5.5.8) is a density on $\boldsymbol{N}_{\mathsf{B}}$ with respect to the σ-finite measure which is Lebesgue measure on each $\boldsymbol{N}_{\mathsf{B}_n}$. This last measure cannot arise as a point process distribution as it does not have total mass 1. If f is defined by this approach then the distribution of N and the 'conditional densities' f_n are determined; the two approaches are indeed equivalent.

The generalization of the Strauss model to a more general pair potential yields the density

$$f(\varphi) = c a^{\varphi(\mathsf{B})} \exp\left(-\sum_{1 \leq i < j \leq \varphi(\mathsf{B})} \theta(\|x_i - x_j\|)\right) \tag{5.5.9}$$

where $x_1, \ldots, x_{\varphi(\mathsf{B})}$ are the constituent points of φ.

A yet more general class of point processes is that of the *Markov point processes*; see Kelly and Ripley (1976), Ripley and Kelly (1977), and Isham (1981). A point process on B is called *Markov* if its conditional distribution on A a subset of B (given its behaviour in $\mathsf{B} \setminus A$) depends only on its behaviour in $E(A) \setminus A$. Here $E(A)$ is the *environment* of A, given by means of a neighbourhood relation $\equiv$ on B; $E(A)$ is defined by

$$E(A) = \{x : x \equiv y \text{ for some } y \text{ in } A\}.$$

An example of a possible rule is,

$$x \equiv y \quad \text{if and only if} \quad \|x - y\| < r$$

where r is fixed.

Ripley and Kelly (1977) determined the general form of Markov point processes; their densities are of the form

$$f(\varphi) = \exp\left(-\sum_{\psi \subset \varphi} V(\psi)\right) \Big/ Z : \varphi \in N_{\mathsf{B}}$$

where V is an *interaction potential.* That is to say V is a function, $V: N_{\mathsf{B}} \to [0, \infty)$, such that $V(\varphi) \neq 0$ implies $x \equiv y$ for all x and y in φ.

Simulation

It is important to be able to simulate Gibbs processes for then statistical techniques such as simulation tests are made available. Most methods in use are based on construction by means of spatial birth-and-death processes; see Ripley (1977); also Lotwick (1982) and Section 5.5.5.

If a process is defined by means of a bounded density f with respect to a Poisson process then it can be obtained (at least in principle) by *rejection sampling*. If the density is not bounded then matters may still be mended by altering the Poisson intensity measure or by fixing attention on samples of a definite total size. The procedure of rejection sampling continually simulates samples of the Poisson process. For each sample the density f is calculated, and compared with a random variable X distributed uniformly over $(0, \sup f)$. The random variable X is resampled independently each time the Poisson process is simulated. The sample is *accepted* as a valid sample if $X < f$, otherwise it is *rejected*. The sequence of accepted samples has the desired Gibbs process distribution.

Problems can arise with the rejection-sampling method because the computation of $\sup f$ and Z is often very complicated. If an upper bound on f is used instead then the sequence of accepted samples still has the desired distribution but more samples than necessary will be rejected. Ripley (1979a) discussed the problem of achieving an economical solution. In practice the use of spatial birth-death processes, as discussed in Section 5.5.5, will often be more practical.

Mase (1984) studied the asymptotic normality of planar Gibbs point processes on lattices, taking the limit as the set B is expanded to fill out the whole lattice. He also discussed the estimation of parameters of parametrized pair potentials.

5.5.3 Gibbs point processes in $\mathbb{R}^d$

The methods of 5.5.2 cannot be used directly in the case when B is replaced by $\mathbb{R}^d$. A more sophisticated approach is required which considers the restrictions of point processes to bounded sets B. Indeed, not only the restrictions must be considered, but the distributions of the restrictions must be conditioned on the behaviour of the point process outside the bounded region B. Formally one considers probabilities of the form $\pi_B(Y \mid \psi)$. This is the probability that the pattern intersected with B belongs to the set Y given that outside B the pattern takes on the configuration ψ. So

$$\pi_B(Y \mid \psi) = \mathsf{P}(\Phi \cap B \in Y \mid \Phi \cap B^c = \psi) \tag{5.5.10}$$

where B is a bounded Borel set, Y is in $\mathcal{N}$, and ψ belongs to the set N_{B^c} where

$$N_{B^c} = \{\varphi \in \mathsf{N} : \varphi(B) = 0\}.$$

These probabilities must satisfy certain measurability and compatibility conditions which need not detain us here. The objective of this section is to show how in many cases the π_B can be expressed elegantly using a weight process and a pair potential.

The mathematical problems of the above formulation were intensively studied in the 1970s, by for example Preston (1976), Glötzl (1980), Kerstan, Matthes, and Mecke (1982), Matthes, Warmuth, and Mecke (1979), and Kallenberg (1983b).

Rather than deriving the π_B from the point process Φ, one can attempt to move in the reverse direction. Indeed this is the more useful procedure for model specification. Suppose the *local specification* of all π_B is given for all bounded Borel B (using the terminology of Preston, 1976). Is there a point process Φ connected with this local specification via (5.5.10)? Is it unique? A point process Φ with distribution P is said to be a *Gibbs point process with respect to the local specification* π_B if for all bounded Borel B

$$P(Y) = \int \pi_B(Y \mid \varphi \cap B^c)\, P(\mathrm{d}\varphi). \tag{5.5.11}$$

The equation (5.5.11) is called the DLR-*equation* after Dobrushin, Lanford, and Ruelle. Note that the solution P to (5.5.11) may not be unique.

The concept of a Gibbs point process is closely connected to the older notion of a *Gibbs distribution* in statistical physics. The corresponding physical question concerns the existence and uniqueness of equilibrium states for dynamic particle systems given local dynamic specifications analogous to (5.5.11). See RUELLE (1969) for a discussion of the general physical theory.

Equation (5.5.2) gives an exponential form for the density of P in the bounded case. By analogy we consider the π_B in exponential form

$$\pi_B(Y \mid \psi) = \int \mathbf{1}_Y\big((\varphi \cap B) \cap \psi\big) \exp\left[-\bar{E}(\varphi \cap B, \psi)\right] Q(\mathrm{d}\varphi)/Z(B, \psi) \quad (5.5.12)$$

Here Q is the distribution of a base or weight process and $Z(B, \psi)$ is a norming constant. The function $\bar{E}: N_f \times N \to (-\infty, \infty]$ is the *conditional energy* (here $N_f = \{\varphi \in \mathsf{N} : \varphi(\mathbb{R}^d) < \infty\}$). The conditional energy $\bar{E}(\chi, \psi)$ can be interpreted as the energy required to add the point set χ to the configuration ψ.

To complete the analogy with (5.5.2) equation (5.5.12) can be derived as the solution to a maximization-of-entropy problem when the energy is held constant.

Two simplifying conditions are imposed in the subsequent discussion.

First, $\bar{E}$ is assumed to have an additive form expressed in terms of the *local energy* E

$$\bar{E}(\chi, \psi) = E(x_1, \psi) + E(x_2, \psi \cup \{x_1\}) + \cdots$$
$$+ E(x_n, \psi \cup \{x_1, \ldots, x_{n-1}\}) \text{ where } \chi = \{x_1, \ldots, x_n\} \quad (5.5.13)$$

(and naturally a consistency condition must be imposed on E to ensure that the sum doesn't depend on the order of the x_i). $E(x, \varphi)$ can be interpreted as the energy required to add the point x to the configuration φ.

Second, the weight process is supposed Poisson with intensity measure Λ_Q. This is an amenable generalization of the 'uniform' weight process used to treat Gibbs processes with non-random finite totals of points as in 5.5.2.

Clearly, it is important to decide when a distribution P leads to a local specification $\{\pi_B\}$ in the form (5.5.12) with a local energy. Some condition resembling absolute continuity is required. The answer is given by a theorem of GLÖTZL (1980).

Theorem 5.1. *A necessary and sufficient condition for the point process distribution P to have local specification $\{\pi_B\}$ in the form of* (5.5.12) *with $\bar{E}$ given by* (5.5.13) *is as follows: the reduced Campbell measure $\mathscr{C}^!$ of P satisfies the absolute continuity condition*

$$\mathscr{C}^! \ll \Lambda_Q \times P. \quad (5.5.14)$$

Here Λ_Q is the intensity measure of the corresponding Poisson weight process. For almost all x and φ (with respect to P)

$$\frac{\mathrm{d}\mathscr{C}^!}{\mathrm{d}(\Lambda_Q \times P)}(x, \varphi) = \exp\big(-E(x, \varphi)\big). \quad (5.5.15)$$

Formula (5.5.15) suggests a plausible argument for the interpretation of $E(x, \varphi)$ as the energy required to add x to φ. For simplicity we consider the case when P and Q are both stationary with intensities λ, λ_Q. Then the Radon–Nikodym derivative on the left of (5.5.15) can be written informally as

$$\frac{d\mathscr{C}^!}{d(\Lambda_Q \times P)}(x, \varphi) = \frac{\mathscr{C}^!(dx \times d\varphi)}{\lambda_Q \nu_d(dx) \, P(d\varphi)}.$$

Using the expression (4.4.4) giving the reduced Palm measure in terms of the reduced Campbell measure the infinitesimal quotient can be written as

$$\frac{\mathscr{C}^!(dx \times d\varphi)}{\lambda_Q \nu_d(dx) \cdot P(d\varphi)} = \frac{P_0{}^!(d\varphi) \, \lambda \nu_d(dx)}{\lambda_Q \nu_d(dx) \, P(d\varphi)}$$

which is proportional to

$$\frac{P_o{}^!(d\varphi)}{P(d\varphi)}.$$

This last quotient is the ratio of the infinitesimal probabilities for Φ under P (in the numerator) and P conditioned to have a point at x (in the denominator). The standard physical interpretation regards the negative logarithm of the quotient as the energy required to make transition from φ to $\varphi \cup \{x\}$.

The question of uniqueness of the process given by the local specification is of great interest and importance in probability theory, but is not of great relevance in spatial statistics. This is because all practical applications involve data observed through bounded windows. The non-uniqueness may possibly exhibit itself in increased sensitivity to boundary conditions. However, it seems that non-uniqueness phenomena really begin to play an important part when the numbers involved achieve thermodynamic proportions of the order of 10^{23} points. On the other hand, when the relatively low numbers of points in statistical problems are being considered, the problem of edge-effects is likely to be of far greater significance (B. D. Ripley, 1984, personal communication).

Together with equation (4.4.4) equation (5.5.15) implies that in the stationary case

$$\lambda P_0{}^!(Y) = \lambda_Q \int_Y \exp\big(-E(o, \varphi)\big) \, P(d\varphi) \tag{5.5.16}$$

which implies that the reduced Palm probability measure $P_0{}^!$ is absolutely continuous with respect to P. This is a way of quickly establishing that a point process is *not* a Gibbs process (with respect to a Poisson process, using local energy). For example, the lattice of points of integral coordinates, translated in both coordinates by random shifts uniform over (0, 1), is a stationary point process which cannot be so expressed. Indeed its reduced Palm probability measure yields simply the lattice with the origin deleted and this is clearly not absolutely continuous with respect to the original law.

An equivalent formulation of (5.5.16) in the stationary case is

$$\lambda \int g(\varphi) \, P_0{}^!(d\varphi) = \lambda_Q \int g(\varphi) \exp\big(-E(o, \varphi)\big) \, P(d\varphi) \tag{5.5.17}$$

for every non-negative measurable $g\colon \mathsf{N} \to \mathbb{R}$.

An important class of general Gibbs processes corresponding to the class of models given by (5.5.12) and (5.5.14) is given by requiring that the local energy

E be expressible in terms of a *pair potential* θ

$$E(x, \varphi) = \alpha + \sum_{x \in \varphi} \theta(\|x - y\|).$$

The constant α is called the *chemical activity.*

If the pair potential $\theta(r)$ is infinite for all $r \leqq h$ and zero for $r > h$ then the corresponding Gibbs process is called a *hard-core Poisson model*; see MÜRMANN (1978).

If θ has a finite *interaction radius* R, that is to say $\theta(r)$ is zero for $r \geqq R$, then in the stationary case the following relationship holds between α, λ_Q, and λ

$$\lambda P_0^!(Y_R) = \lambda_Q P(Y_R)\, e^{-\alpha} \tag{5.5.18}$$

where $Y_R = \{\varphi \in N : \varphi(b(o, R)) = 0\}$.

For the rest of the chapter, it is assumed that $\lambda_Q = 1$; if λ, $P_0^!(Y_R)$, and $P(Y_R)$ are fixed then this constraint is balanced by the freedom to choose α.

5.5.4 Statistical estimation of the pair potential of a planar stationary Gibbs process

Several methods have been developed in order to estimate the pair potential and the chemical activity. We discuss first two approaches to the estimation of the pair potential $\theta_{h,b,R}$ described above.

A reasonable estimate for h in $\theta_{h,b,R}$ is the minimum interpoint distance. RIPLEY (1977) used a trial-and-error method to estimate b and R. The procedure is to choose repeatedly trial pairs (b_i, R_i) and to use Monte Carlo tests to evaluate the significance levels of the successive hypotheses that the pattern can be modelled via the pair potential $\theta_{h,b,R}$. The estimate for (b, R) is the first pair (b_i, R_i) for which the hypothesis is acceptable.

HANISCH and STOYAN (1983) proposed a 'cusp' method, exploiting the fact that the discontinuities of $\theta_{h,b,R}$ at $r = h$, $r = R$ are reflected in cusps of the reduced second moment function K. The procedure is to estimate K and then to look for a clear cusp of its graph (apart from the one at $(h, 0)$). The corresponding r-coordinate estimates R. Moreover, b may be estimated by using the fact

$$\frac{\lim_{r \uparrow R} K'(r)}{\lim_{r \downarrow R} K'(r)} = e^b.$$

Example 5.5 continued.

Consider the interpretation of the point pattern of Figure 5.9 as part of a sample of a stationary Gibbs process with pair potential $\theta_{h,b,R}$. Figure 5.10 graphs the corresponding estimate of K. From this curve the model parameters are estimated by

$$\hat{h} = 15 \text{ m}$$
$$\hat{R} = 100 \text{ m}$$
$$\hat{b} = 2.0 \text{ m}$$

as described above.

We now consider the estimation of more general pair potentials. A method suggested by Glötzl and Rauchenschwandtner (1981) is based on approximation of the differential quotients of (5.5.15). Preferable to this is a method of Takacs (1983; 1986) which assumes the pair potential θ to be piecewise constant. Its hard-core is h, its radius of interaction is $h + n\delta$, and it takes a different constant value in each interval $(h + (j - 1)\,\delta, h + j\delta]$;

$$\theta(r) = a_j \quad \text{if } h + (j-1)\,\delta < r \leqq h + j\delta, \text{ for } j = 1, \ldots, n.$$

Assuming this general form the equation (5.5.17) can be used to derive estimating equations for the parameters a_j. This is because for such θ the local energy $E(o, \varphi)$ can be written as

$$E(o, \varphi) = \begin{cases} \alpha + \sum_{j=1}^{n} a_j \varphi(K_j) & \text{if } \varphi\big(b(o, h)\big) = 0 \\ \infty \text{ otherwise.} \end{cases}$$

Here K_j is the annulus

$$K_j = \{x \in \mathbb{R}^2 : h + (j-1)\,\delta < \|x\| \leqq h + j\delta\}.$$

Consequently if $g_i(\varphi) = \varphi(K_i)$ then (5.5.17) takes the form

$$\lambda \int g_i(\varphi)\, P_o^!(\mathrm{d}\varphi) = \int g_i(\varphi) \exp\left(-\alpha - \sum_{j=1}^{n} a_j g_j(\varphi)\right) P(\mathrm{d}\varphi) \qquad (5.5.19)$$

integrating over $\{\varphi \in \mathsf{N} : \varphi\big(b(o, h)\big) = 0\}$. The mean values in (5.5.19) can be estimated easily; estimation of the mean value on the left-hand side is similar to that of $\lambda K(r)$ even to the extent of edge-corrections. Let l_i be this estimate and let $r_i(a_1, \ldots, a_n; \alpha)$ be the expression produced by estimating the right-hand side as a function of the unknown a_i's and α. Furthermore, let $r_0(a_1, \ldots, a_n; \alpha)$ be the expression in the a_i's and α produced by estimating

$$\int \exp\left(-\alpha - \sum_{j=1}^{n} a_j g_j(\varphi)\right) P(\mathrm{d}\varphi).$$

By (5.5.18) the quantity $r_0(a_1, \ldots, a_n; \alpha)$ estimates λ. So the system of $n + 1$ equations

$$l_i = r_i(a_1, \ldots, a_n; \alpha)$$

$$\lambda = r_0(a_1, \ldots, a_n; \alpha)$$

contains $n + 1$ unknowns $a_1, \ldots, a_n, \alpha$; a solution for this system gives estimates for the unknown parameters.

Takacs suggested that higher moments such as

$$\lambda \int g_i(\varphi)\, g_j(\varphi)\, P_o^!(\mathrm{d}\varphi)$$

could also be used. Formulae similar to (5.5.9) would then be derived leading to a system of equations containing more equations than unknowns, allowing the possibility of finding estimates by minimising the sum of squared deviations from equality in the system.

FIKSEL (1984a) discussed a similar method for use when the pair potential θ_β depends on a vector parameter $\boldsymbol{\beta}$. Again suitable functions $g: \mathsf{N} \to \mathbb{R}$ are chosen and estimates are made of the mean values involved in such expressions as

$$\lambda \int g(\varphi)\, P_o^{!}(\mathrm{d}\varphi)$$

and

$$\int g(\varphi) \exp\left(-E(o, \varphi)\right) P(\mathrm{d}\varphi) = \int g(\varphi) \exp\left(-\alpha - \sum_{x \in \varphi} \theta_\beta(\|x\|)\right) P(\mathrm{d}\varphi)$$

as a function of $\boldsymbol{\beta}$. The criterion to determine $\boldsymbol{\beta}$ is to achieve the best possible agreement between the corresponding estimated quantities.

In the particular case when $\boldsymbol{\beta}$ is just a real number β, one can put $g \equiv 1$ and so obtain

$$\begin{aligned}
\lambda &= \mathrm{e}^{-\alpha} \int \exp\left(-\sum_{x \in \varphi} \theta_\beta(\|x\|)\right) P(\mathrm{d}\varphi) \\
&= \mathrm{e}^{-\alpha} \int \exp\left(-\int \theta_\beta(\|x\|)\, \varphi(\mathrm{d}x)\right) P(\mathrm{d}\varphi) \\
&= \mathrm{e}^{-\alpha} L_P(u)
\end{aligned}$$

where $u(x) = \theta_\beta(\|x\|)$. Here L_P is the Laplace functional of the point process distribution P.

These parametric procedures tacitly assume that the radius of interaction of the pair potential is finite.

Example 5.6. *Pair potential of trees in a sprnce forest.*

The point pattern of this example is to be found in Figure 5.11. It gives tree positions in a forest of pine trees. Fiksel studied these data under the assumption that they form a portion of a sample of a stationary Gibbs process. He considered two families of pair potentials

$$\theta_1(r) = \theta_{h_1, b_1, R_1}(r) \quad \text{as on p. 152}$$

and

$$\theta_2(r) = \begin{cases} \infty & \text{if } r \leq h_2 \\ a\, \mathrm{e}^{-\zeta r} & \text{if } h_2 < r \leq R_2 \\ 0 & \text{if } r \geq R_2 . \end{cases}$$

Estimates for the parameters are

$$\begin{array}{ll}
\hat{h}_1 = 1 \text{ m} & \hat{h}_2 = 1 \text{ m} \\
\hat{R}_1 = 2.7 \text{ m} & \hat{R}_2 = 3.5 \text{ m} \\
\hat{b}_1 = -0.8 & \hat{a} = 6.0 \\
 & \hat{\zeta} = 1.0 \text{ m}^{-1}
\end{array}$$

and the chemical activities are estimated by

$$\hat{\alpha}_1 = 2.0, \qquad \hat{\alpha}_2 = 1.9 .$$

These estimates suggest inhibition between trees up to a distance of approximately 3 m.

The values were obtained by the use of the least-squares method referred to above using $g(\varphi) = \varphi(b(o, r))$ for values of r

$$r = 1.25, 1.50, \ldots, 5.00.$$

It was found that θ_2 fitted the data better than θ_1, at least insofar as it yielded a lower sum of squares.

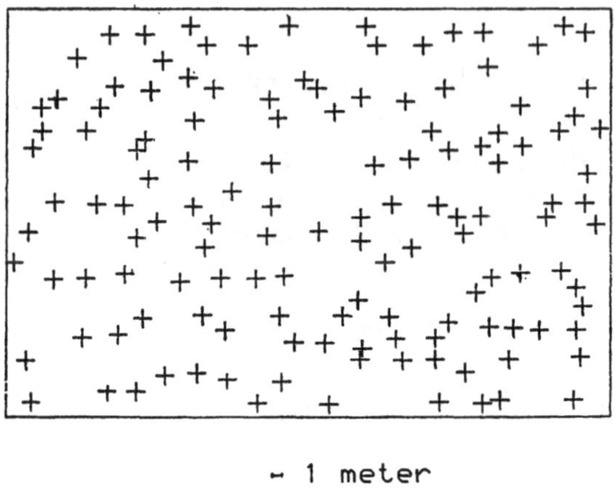

Fig. 5.11. Positions of 134 trees in a spruce forest

Marked Gibbs processes have also been studied by FIKSEL (1987b) and by OGATA and TANEMURA (1983). In the case of the trees example, marks might be stem diameters or stem volumes.

WESTCOTT (1982) studied the hard-core Poisson model on the surface of the sphere.

5.5.5 Spatial birth-death processes

A spatial birth-death process is a continuous-time Markov process with state-space N the family of point patterns; see PRESTON (1976) and GLÖTZL (1981). These processes provide models of spatio-temporal point processes (mentioned on page 96). Their relevance here lies in their close relationship to Gibbs processes and especially in the way in which they provide a means to simulate Gibbs processes as suggested by RIPLEY (1977).

A brief description of a spatial birth-death process is that at any one time it is a point process in space which changes at isolated instants of time either by the *birth* of a new point, which is added to the pattern, or by the *death* of an old point, which is deleted. Moreover, it has a *Markov* property in time; the probability of a change depends only on the current configuration of the process. Births are controlled by a *birth rate* $\boldsymbol{b}$, a positive measurable function

$$\boldsymbol{b} : \mathbb{R}^d \times \mathsf{N} \to \mathbb{R}$$

with

$$\int_B \boldsymbol{b}(x, \varphi)\, \mathrm{d}x < \infty$$

for all bounded Borel sets B and all φ in N. The probability that a birth occurs within the set B in the time interval $[t, t + s)$, given that the process has configuration φ at time t, is given by

$$s \cdot \int_B \boldsymbol{b}(x, \varphi)\, \mathrm{d}x + o(s).$$

The occurrence of a death is controlled by a *death rate* $\boldsymbol{d}$, a positive measurable function

$$\boldsymbol{d}\colon \mathbb{R}^d \times \mathsf{N} \to \mathbb{R}.$$

The probability that the point x is deleted from the pattern φ in the time interval $[t, t + s)$, given that φ is the state of the birth-death process at time t and that $x \in \varphi$, is given by

$$s \cdot \boldsymbol{d}(x, \varphi) + o(s).$$

The functions $\boldsymbol{b}$ and $\boldsymbol{d}$ characterize the birth-death process. It is of course necessary to impose conditions on $\boldsymbol{b}$ and $\boldsymbol{d}$ to eliminate various pathological properties; suitable conditions as well as a formal construction can be found in PRESTON (1976).

The *equilibrium* or time-stationary states of the process are of particular interest. A point process distribution P is an *equilibrium state* of the birth-death process if when the birth-death process is started with distribution P at time 0 it remains having distribution P for all time (the actual configuration may change but the statistics of the configuration remain of the same distribution). An equilibrium state is a *time-reversible* state of the birth-death process if in addition a realization of the ensuing evolution cannot be statistically distinguished from its reversal in time. Technically for P to be reversible it is required that

$$\int g(\varphi)\, \Omega h(\varphi)\, P(\mathrm{d}\varphi) = \int h(\varphi)\, \Omega g(\varphi)\, P(\mathrm{d}\varphi). \tag{5.5.21}$$

Here Ω is the closure of the infinitesimal generator of the birth-death process, defined on suitable g by

$$\begin{aligned}\Omega g(\varphi) = &\int \boldsymbol{b}(x, \varphi)\, [g(\varphi \cup \{x\}) - g(\varphi)]\, \varphi(\mathrm{d}x)\\ &+ \int \boldsymbol{d}(x, \varphi \setminus \{x\})\, [g(\varphi \setminus \{x\}) - g(\varphi)]\, \varphi(\mathrm{d}x).\end{aligned} \tag{5.5.22}$$

(For these definitions to make sense it is necessary to assume $\boldsymbol{b}$ and $\boldsymbol{d}$ are well behaved; in particular that they generate a uniquely defined Markov process.)

If the birth-death process remains confined to a bounded set then its distribution may be represented by a density as in 5.5.2 with random total number of points. In case of reversibility the (uniquely determined) reversible density f satisfies the following *detailed balance conditions*

$$\boldsymbol{b}(x, \varphi)\, f(\varphi) = \boldsymbol{d}(x, \varphi)\, f(\varphi \cup \{x\}) \tag{5.5.23}$$

assuming that $f(\varphi \cup \{x\}) > 0$.

Under suitable conditions the spatial birth-death process will converge to a statistical equilibrium which is the time-reversible distribution P. PRESTON

(1976, Theorem 7.1) and LOTWICK and SILVERMAN (1981) demonstrated this using coupling arguments. This fact together with Theorem 5.2 below forms the basis for a practical simulation procedure for Gibbs processes.

If the birth-death process has a time-reversible equilibrium distribution P then whatever initial distribution it is given it converges towards P. This fact forms the basis for simulation of stationary Gibbs processes as described below.

GLÖTZL (1981) demonstrated the relationship between time-reversible birth-death process and Gibbs processes

Theorem 5.2. *Let the rates* $\boldsymbol{b}$ *and* $\boldsymbol{d}$ *satisfy*

$$\frac{\boldsymbol{b}(x, \varphi)\, \boldsymbol{b}(y, \varphi \cup \{x\})}{\boldsymbol{d}(x, \varphi)\, \boldsymbol{d}(y, \varphi \cup \{x\})} = \frac{\boldsymbol{b}(y, \varphi)\, \boldsymbol{b}(x, \varphi \cup \{y\})}{\boldsymbol{d}(y, \varphi)\, \boldsymbol{d}(x, \varphi \cup \{y\})}$$

for all x *and* y *in* $\mathbb{R}^d$ *and* φ *in* N.

Then the distribution P *is a time-reversible distribution of the spatial birth-death process with rates* $\boldsymbol{b}$ *and* $\boldsymbol{d}$ *if and only if* P *is the distribution of a Gibbs process based on a stationary Poisson weight process with intensity* 1 *and local energy*

$$E(x, \varphi) = -\ln \left[\frac{\boldsymbol{b}(x, \varphi)}{\boldsymbol{d}(x, \varphi)}\right]$$

for x *in* $\mathbb{R}^d$ *and* φ *in* N.

Using these results a method of simulating Gibbs processes can be devised; see RIPLEY (1977, 1979a) and LOTWICK (1982). Of course for practical reasons only point processes in bounded regions can be generated; in the following we take the region to be $\mathsf{B} = [0, 1]^2$ the unit square.

The simulation method is as follows:

(1) To choose a starting configuration;

(2) To choose an environment, $\boldsymbol{b}$ and $\boldsymbol{d}$;

(3) To simulate the birth-death process with these rates till it is considered sufficiently close to equilibrium. Then the theorem above is applied to assert that the distribution of the final configuration is approximately that of the desired Gibbs process.

The choice (1) can be of n points generated by the uniform distribution on B. In order to obtain a stationary pattern it is necessary to suppose this and all subsequent patterns to be repeated periodically outside of B; or equivalently that by suitable identification of edges B is in fact a torus. Figure 5.12 illustrates the periodic continuation.

The choice (2) can take many forms. One possibility is

$\boldsymbol{b}(x, \varphi) = 1$, which simplifies the simulation of births,

$\boldsymbol{d}(x, \varphi) = \mathrm{e}^{E(x,\varphi)}$.

The simulation (3) takes place in continuous time running in variable time-steps from one incident of birth or death to the next. It can be described inductively as follows.

Suppose at time t a birth or a death has just happened and the configuration in B is of k points $x_1, \ldots, x_k$ (it is continued periodically outside B). Then it must be determined whether the next change of state is to be a birth or a death, when it is to happen, and if a birth where it occurs, and if a death which point is removed. This is done by simulating $k + 1$ independent exponential random variables $X_1, \ldots, X_{k+1}$, with means $\mu_1^{-1}, \ldots, \mu_{k+1}^{-1}$, where

$$\mu_1 = \boldsymbol{d}(x_1, \varphi)$$

$$\cdots\cdots\cdots\cdots$$

$$\mu_k = \boldsymbol{d}(x_k, \varphi)$$

$$\mu_{k+1} = 1 .$$

The random variables $X_1, \ldots, X_k$ may be thought of as the 'virtual residual lifetimes' of the points $x_1, \ldots, x_k$; while X_{k+1} is the 'virtual residual time till birth'.

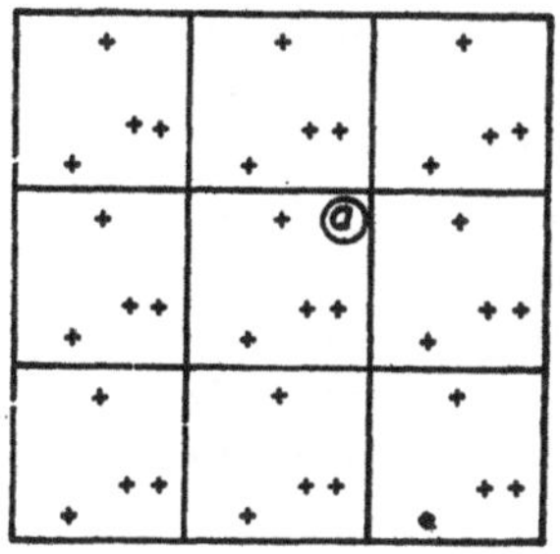

Fig. 5.12. A point pattern in $[0, 1]^2$ and its periodic continuation

The next change of state occurs at the time

$$t + T^*$$

where

$$T^* = \min \{X_1, \ldots, X_{k+1}\} .$$

The specific change depends on precisely which of the X_i achieves the minimum. If

$$T^* = X_i \quad \text{for some } i \text{ in } \{1, \ldots, k\}$$

then the point x_i is deleted; if

$$T^* = X_{k+1}$$

then a birth occurs; a new point x_{k+1} is added to the configuration chosen at random according to the uniform distribution on B.

Finally the points are re-ordered if a death has occurred, time is updated to $t + T^*$, the configuration is repeated periodically outside B, and the whole process is repeated.

Of course each iteration must generate its own random variables independent from all previous iterations.

Remarks. If a hard-core process is simulated then the procedure must be modified. The form of $\boldsymbol{d}$ will then ensure that pairs of points that are too close together die immediately. The simplest remedy is to delete them both, and throughout the simulation to employ rejection sampling when locating newborn points to ensure that they do not violate the hard-cores of other points.

It is not necessary to generate all the X_i's. Straightforward calculation with exponential distributions shows that T^* is exponential with mean $\mu^{-1} = \{\mu_1 + \cdots + \mu_{k+1}\}^{-1}$. Moreover, the probability that $X_i = T^*$ is given by $p_1 = \mu_1/\mu$. Thus it is more efficient to generate one random variable Z, uniform on (0, 1), to kill the ith point of the process if

$$p_0 + p_1 + \cdots p_{i-1} < Z \leqq p_0 + p_0 + \cdots + p_i, \quad \text{taking } p_0 = 0,$$

and to create a new point uniformly at random on B if

$$Z > p_0 + p_1 + \cdots + p_k.$$

This is derived from the jump chain construction of a Markov process.

The Gibbs process is theoretically only achieved at the limit after infinitely long time. In practice it suffices to continue the simulation until statistical equilibrium is judged to be satisfactory. Ripley (1977) made suggestions on the number of iterations required. (But note the work of Gates and Vestcott (1986), which suggests convergence may be very show in the case when the pair potential is allowed to take negative values.

In conclusion, we note the use of spatial birth-death processes in Hanisch and Stoyan (1983), Fiksel and Stoyan (1983), and Fiksel (1984b), for modelling evolution of time-dependent point patterns such as earthquake epicentres and sink-holes.

6. Random closed sets II – The general case

6.1 Basic properties

Introduction

Random closed sets serve as mathematical models for irregular random patterns. In the planar case they are sometimes called *random area patterns*. They play an important role in stochastic geometry. In Chapter 3 we studied an important special case of a random set: the case of the Boolean model. We now consider the more general theory.

What today are known as random sets occur relatively early in the literature of probability theory. Systematic studies can be found as early as in the decade of the thirties; the famous book of Kolmogoroff (1933, p. 41) contains the phrase 'region ... whose shape depends on chance', and Kolmogoroff gives there a mean-value formula closely related to formula (6.2.1) below. Studies of particular random set models can be found in the literature of the forties and fifties; see the references in Section 3.1.

Foundations of the modern theory of random closed sets were laid by Choquet (1953/54) in his work on *capacities*. Matheron (1967, 1975) studied random closed sets in locally compact separable Hausdorf spaces (LCS spaces). A more general theory has been published by D. G. Kendall (1974); this derives from work in the sixties in which Davidson played an important part (see the remarks in Kendall, 1974).

In this chapter, the theory of Matheron (1975) will be discussed, but only for the particular case in which the random sets lie in $\mathbb{R}^d$. (A reader familiar with the theory of LCS spaces will find that the generalization is straightforward.) As a matter of convention the random sets considered here are *closed*; boundaries of sets are included in the sets themselves.

The σ-algebra $\mathcal{F}$

Random sets can be considered and characterized in many ways. It is natural to consider some system of test sets and to analyze intersections of random sets with various of these test sets. Thus two choices must be made: of the system of test sets, and then of the nature of measurement of the intersection. For example, if the random set Z might be infinite in extent then it is best that the test sets are bounded. On the other hand, as Kendall (1974) points out, some measurements (such as the Lebesgue measure of the intersection $Z \cap T$ for test sets T) are not always appropriate. Rather, to quote Kendall, 'instead of working with some version of the "size" of $Z \cap T$ we should merely note whether $Z \cap T$ is empty or not'. This means of measurement, the mere observation of whether the intersection is void, is fundamental to the modern random set theories as developed by Matheron and Kendall. The σ-algebra $\mathcal{F}$ is closely related to this kind of measurement.

Suppose, as in Section 1.2, that $\mathbf{F}$ is the family of all closed subsets of R^d. Then $\mathcal{F}$ denotes the smallest σ-algebra of subsets of $\mathbf{F}$ that contains all the 'hitting sets'

$$\boldsymbol{F}_K = \{F \in \mathbf{F} : F \cap K \text{ is not empty}\}$$

(K is any compact subset of $\mathbb{R}^d$). Thus the system of test sets is the system of all compact subsets of $\mathbb{R}^d$; and the measurement of the intersection is simply an observation of whether or not the intersection is void. The σ-algebra $\mathcal{F}$ corresponds to a family of assertions about whether or not the random set hits particular countable families of compact subsets, and fails to hit other particular countable families of compact subsets.

Important properties of $\mathcal{F}$ can be found in MATHERON (1975); $\mathcal{F}$ is the Borel σ-algebra with respect to a suitable topology on $\mathbf{F}$.

Random closed sets

A random closed set Ξ is a random variable taking values in $(\mathbf{F}, \mathcal{F})$. So Ξ is an $(\mathcal{A}, \mathcal{F})$-measurable mapping of a probability space $[\Omega, \mathcal{A}, \mathsf{P}]$ into $(\mathbf{F}, \mathcal{F})$. This means that probabilities can be assigned to statements about Ξ when the statements correspond to assertions that Ξ hits each of a countable family of compact sets and fails to hit any of another countable family of compact sets. Thus Ξ generates a distribution P on $(\mathbf{F}, \mathcal{F})$ called the ***distribution*** of Ξ (so $P(\boldsymbol{A}) = \mathsf{P}(\Xi \in \boldsymbol{A})$).

Simple examples of random sets

The full sophistication of the formal definition is not needed in order to appreciate many examples of random sets. Here are some simple examples of random sets:

random points;
closed spheres of random midpoints and radii;
closed cubes with random corner positions and random edge lengths.

More complicated random sets can be generated by using set-theoretic operations such as $\cup$, $\cap$, $\oplus$. MATHERON (1975) verified that these operations have suitable measurability properties so that the resulting structures are indeed still random closed sets. An important example of a random closed set generated by such means is of course the Boolean model which uses the operations of set union and translation.

Capacity functional

An effective means of characterizing the distribution P of a random set Ξ is given by the *capacity functional* T_Ξ. It is defined by

$$T_\Xi(K) = \mathsf{P}(\Xi \cap K \text{ is non-empty})$$
$$= \mathsf{P}(F_K)$$

where K is any compact subset of $\mathbb{R}^d$. Considered as a function on $\mathbb{K}$ the space of compact sets the capacity functional is actually an *alternating Choquet capacity of infinite order* (MATHERON, 1975). That is to say, T_Ξ satisfies

(a) if $K_n \downarrow K$ then $T_\Xi(K_n) \downarrow T_\Xi(K)$;

(b) $S_n(K; K_1, \ldots, K_n) \geqq 0$

where

$$S_n(K; K_1, \ldots, K_n)$$
$$= S_{n-1}(K; K_1, \ldots, K_{n-1}) - S_{n-1}(K \cup K_n; K_1, \ldots, K_{n-1})$$

and $S_0(K) = 1 - T_\Xi(K)$; here K and $K_1, \ldots, K_n$ are compact subsets of $\mathbb{R}^d$. Note that $S_n(K; K_1, \ldots, K_n)$ is equal to

$$\mathsf{P}(\Xi \cap K \text{ is empty but } \Xi \cap K_1, \ldots, \Xi \cap K_n \text{ are non-empty})$$

and that

$$T_\Xi(\emptyset) = 0$$

and $1 \geqq T_\Xi(K) \geqq 0$ if K is compact.

The Choquet theorem

The Choquet theorem on capacities gives a precise characterization of the distribution of a random closed set; the distribution is exactly determined by knowledge of the capacity functional.

Theorem. *Let T be a functional on $\mathbb{K}$. Then there is a (necessarily unique) distribution P on $\mathcal{F}$ with*

$$P(F_K) = T(K)$$

if and only if T is an alternating Choquet capacity of infinite order and such that

$$1 \geqq T(K) \geqq 0$$

for K compact and $T(\emptyset) = 0$.

The necessity of these conditions is clear from the discussion above. A proof of the sufficiency can be found in MATHERON (1975). Its basis is an extension theorem for set functions defined on semi-algebras. Note that the theorem is plausible, since the hitting sets F_K generate $\mathcal{F}$. If the random closed set satisfies additional properties then it may be possible to determine its distribution using a smaller family of hitting sets; see CRESSIE and LASLETT (1986).

For example, the distribution of a *convex* random closed set Ξ is determined by the probabilities

$$\mathsf{P}(\Xi \in K)$$

for all convex compact K; see Vitale (1983).

This theorem shows that capacity functionals play the same defining role for random closed sets as do distribution functions for random variables. Several distributional properties of random closed sets can be succinctly expressed in terms of their capacity functionals. Also convergence in distribution is ensured iff the corresponding capacity functionals are convergent; see Norberg (1984).

Independence

Two random closed sets Ξ_1 and Ξ_2 are said to be *independent* if for arbitrary $\boldsymbol{A}_1$ and $\boldsymbol{A}_2$ in $\mathcal{F}$

$$\mathsf{P}(\Xi_1 \in \boldsymbol{A}_1, \Xi_2 \in \boldsymbol{A}_2) = \mathsf{P}(\Xi_1 \in \boldsymbol{A}_1)\, P(\Xi_2 \in \boldsymbol{A}_2).$$

From Choquet's theorem this is equivalent to the condition

$$\mathsf{P}(\Xi_1 \cap K_1, \Xi_2 \cap K_2 \text{ are both non-empty}) = T_{\Xi_1}(K_1)\, T_{\Xi_2}(K_2)$$

for all compact K_1 and K_2.

Stationarity and isotropy

A random closed set Ξ is said to be *stationary* if Ξ and the translated sets $\Xi_x = \Xi + x$ all have the same distribution, whatever x in $\mathbb{R}^d$ may be. Again Choquet's theorem shows that this holds if and only if the capacity functional is translation-invariant

$$T_\Xi(K) = T_\Xi(K_y)$$

for every compact K and every y in $\mathbb{R}^d$. The proof of this uses the equalities

$$\begin{aligned} T_\Xi(K) &= \mathsf{P}(\Xi \cap K \text{ is non-empty}), \\ &= \mathsf{P}\big((\Xi - y) \cap K \text{ is non-empty}\big), \\ &= \mathsf{P}(\Xi \cap (K + y) \text{ is non-empty}) = T_\Xi(K_y). \end{aligned}$$

A random closed set is said to be *stationary and isotropic* (or *motion-invariant*) if for every rigid motion $\mathfrak{m}$ the distributions of Ξ and of $\mathfrak{m}\Xi$ are the same. As before an equivalent condition can be given using capacity functionals:

$$T_\Xi(K) = T_\Xi(\mathfrak{m}K)$$

for all such $\mathfrak{m}$ and compact K.

Ergodicity

Consider the translated family of sets

$$\boldsymbol{A}_x = \{F_x : F \in \boldsymbol{A}\}$$

for $\boldsymbol{A} \in \mathcal{F}$ and x in $\mathbb{R}^d$. A random closed set Ξ (or its distribution P) is said to be *ergodic* if the condition on $\boldsymbol{A}$,

$$P(\boldsymbol{A} \setminus \boldsymbol{A}_x \cup \boldsymbol{A}_x \setminus \boldsymbol{A}) = 0$$

always implies that $P(\boldsymbol{A})$ is zero or one. The condition on $\boldsymbol{A}$ stated here means (if we neglect events of probability zero) that Ξ belongs or fails to belong to $\boldsymbol{A}$ and to $\boldsymbol{A}_x$ simultaneously, so that the family of sets $\boldsymbol{A}$ is translation-invariant. Consequently if Ξ is ergodic then assertions that are translation-invariant (make no explicit or implicit reference to the location of Ξ) are either almost certainly true or almost certainly false.

A sufficient condition for ergodicity is the *mixing property*:

$$P(\boldsymbol{A} \cap \boldsymbol{B}_x) \to P(\boldsymbol{A})\, P(\boldsymbol{B})$$

as $\|x\| \to \infty$, for all $\boldsymbol{A}$, $\boldsymbol{B}$ in $\mathcal{F}$.

Boolean models with convex compact grains are ergodic random sets; see Nguyen and Zessin (1979) and Weil (1984).

The use of the property of ergodicity may be summarized as follows: statistical averages can be expressed by limits of arithmetic or spatial averages. The classical Strong Law of Large Numbers is typical of such results. For example, the classical ergodic theorem gives conditions to ensure that arithmetic averages

$$\left(f(Tx) + \cdots + f(T^n x)\right)$$

converge to $\mathsf{E}f(X)$, when f is a bounded measurable function, T^k is the kth iterate of a measure-preserving transformation, and X is a random variable. Fritz (1970), Nguyen and Zessin (1979), and Nguyen (1979) prove ergodic theorems for 'spatial processes', in which spatial averages replace arithmetic averages.

A simple case of such results concerns spatial processes of the form

$$X_B = h(\Xi \cap B) \quad \text{for } B \text{ a bounded Borel subset of } \mathbb{R}^d,$$

where h is a suitable function. Such a case arises in the study of statistics of random sets and in the definition of 'densities' of random sets.

In order to discuss this, notation must be introduced to describe the kind of spatial averaging being used. Let C_0 denote the semi-open cube,

$$C_0 = \{(x_1, \ldots, x_d) \in \mathbb{R}^d : 1/2 > x_i \geqq -1/2\}.$$

For any bounded Borel subset B of $\mathbb{R}^d$ let $r(B)$ be the supremum of the radii of spheres lying in B. Finally let $\{K_n\}$ be a sequence of nonempty convex compact sets with

$$K_m \text{ contained in } K_n \text{ whenever } m < n \quad \text{and} \quad r(K_n) \to \infty \text{ as } n \to \infty.$$

Theorem. *Suppose the mapping $h : \mathcal{B}_0{}^d \to \mathbb{R}$ is a measurable, translation-invariant, additive set-function defined on bounded Borel sets. Suppose, moreover, that Ξ is a stationary ergodic random closed set.*

(1) *For every $\varepsilon > 0$ there is a $\eta = \eta(\varepsilon) > 0$ such that if sup $\{\mathsf{E}\,|h(\Xi \cap K)| :$ K non-empty and in $C_0\} < \infty$ then*

$$\mathsf{E}\left|\frac{h(\Xi \cap K)}{\nu_d(K)} - \mathsf{E}h(\Xi \cap C_0)\right| < \varepsilon$$

for compact convex K with $r(K) > \eta$.

(2) *Suppose there is a non-negative random variable ξ of finite mean such that*

$$|h(\Xi \cap K)| < \xi \text{ almost surely for non-empty } K \text{ in } C_0.$$

Then

$$\lim_{n\to\infty} \frac{h(\Xi \cap K_n)}{\nu_d(K_n)} = \mathsf{E}h(\Xi \cap C_0) \quad \text{almost surely.}$$

Thus the spatial average $h(\Xi \cap K)/\nu_d(K)$ converges to the statistical average $\mathsf{E}h(\Xi \cap C_0)$ under conditions of stationarity and ergodicity. Case (1) of the theorem provides a weaker form of convergence than case (2).

Many authors have established Central Limit Theorems, Laws of Large Numbers, and large-deviation results for bounded and unbounded random closed sets. See Artstein (1984), Artstein and Hart (1981), Artstein and Vitale (1982), Lyashenko (1982, 1983), Mase (1982a, b) and Weil (1982b). Surveys are provided by Vitale (1983) and Cressie and Laslett (1986).

Random fields

From any random closed set Ξ a random field $\{X(x) : x \in \mathbb{R}^d\}$ can be constructed as the indicator function of the set Ξ. Thus

$$X(x) = \mathbf{1}_\Xi(x) \quad \text{for } x \text{ in } \mathbb{R}^d.$$

This correspondence is of use in applying methods of the statistics of random fields to the statistical analysis of random sets.

Mean value formulae for convex random sets

In this section the random closed set Ξ is taken to be convex and compact with probability one. Thus the Minkowski functionals $W_k(\Xi)$ can be considered. It is assumed that their first moments are finite:

$$W_k = \mathsf{E}W_k(\Xi) < \infty \quad \text{for } k = 0, 1, \ldots, d.$$

The *Steiner, Crofton*, and other integral-geometric *formulae* will then be satisfied for Ξ almost surely and analogous formulae will hold for the mean values.

For example corresponding to the Steiner formula there is

$$\mathsf{E}\nu_d(\Xi \oplus b(o, r)) = \sum_{k=0}^{d} \binom{d}{k} \overline{W}_k r^k \quad \text{for } r \geqq 0.$$

If Ξ has distribution invariant with respect to rotations about the origin o then the so-called *generalized Steiner formula* holds for every compact convex K

$$\mathsf{E}\nu_d(\Xi \oplus K) = \frac{1}{\omega_d} \sum_{k=0}^{d} \binom{d}{k} \overline{W}_k W_{d-k}(K). \tag{6.1.1}$$

This is of considerable importance. Its proof rests on the Hadwiger characterization theorem; the functional h defined by

$$h(K) = \mathsf{E}\nu_d(\Xi \oplus K)$$

can be shown to be motion-invariant, C-additive, and monotonic in the compact set K. Thus from Hadwiger's theorem h is a linear combination of the $W_k(K)$ and its coefficients can be derived by setting $K = b(o, r)$.

In the important planar and spatial cases the formula (6.1.1) takes on the forms

Planar ($d = 2$)

$$\mathsf{E}A(\Xi \oplus K) = \overline{A} + \frac{1}{2\pi} \overline{U} U(K) + A(K). \tag{6.1.2}$$

Spatial ($d = 3$)

$$\mathsf{E}V(\Xi \oplus K) = \overline{V} + \frac{1}{2} \overline{\overline{b}} S(K) + \frac{1}{2} \overline{S} \overline{b}(K) + V(K). \tag{6.1.3}$$

A similar generalization can be carried out on the formulae (1.6.10) for the Minkowski functionals, and also for the generalized Minkowski functionals of sets in the convex ring $\mathcal{R}$. In this last case W_d is replaced by $\overline{\chi}$.

Comparison of random closed sets

Stoyan and Stoyan (1980a) considered some partial orderings of random closed sets. These orderings express ways in which one set is larger than another. In the case of two sets of equal shape the orderings all reduce to ordering by size. One example enables the comparison of random sets with the same mean volume.

6.2 Characteristics for stationary and isotropic random closed sets

The volume fraction

The volume fraction p of a stationary random closed set plays a role that is similar to that of the mean of a random variable. The volume fraction is the mean of the volume of Ξ in the unit cube C_0

$$p = \mathsf{E}\nu_d(\Xi \cap C_0).$$

It satisfies

$$p = \mathsf{P}(o \in \Xi) = \mathsf{P}(x \in \Xi) = T_\Xi(\{o\}) \tag{6.2.1}$$

where x belongs to $\mathbb{R}^d$. These relations follow from the stationarity of Ξ by the following chain of equalities

$$\begin{aligned}\mathsf{E}\nu_d(\Xi \cap C_0) &= \mathsf{E}\int_{\mathbb{R}^d} \mathbf{1}_{\Xi \cap C_0}(x)\,\mathrm{d}x \\ &= \mathsf{E}\int_{C_0} \mathbf{1}_\Xi(x)\,\mathrm{d}x = \int_{C_0} \mathsf{E}\mathbf{1}_\Xi(x)\,\mathrm{d}x \\ &= \int_{C_0} \mathsf{P}(x \in \Xi)\,\mathrm{d}x = \mathsf{P}(o \in \Xi)\end{aligned}$$

since by stationarity $\mathsf{P}(x \in \Xi) = \mathsf{P}(o \in \Xi)$ for all x in $\mathbb{R}^d$.

The volume fraction can arise in other ways.

(a) If the volume measure V_Ξ is defined on $\mathbb{R}^d$ by

$$V_\Xi(B) = \nu_d(\Xi \cap B) \quad \text{for all Borel sets } B$$

then V_Ξ is a stationary random measure. Its *intensity* is clearly p.

(b) Consider v a translation-invariant measure $\mathbb{R}^d$ defined by

$$v(B) = \mathsf{E}\nu_d(\Xi \cap B) \quad \text{for all Borel sets } B.$$

It is finite on bounded Borel sets. Suppose that K is a convex body with inner points. Then the corresponding *density*

$$D_V(\Xi) = \lim_{r\to\infty} \frac{v(rK)}{\nu_d(rK)}$$

exist and equals p.

(c) Suppose that Ξ is ergodic. Then by the ergodic theorem given above as $r \to \infty$ so the quantity

$$\frac{\nu_d\big(\Xi \cap (rK)\big)}{\nu_d(rK)}$$

converges to p in some sense (the theorem gives precise statements of possible forms of convergence).

The covariance

Just as p describes the first-order structure of Ξ so the covariance measures the second-order structure. However, for this to be effective p must be assumed positive. Some random sets, such as point or fibre processes, have $p = 0$; their second-order structure is described by moment measures.

The covariance C is given by

$$C(\boldsymbol{r}) = \mathsf{P}(o \in \Xi, \boldsymbol{r} \in \Xi) \quad \text{for } \boldsymbol{r} \text{ in } \mathbb{R}^d.$$

It satisfies the equation

$$C(\boldsymbol{r}) = \mathsf{P}(o \notin \Xi, \boldsymbol{r} \notin \Xi) + 2p - 1 = \mathsf{P}(o \in \Xi \ominus \check{B}) \tag{6.2.2}$$

where $B = \{o, \boldsymbol{r}\}$ and $\check{B} = -B$ as in section 1.3. Thus $C(\boldsymbol{r})$ is the volume fraction of the set Ξ eroded by the structuring element $B = \{o, \boldsymbol{r}\}$ (so B is a pair of points). This formula (6.2.2) forms the basis for statistical estimation of $C(\boldsymbol{r})$ by means of image analyzers, and is a simple consequence of the definition of the operation of erosion:

$$o \in \Xi \ominus \check{B} \quad \text{if and only if } o \in \Xi \text{ and } \boldsymbol{r} \in \Xi.$$

There is a close relation between C and the *covariance function* of the random field associated with Ξ; the mean of the stationary random field

$$\{\mathbf{1}_\Xi(x) : x \in \mathbb{R}^d\}$$

is given by $\mathsf{E}\mathbf{1}_\Xi(x) = p$, and its covariance function k is

$$\begin{aligned} k(\boldsymbol{r}) &= \mathsf{E}\big(\mathbf{1}_\Xi(o) - p\big)\big(\mathbf{1}_\Xi(\boldsymbol{r}) - p\big) \\ &= C(\boldsymbol{r}) - p^2 \end{aligned}$$

as can be seen on expanding the product.

The covariance function is often normalized by dividing by $k(o) = p(1 - p)$, which gives the *correlation function*

$$\varkappa(\boldsymbol{r}) = k(\boldsymbol{r})/k(o) \quad \text{for } \boldsymbol{r} \in \mathbb{R}^d.$$

If Ξ is not only stationary but also isotropic then the functions C, k, $\varkappa$ depend only on the distance $r = \|\boldsymbol{r}\|$. For simplicity the argument is then written as a scalar r. For example,

$$C(\boldsymbol{r}) = C(\|\boldsymbol{r}\|) = C(r).$$

Analysis of the three functions C, k, $\varkappa$ allows for information to be gained about the structure of Ξ even in the case where no image of Ξ is available (as for example in X-ray scattering analysis, or in geological exploration by means of boreholes). Figure 6.1 illustrates this by giving three typical images of samples of planar random sets together with corresponding empirical covariances (determined by image analyzer). Thus the functions can be used in a descriptive analysis. They also have theoretical uses in determining the accuracy of statistical estimators, as is covered in 6.3. Finally they can themselves be used directly in estimation of model parameters, for example in the Boolean model as demonstrated in Example 3.3.

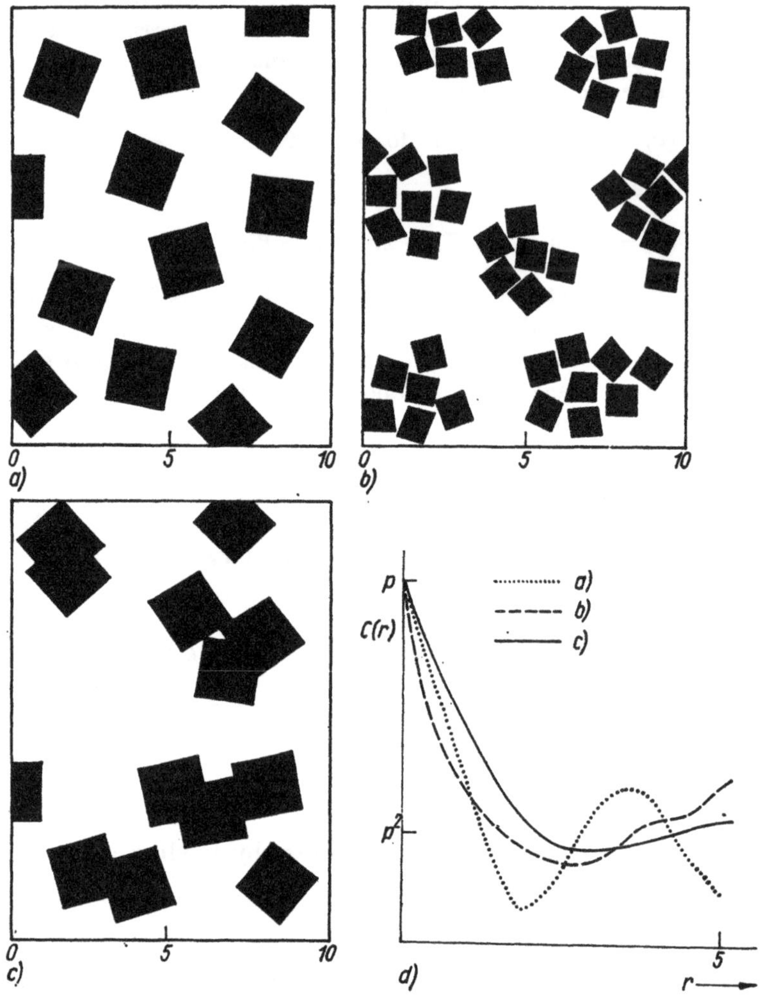

Fig. 6.1. Forms of the covariance $C(r)$ for realizations of three different random closed sets sharing the same volume fraction p. The form of $C(r)$ allows conclusions to be drawn concerning the properties of the corresponding random sets, such as the inter-particle distances or the tendency to clustering. See also Figure 6.3

If Ξ is isotropic then the covariance satisfies some basic properties:

$$C(0) = p$$

and if Ξ satisfies the mixing property then also

$$C(\infty) = p^2.$$

If the random set Ξ is suitably regular then the specific surface content $S_{V^{(d)}}$ will be finite (see page 200 for a definition of specific surface area). Suppose in addition to this that for every convex compact K

$$\nu_{d-1}(\partial\Xi \cap K) = \nu_{d-1}\big(\partial(\Xi \setminus \partial\Xi) \cap K\big)$$

that is to say Ξ contains no flat sheets of zero thickness. Then the derivative $C'(0)$ is related to $S_{V^{(d)}}$

$$S_{V^{(d)}} = -2\sqrt{\pi}\,\Gamma\big((d+1)/2\big)\,C'(0)/\Gamma(d/2). \tag{6.2.3}$$

In the particular cases of $d = 1, 2,$ or 3 the coefficient of $-C'(0)$ is $2, \pi$, or 4 respectively.

For $d = 3$ the formula (6.2.3) can be established as follows.

Let g be an oriented line through o. Then the set $\partial\Xi \cap g$ is a point process of finite intensity P_L, where P_L is connected to $S_V = S_{V^{(3)}}$ by

$$P_L = S_V/2.$$

This is a stereological formula; see Section 11.2.2.

If $\{x_n\}$ is the point process of left end-points of $\partial\Xi \cap g$ (recall that g is oriented, so 'left' has a meaning along g) then a mark l_n is attached to each x_n by defining l_n to be the length of the subsequent interval in Ξ. The marked point process $\Psi = \{[x_n; l_n]\}$ inherits stationarity from Ξ and has intensity $\lambda = S_V/4$. Let M be the mark distribution function of Ψ. Then M is the chord-length distribution function of $\Xi \cap g$. Let L be the length distribution function of a typical chord of $g \setminus \Xi$. The volume fraction $V_V = p$ satisfies

$$p = \lambda \int_0^\infty x \,\mathrm{d}M(x)$$

and the covariance $C(r)$ fulfils the inequality

$$\lambda\left[\int_0^\infty (x-r)_+ \,\mathrm{d}M(x) + \int_0^\infty (x-r)_+ \,\mathrm{d}L(x)\right] \geqq C(r) \geqq \lambda \int_0^\infty (x-r)_+ \,\mathrm{d}M(x).$$

This yields

$$-\lambda\left\{\int_0^r x \,\mathrm{d}M(x) + r[1-M(r)] + \int_0^r (r-x)\,\mathrm{d}L(x)\right\} \geqq C(r) - C(0)$$

$$\geqq -\lambda\left\{\int_0^r x\,\mathrm{d}M(x) + r[1-M(r)]\right\}.$$

But

$$\lim_{r\to\infty} \frac{\int_0^r [1-M(x)]\,\mathrm{d}x}{r} = -\lim_{r\to 0}\int_0^r \left(\frac{x}{r}\right)\mathrm{d}M(x) - [1-M(0+)]$$

$$= [1-M(0+)]$$

and

$$\lim_{r\downarrow 0}\int_0^r (r-x)\,\mathrm{d}L(x) = 0$$

and the boundary assumption stated above (6.2.3) implies that $M(0+) = 0$. Therefore the right-hand derivative $C'(0)$ of C at 0 exists and satisfies $\lambda = S_V/4$.

It has proved useful in many applications to approximate complicated covariances by a simpler form, the so-called *exponential covariance* C_e:

$$C_e(r) = p(1-p)\,\mathrm{e}^{-\alpha r} + p^2 \quad \text{for } r \geqq 0, \qquad \alpha \geqq 0.$$

The corresponding covariance function for the random field and correlation function are given by

$$\left.\begin{aligned} k_e(r) &= p(1-p)\,\mathrm{e}^{-\alpha r} \\ \varkappa_e(r) &= \mathrm{e}^{-\alpha r} \end{aligned}\right\} \quad \text{both for } r \geqq 0.$$

In the spatial case ($d = 3$) it can be shown that $\alpha = S_V/\big(4 \cdot V_V \cdot (1 - V_V)\big)$. This is a useful expression of the theoretical quantity α in terms of quantities that can be estimated.

A planar random set with exponential covariance can be constructed as follows. Consider a Poisson line mosaic with parameter ϱ as described in 10.5.1. Let each cell of the mosaic be painted independently of the others, painted black with probability p and white with probability $1 - p$. Then the closure of the union of all black regions is a random closed set with exponential covariance C_e, with $\alpha = \varrho$. Oganian (1973) has given examples of random sets with exponential covariances but with curved boundaries.

Contact distribution functions

It is frequently the case that random sets must be considered for which identification of single particles or pores is inappropriate. Nevertheless, some kind of measurement of 'size' would be most useful. The *contact distribution functions* H_B provide a useful tool for this. Delfiner (1972) and Serra (1982, Ch. XIII) discuss the general need for such a concept.

A contact distribution function depends on the choice of *structuring element* B. If B is compact and $o \in B$ then H_B is defined by

$$\begin{aligned} H_B(r) &= 1 - \mathsf{P}(o \notin \Xi \oplus r\check{B} \mid o \notin \Xi) \\ &= 1 - \mathsf{P}(o \notin \Xi \oplus r\check{B})/(1-p) \quad \text{for } r \geqq 0. \end{aligned}$$

The cases of particular practical importance are those of the two extremes: the *linear contact distribution function* H_l (for $B = l$ a segment of unit length — in the case of anisotropy the orientation of this segment is of importance; see page 181); and the *spherical contact distribution function* H_s (for $B = s$ a unit sphere in the spatial case, or unit disc in the planar case).

The quantity $1 - H_s(r)$ is precisely the conditional probability that a fixed point is centre of a sphere of radius r lying completely outside Ξ, given that the fixed point in question does not belong to Ξ. Thus the spherical contact distribution function H_s is also referred to as 'the law of first contact'.

For isotropic Ξ of sufficiently smooth boundary (see page 187) the spherical contact distribution function is related to the surface intensity by

$$S_V^{(d)} = (1 - p)\, H_s'(0). \tag{6.2.4}$$

Analogous quantities can be defined using the closure of complement $(\Xi^c)^{cl}$ rather than Ξ. Because in our approach all random sets are closed, the closure of the complement is the appropriate representation of the exterior of Ξ.

Chord length distributions

The linear contact distribution function H_l characterizes the distribution of chord lengths of the complement of Ξ. In fact, $H_l(r)$ is the distribution function of residual chord length in the direction of a test line segment begun at a fixed point o, under the condition that o does not belong to Ξ. This is illustrated in Figure 6.2.

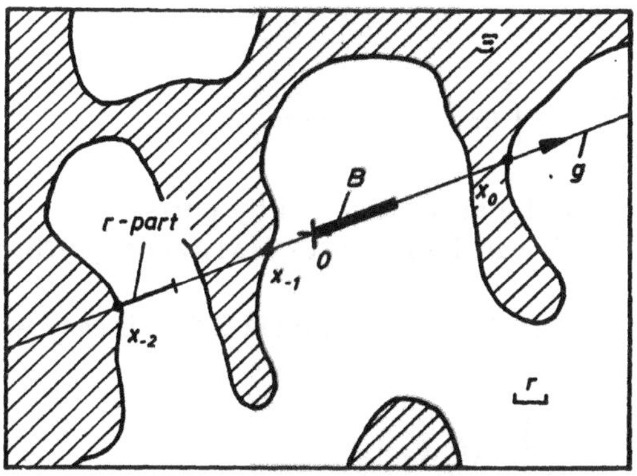

Fig. 6.2. Intersection of a closed random set Ξ and a line g (the direction of the line being defined by a structuring element B that is a segment). The left endpoints $\ldots, x_{-1}, x_0, x_1, \ldots$ form a point process. The r-part of a chord is the set of all points of distance more than r from the right endpoint.

This interpretation must be considered in the light of the difference between 'length-biased' and 'unbiased' sampling of chord-lengths. For suppose that g is a directed line containing the test segment and consider the stationary point process $g \cap \partial\Xi$. We may consider $g \cap \partial\Xi$ to be a point process on $\mathbb{R}$; consider this point process under the additional assumption that its intensity is finite.

Then from $g \cap \partial\Xi$ we may construct a marked point process $\{[x_n; r_n]\}$. Here $\{x_n\}$ is an enumeration of the sequence of left end-points of chords of $(\Xi^c)^{cl}$ and r_n is the length of the chord whose left end-point is x_n. The mark distribution function L is the length distribution function of a 'typical' chord of the complement of Ξ, selected in a fashion that is not length-biased. The distribution functions L and H_l are linked by

$$H_l(r) = \int_0^r \big(1 - L(x)\big)\,\mathrm{d}x/m, \tag{6.2.5}$$

where m is the mean typical chord length

$$m = \int_0^\infty x\,\mathrm{d}L(x)$$

with $0 < m < \infty$. Formula (6.2.5) can be shown as follows. For a chord $[x_n, r_n]$ of length greater than r let the 'r-part' be the subset of points in the chord further than r from the right end-point. Thus the r-part has length $r_n - r$ (see Figure 6.2). $1 - H_l(r)$ is the probability that o lies in an r-part given that o does not lie in Ξ. Hence $1 - H_l(r)$ is the quotient of $L_L(r)$ (the length fraction of the union of all r-parts) by L_L (the length fraction of $\boldsymbol{g} \cap \Xi$). If λ is the intensity of the process $\{x_n\}$ of left end-points then

$$L_L = \lambda m$$

$$L_L(r) = \lambda \int_r^\infty (x - r)\,\mathrm{d}L(x)$$

so

$$1 - H_l(r) = \frac{L_L(r)}{L_L} = \int_r^\infty (x - r)\,\mathrm{d}L(x)/m$$

and the relation (6.2.5) follows by an integration by parts.

Further discussion of chords can be found in Matheron (1975, p. 53) and Serra (1982, Ch. X).

By means of stereological formulae (9.3.7) and (11.2.3) it is easy to show that the mean chord length m satisfies

$$\left.\begin{aligned} m &= \pi(1 - A_A)/L_A \\ A_A &= p \end{aligned}\right\} \quad (\text{if } d = 2)$$

$$\left.\begin{aligned} m &= 4(1 - V_V)/S_V \\ V_V &= p \end{aligned}\right\} \quad (\text{if } d = 3).$$

Following the duality between Ξ and $(\Xi^c)^{cl}$ mentioned above, lengths of chords *inside* $\Xi \cap \boldsymbol{g}$ can be studied as well.

If Ξ is a Boolean model of convex grains, or a semi-Markovian random set in the sense of Matheron (1975), then the chords of the complement of Ξ have an exponential distribution. Ambartzumian (1982a) considered the question of which other random sets have this property.

Orientation analysis of random closed sets

Many area patterns, studied using the theory of random sets, exhibit anisotropy. Such anisotropy can be quantified by analysis of $C(\boldsymbol{r})$ and $H_l(r)$, by means of varying the direction of $\boldsymbol{r}$ and of the linear segment used in H_l.

Figure 6.3(a) shows a pattern exhibiting clear anisotropy, and Figures 6.3(b), (c) show $C(\boldsymbol{r})$ and $H_l(r)$ measured for $\overline{\Xi^c}$ in two perpendicular directions. One direction is parallel to the bottom line of Figure 6.3(a), the other perpendicular. The reader can see the differences between the two directions that result from anisotropy.

a)

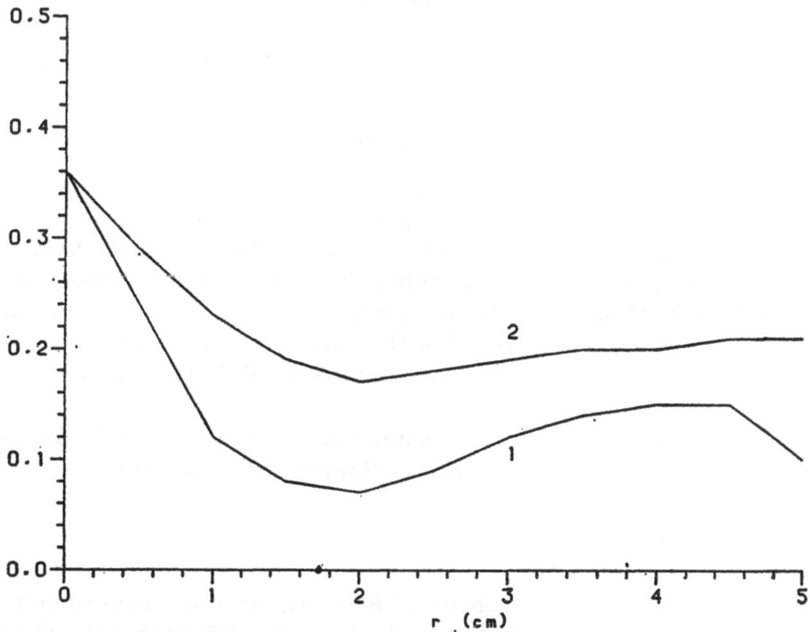

b)

Fig. 6.3 (a). An anisotropic area pattern. (b) Estimate of $C(\boldsymbol{r})$ for two directions (vertically and horizontally with respect to the page). (c) Estimate of $H_l(r)$ for the complement $(\Xi^c)^{cl}$ using the same directions as in (b). Horizontal direction (→), 1; Vertical direction (↑), 2; see p. 182.

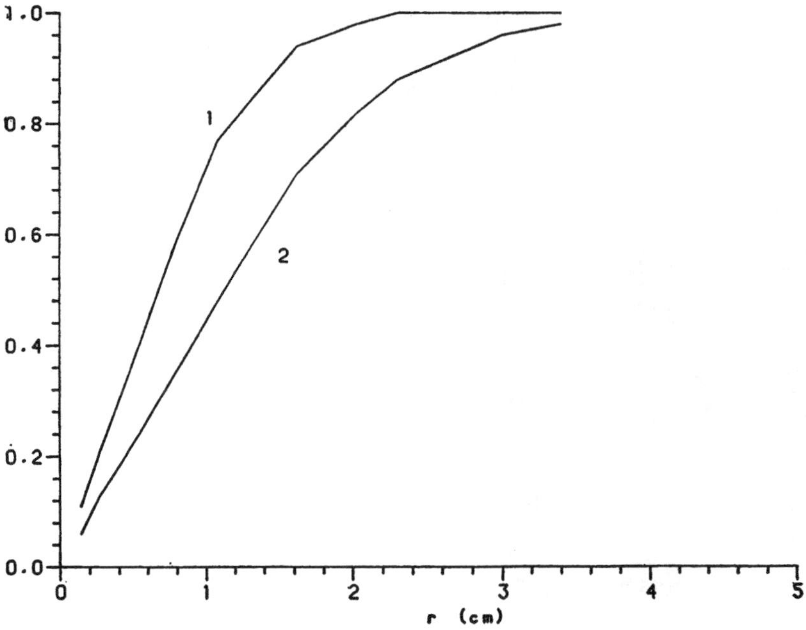

Fig. 6.3 (c)

Intensities and densities of random closed sets

Various random measures are associated to a random closed set Ξ. One example is the *volume measure* V_Ξ which is defined by

$$V_\Xi(B) = \nu_d(\Xi \cap B) \text{ for } B \text{ a bounded Borel set.}$$

Other examples include the *Minkowski measures* discussed in 7.3.

If such an associated random measure μ_Ξ is stationary then it has an *intensity* $\lambda(\mu_\Xi)$. When $\lambda(\mu_\Xi)$ is finite and positive it is an instructive characteristic of Ξ. The *specific surface* $S_V^{(d)}$ arises in this fashion as the mean surface area of Ξ per unit volume. If $S_V^{(d)}$ is to be well-defined then either assumptions on the smoothness of $\partial\Xi$ must be imposed or the notion of surface area must be modified. Zähle (1982, 1983a) discussed suitable modifications of surface area as noted in page 231.

Generally let h be an additive translation-invariant functional on the Borel sets or on some sub-system of the Borel sets. Suppose that the limit

$$D_h(\Xi) = \lim_{r\to\infty} \frac{\mathsf{E}h(\Xi \cap rK)}{\nu_d(rK)}$$

exists and does not depend on the choice of K as long as K is a convex body with inner points. Then $D_h(\Xi)$ is said to be the *h-density* of Ξ. Such h-densities have been studied in Weil (1984) and in Weil and Wieacker (1983).

6.3 Statistics for random closed sets

This section discusses methods for the estimation of p, C, and H_B and the problem of determining the accuracy of estimation for p. Throughout the section the random closed sets under construction are assumed to be stationary and isotropic. Statistical methods for the particular case of germ-grain models are also discussed in 6.4.

There is some literature on the methods for determining the volume fraction of a phase in a bounded inhomogeneous body such as an animal organ. See, for example, Cruz-Orive and Myking (1981), Kroustrup and Gundersen (1983), and Leder (1979). Obviously in such cases the stationarity assumption noted above fails. These cases are not considered here.

Estimation of p

The methods described here apply to planar sets or to planar sections of spatial sets (for the case of projections of thick sections, see Section 11.2.3). There are three methods, corresponding to sampling the image on a finite grid of isolated points, or along parallel lines, or actually over the whole of a bounded region.

The point-count method (Thompson, 1930, and Glagolev, 1933)

A grid of points $x_1, \ldots, x_n$ is selected from the image and p is estimated by

$$\hat{p}_p = \frac{1}{n}\left[\sum_{i=1}^{n} \mathbf{1}_\Xi(x_i)\right]$$

the fraction of grid points lying in Ξ. The estimator is unbiased since

$$\mathsf{E}\hat{p}_p = \frac{1}{n}\sum_{i=1}^{n} \mathsf{E}\mathbf{1}_\Xi(x_i) = \frac{1}{n}\,np.$$

Its variance σ_p^2 is given by

$$\sigma_p^2 = \frac{1}{n^2}\,\mathsf{E}\left[np - \sum_{i=1}^{n} \mathbf{1}_\Xi(x_i)\right]^2$$

$$= \frac{1}{n^2}\left[np(1-p) + 2\sum_{i \neq j} k(r_{ij})\right] \tag{6.3.1}$$

where $r_{ij} = \|x_i - x_j\|$ and k is the covariance function of Ξ. The formula (6.3.1) follows from expanding the square

$$\left[np - \sum_{i=1}^{n} \mathbf{1}_\Xi(x_i)\right]^2 = \left[\sum_{i=1}^{n} (p - \mathbf{1}_\Xi(x_i))\right]^2.$$

The variance formula (6.3.1) depends both on the grid geometry and the form of the covariance function k. If n is large and k decays to zero quickly then approximations are available. Many authors have studied point-count methods. See, for example, Matérn (1960) and Diggle and ter Braak (1982).

Example. Suppose the grid is rectangular of mesh width Δ and k is of exponential form with $\alpha\Delta \gg 1$. Neglecting edge effects and only considering up to second-nearest neighbours in the grid,

$$\sum_{i \neq j} k(r_{ij}) = 4np(1-p) \exp(-\alpha\Delta) \left[1 + \exp\left(-\left(\sqrt{2}-1\right) \alpha\Delta\right)\right]$$

thus

$$\sigma_p^2 \simeq \frac{1}{n} \left\{p(1-p) + 8p(1-p) \exp(-\alpha\Delta) \left[1 + \exp\left(-\left(\sqrt{2}-1\right)\alpha\Delta\right)\right]\right\}. \tag{6.3.2}$$

The lineal method (Rosiwal, 1898)

An array of N parallel line segments, each of length l, is laid down on the image. The volume fraction p is estimated by

$$\hat{p}_l = \frac{L}{Nl} \tag{6.3.3}$$

where L is the (random) total length of line segment intersecting Ξ. As in the point-count method this estimator is unbiased. Its accuracy has been studied in Stoyan and Steyer (1979) and Stoyan (1979b), with reference also to the interline distance. The particular models used in these studies were either of exponential covariance or formed by a Boolean model of spherical primary grains. In the case of exponential covariance with large α the estimation variance σ_l^2 is approximately

$$\sigma_l^2 \simeq \frac{2p(1-p)\left(1 - 1/(\alpha l)\right)}{N\alpha l}. \tag{6.3.3a}$$

In the case of a planar sample from a spatial set σ_l^2 can be expressed in terms of V_V and S_V

$$\sigma_l^2 \simeq \frac{8V_V^2(1-V_V)^2}{NlS_V}.$$

The area method

It is possible to measure the area fraction p_A directly, for example by using an image-analyzer. Measurement errors occur only through the bounded nature of the window (if discretization errors are ignored). The estimation variance σ_A^2 is given by

$$\sigma_A^2 = \mathsf{E}(\hat{p}_A - p)^2 = \frac{1}{A^2} \int_W \int_W k(\|x-y\|)\, dx\, dy,$$

where A is the area of the window W.

If the window is suitably large then an approximate formula is available in the case of an exponential covariance. Here W must be large in the sense of extending far in all directions, as for example in the case of a disc of large radius. Then

$$\sigma_A^2 \simeq \frac{2\pi p(1-p)}{\alpha^2 A}. \tag{6.3.4}$$

See SERRA (1982) and HANISCH, OBERSTEDT, REINSCH, and STOYAN (1978). An alternative formula is

$$\sigma_A^2 \simeq 100 V_V^3(1 - V_V)^3/(AS_V^2).$$

The estimation variance has been determined numerically and presented in nomograms or tables for the case of exponential covariance: see JOURNEL and HUIJBREGTS (1978) and AGTERBERG (1981). OHSER (1985) calculated correction-factors for formula (6.3.4) for rectangular windows W.

Example 6.1. *Comparison of estimation variances of the point-count, lineal, and area methods.*

To illustrate the methods discussed above consider a random closed set of exponential covariance with $\alpha = 5$, observed through a square sampling window W of dimension 30×30 in suitable units. For the point-count method a rectangular grid is used with $31 \times 31 = 961$ points with mesh width $\Delta = 1$. For the lineal method 31 parallel lines (parallel also to a window side) are employed, each of length 30 and with separation distance 1.

In the case of specific area $p = 0.10$ the equations (6.3.2, 6.3.3a, 6.3.4) yield approximate estimation variances given below and compared with exact values (given in brackets). The exact values were obtained by numerical integration.

$$\sigma_p^2 = 0.010\ (0.0098)$$
$$\sigma_l^2 = 0.006\ (0.0059)$$
$$\sigma_A^2 = 0.005\ (0.0053).$$

Naturally, greater use of information contained in the window is reflected in lower estimation variance; this has to be set against any increased work that might be entailed.

It should be noted that in case of anisotropy the lineal method, based on parallel lines, may be affected adversely by the relationship between axes of anisotropy and the common direction of the sampling lines.

Estimation of the covariance

Using (6.2.2) the covariance $C(\boldsymbol{r})$ can be estimated as the area fraction of the set $\Xi \ominus \{o, \boldsymbol{r}\}$. Under the assumption of isotropy the same estimator can be used for $C(r)$ with $\|\boldsymbol{r}\| = r$. With an image analyzer the measurement is easily made. Manual measurements can be carried out with the aid of a grid; by counting the number $n_{\boldsymbol{r}}$ of grid points x such that $x + \boldsymbol{r}$ is also a grid point and then setting

$$\hat{C}(\boldsymbol{r}) = (\text{grid point pairs } (x, x + \boldsymbol{r}) \text{ which are both in } \Xi)/n_{\boldsymbol{r}}.$$

The estimator $\hat{C}(\boldsymbol{r})$ is unbiased for $C(\boldsymbol{r})$.

Estimation of $H_B(r)$

Since $1 - H_B(r)$ is the area fraction of $\Xi \oplus r\check{B}$ methods similar to those of the previous section are suitable. An image analyzer can be used to measure the appropriate area fraction directly. Manual measurements use a grid, counting n_B the number of grid points outside Ξ and setting

$$1 - \hat{H}_B(r) = \frac{1}{n_B} \# \begin{bmatrix} \text{grid points } x \text{ outside } \Xi \\ \text{such that } x + rB,\ \Xi \text{ are disjoint} \end{bmatrix}.$$

Little is known of the theoretical properties of the estimators given above. If Ξ is ergodic then they are consistent. Under certain mixing conditions the estimator $\hat{p}_A$ can be shown to be asymptotically normal (BADDELEY, 1980a; MASE, 1982a). In MASE (1982b) sufficient conditions are given for these mixing assumptions to hold. MASE (1982a) contains approximations for the asymptotic variance.

6.4 Germ-grain and grain models

Basic facts

In a natural generalization of the Boolean model the Poisson point process of germs is replaced by a general point process. This leads to the class of *germ-grain models*. In particular, the generalization allows for treatment of collections of non-overlapping grains; in the Boolean model the possibility of overlapping cannot be excluded.

A formal definition starts with a marked point process $\Psi = \{[x_n; \Xi_n]\}$ where the x_n are points of $\mathbb{R}^d$ and the Ξ_n are compact subsets of $\mathbb{R}^d$. A *germ-grain model* Ξ as defined by HANISCH (1981) arises from such a point process as a union

$$\Xi = \bigcup_n (\Xi_n + x_n) = (\Xi_1 + x_1) \cup (\Xi_2 + x_2) \cup \cdots$$

The x_n are the *germs* and the Ξ_n are the *primary grains* of the germ-grain model. In the case when Ξ is stationary it is useful to introduce the 'typical' grain Ξ_0, which is a random closed set of the same distribution as the sets in the sequence of primary grains $\{\Xi_n\}$ but independent both of them and of the germs (to make complete sense of this it is necessary to appeal to the theory of Palm distributions). The distribution M of the 'typical' grain is the mark distribution of the marked point process Ψ; it is a distribution on the space $\mathbf{K}$ of compact sets endowed with the σ-algebra that is the trace of $\mathcal{F}$ on $\mathbf{K}$. The intensity of the germ process is denoted by λ.

When the grains Ξ_n are convex the germ-grain model can be represented as a *random $\mathcal{S}$-set* or *grain model*. A random $\mathcal{S}$-set Ξ has the property that almost surely for every compact convex set K

$$\Xi \cap K \in \mathcal{R}.$$

Here $\mathcal{R}$ is the *convex ring*, the family of all sets that can be expressed as finite unions of compact convex sets (see Section 1.7). So a random $\mathcal{S}$-set is locally a finite union of compact convex grains. WEIL and WIEACKER (1986) showed that for every random $\mathcal{S}$-set Ξ there is a point process in $C(\mathbf{K})$ such that the union of its 'points' (which are of course really compact convex sets) is equal to Ξ. Since this point process in $C(\mathbf{K})$ can be represented as a marked point process Ψ as above, random $\mathcal{S}$-sets are essentially the same as germ-grain models with convex grains. In particular, if the $\mathcal{S}$-set is stationary then the germ-grain model can be chosen to be stationary as well.

Random $\mathscr{S}$-sets Ξ have boundaries that are smooth enough for the definition of random Minkowski measures as in Chapter 7. These are examples of random measures accompanying Ξ as described in 6.2. When Ξ is stationary then so are the Minkowski measures.

Formulae for p and C

Explicit formulae for general germ-grain models are complicated. The capacity functional T_Ξ can be expressed in terms of the generating functional G of the marked point process Ψ

$$T_\Xi(K) = \mathsf{P}(\Xi \cap K \text{ is non-empty}) = 1 - G(v_K)$$

where K is compact and

$$v_K(x, C) = 1 - \mathbf{1}_{\check{C} \oplus K}(x) \quad \text{for compact } C \text{ and } x \text{ in } \mathbb{R}^d.$$

When the grains fail to overlap, which is to say when

$$\nu_d[(\Xi_i + x_i) \cap (\Xi_j + x_j)] = 0 \quad \text{for all } i \neq j$$

with probability one, and when, moreover, Ξ is stationary then the volume fraction is given by

$$p = \lambda \overline{V},$$

where $\overline{V}$ is the mean volume of a typical grain Ξ_0.

In this case of non-overlapping and stationarity there are also formulae for the intensities of the Minkowski measures

$$w_k = \lambda \overline{W}_k,$$

where $\overline{W}_k$ is the mean of the kth Minkowski functional of a typical grain Ξ_0.

HANISCH (1983c, 1984b) has given formulae for the covariance function $C(r)$ in the case of non-overlapping independent primary grains of positive volume. For example,

$$C(r) = \lambda \int_{\mathbb{R}^d} c_0(x, x - r)\, dx + \int_{\mathbb{R}^d} \int_{\mathbb{R}^d} p(x)\, p(x - r + z)\, \varrho^{2}(z)\, dx\, dz$$

Here λ and ϱ^2 are the intensity and second-order product intensity of the (stationary) germ-process. The functions c_0 and p describe the form of a typical grain Ξ_0 in the fashion given below: if M is the mark distribution of Ψ then

$$\begin{aligned} p(x) &= M(\{K : x \in \check{K}\}) \\ &= \mathsf{P}(x \in \check{\Xi}) \end{aligned}$$

and

$$\begin{aligned} c_0(x, y) &= M\{(K : x, y \in \check{K}\}) \\ &= \mathsf{P}(x \in \check{\Xi},\, y \in \check{\Xi}). \end{aligned}$$

Weighted grain distributions

The mark distribution M describes the shape and size of a 'typical' grain of a germ-grain model. It is sometimes useful to weight the grain by a size parameter and to study the weighted mark distribution of the grain. The *volume-weighted mark distribution* $M_{(V)}$ is an important special case. It is defined by

$$\int f(K)\, M_{(V)}(\mathrm{d}K) = \int f(K)\, \nu_d(K)\, M(\mathrm{d}K)/\overline{V} \tag{6.4.1}$$

for non-negative measurable functions f on $\mathbb{K}$, where $\overline{V}$ is the mean volume of the 'typical' (unweighted) grain. A small number of large grains may contribute as much to this distribution as a large number of small grains.

Such volume-weighted mark distributions can be of more interest than the unmodified or *number-weighted* mark distribution M. In the case of spherical particles the term 'sieving distribution' is often used, since a series of sieving operations using sieves of different mesh sizes will produce an approximation to a volume-weighted distribution of spheres.

Weighting can be carried out by using other functionals such as surface area rather than volume; see Gundersen and Jensen (1983), Jensen (1985), and Nagel (1985).

A series of moment formulae are implied by (6.4.1). In particular

$$\int \nu_d(K)^n\, M_{(V)}(\mathrm{d}K) = \int \nu_d(K)^{n+1}\, M(\mathrm{d}K)/\overline{V}. \tag{6.4.2}$$

Intersection formulae

Measurements made by image analysis tend to be of the form $h(\Xi \cap T)$ for a given test set T and suitable functionals h. So formulae for the expectations of such quantities are important in statistical applications. Typically, T is a (planar) sampling window, and $h(B)$ yields, for example, the boundary length of B. In such a case $\mathsf{E}h(\Xi \cap B)$ equals the mean length of $(\partial\Xi) \cap T$ (the boundary length of Ξ contained in T) plus the mean length of $\Xi \cap \partial T$. Formulae for such means in the case of stationary random $\mathcal{S}$-sets are given by Davy (1978) and by Weil and Wieacker (1983).

There are two important special cases:

$$\mathsf{E}V(\Xi \cap T) = pV(T) \tag{6.4.3}$$

and

$$\mathsf{E}S(\Xi \cap T) = pS(T) + S_V{}^{(d)}\, V(T). \tag{6.4.4}$$

Formula (6.4.3) is clearly true for general stationary random closed sets under the sole assumption that $V(T)$ is finite. Formula (6.4.4) concerns the case described above; S gives boundary length (surface area in the spatial case) while $S_V{}^{(d)}$ gives the intensity of the boundary process formed by $\partial\Xi$, the mean boundary length per unit area (mean surface area per unit volume in the spatial case). The first term of (6.4.4) derives from the intersection of Ξ with the boundary of T.

If the random set Ξ is motion-invariant, and T is convex, then similar formulae hold for all generalized Minkowski functions W_k of $\Xi \cap T$; see Davy

(1978) and WEIL (1983, 1984). Generally

$$\mathsf{E}W_k(\Xi \cap T) = \frac{\omega_k}{\omega_d\,\omega_{d-k}} \sum_{i=0}^{k} \frac{\omega_{d-i}\omega_{d+i-k}}{\omega_i\omega_{k-i}} \binom{k}{i} w_{k-i}\, W_i(T) \quad \text{for } k = 0, 1, \ldots, d \tag{6.4.5}$$

where w_k is the density of the kth Minkowski functional (that is, the intensity of the corresponding signed Minkowski measure; see Section 7.3.4). In stereological notation, as in Section 11.1,

$$w_0 = A_A, \quad w_1 = L_A/2, \quad w_2 = \pi N_A, \quad \text{if } d = 2,$$

$$w_0 = V_V, \quad w_1 = S_V/3, \quad w_2 = M_V/3, \quad w_3 = 4\pi N_V/3, \quad \text{if } d = 3.$$

Formula (6.4.5) follows from the Hadwiger characterization theorem, part (b), since the functionals $h_k(\cdot) = \mathsf{E}W_k(\Xi \cap \cdot)$ are finite, continuous, C-additive, and motion-invariant.

In the planar case of $d = 2$ a particularly important version of (6.4.5) is that of $k = 2$. In that case

$$\begin{aligned} \mathsf{E}\chi(\Xi \cap T) &= \mathsf{E}W_2(\Xi \cap T)/\pi \\ &= [w_2 W_0(T) + 2w_1 W_1(T) + w_0 W_2(T)]/\pi \\ &= A_A + L_A U(T)/(2\pi) + N_A A(T), \end{aligned} \tag{6.4.6}$$

where χ denotes the connectivity number, $A_A = p$, L_A is mean boundary length per unit area, and N_A is the specific connectivity number.

Let us consider the particular case of a germ-grain model with convex disjoint grains. The number of primary grains to hit T is given by $\chi(\Xi \cap T)$. If the marked point process Ψ is stationary then the Campbell theorem yields

$$\mathsf{E}\chi(\Xi \cap T) = \lambda \mathsf{E}\nu_d(\Xi_0 \oplus T) \tag{6.4.7}$$

where Ξ_0 denotes the typical grain and λ is the intensity of the germ process (so $\lambda = N_A$). The expectation on the right-hand side can be elaborated to show

$$\begin{aligned} \mathsf{E}\chi(\Xi \cap T) &= \lambda \int_T \int_{\mathbb{R}^d} \mathbf{1}_{H(T)}(x, K)\, \mathrm{d}x M(\mathrm{d}K) \\ &= \lambda \int_T \int_{T \oplus \check{K}} \mathrm{d}x M(\mathrm{d}K) \\ &= \lambda \int_T \nu_d(T \oplus \check{K})\, M(\mathrm{d}K) \end{aligned}$$

where $H(T) = \{(x, K) \in \mathbb{R}^d \times C(\mathbb{K}) : (x + K) \cap T \text{ is not empty}\}$ and K is a dummy variable of integration ranging over the system of compact convex sets $C(\mathbb{K})$.

The volume $\nu_d(\Xi_0 \oplus T)$ can be calculated by methods from integral geometry. If Ξ_0 is rotation-invariant then the generalized Steiner formula can be used. In the case $d = 2$ the formula (6.4.6) follows with $N_A = \lambda$.

It should be noted that the entire germ-grain model need not be motion-invariant in order for the above reasoning to hold good. All that is required is that Ψ is stationary and that the typical grain has a distribution which is rotation-invariant.

Statistics for planar motion-invariant germ-grain and grain models

Practical measurement procedures generally involve a planar sample of a germ-grain model Ξ only via a convex compact window W. Here Ξ need not have been planar originally; the sample may be constituted of a plane section or a projection of a thin section.

The fundamental statistical problems are the determination of N_A and the description of the grains. These problems are made non-trivial because of edge-effects; some grains will only be partially observed as they do not lie completely in the window W. Measurements on these grains will be difficult or impossible; the possibility of such grains introduces potential bias in counting procedures.

Determination of N_A and mean values for grain characteristics

Let $U(W)$, $A(W)$ be the perimeter and area of W. Generally the following quantities can be measured

$A(\Xi \cap W)$	the area of Ξ in W,
$L(\Xi \cap W)$	the boundary length of Ξ plus the length of ∂W lying in Ξ,
$\chi(\Xi \cap W)$	the connectivity number of $\Xi \cap W$.

These satisfy the system of equations

$$\mathsf{E}A(\Xi \cap W) = A_A A(W)$$

$$\mathsf{E}L(\Xi \cap W) = A_A U(W) + L_A A(W)$$

$$\mathsf{E}\chi(\Xi \cap W) = A_A + \frac{L_A}{2\pi} U(W) + N_A A(W)$$

which follow from (6.4.1), (6.4.2), and (6.4.4) by putting $T = W$. If the expectations on the left-hand sides are replaced by the corresponding measurable random quantities $A(\Xi \cap W)$, $L(\Xi \cap W)$, $\chi(\Xi \cap W)$ then solving for A_A, L_A, N_A produces unbiased estimators for these fractions.

Schwandtke, Ohser, and Stoyan (1987) study the estimation variances of these and other estimators.

Gundersen (1978) described simple counting procedures for the determination of N_A for germ-grain-models. The most natural procedure assigns to each grain a unique point, such as the extremal point in a fixed direction or the centre of gravity. An unbiased estimator of N_A is given by the number of these points in W divided by $\nu_2(W)$. Such procedures often require information from outside of the sampling window W.

Example 6.2. *Determination of N_A, A_A, L_A for the pattern of Figure* 6.4.

Figure 6.4 shows an area pattern Ξ which for the purposes of this example is interpreted as a germ-grain model observed through the window W. The measurements made only through the window can be compared with others obtained by observing all of the grains that happen to intersect the window.

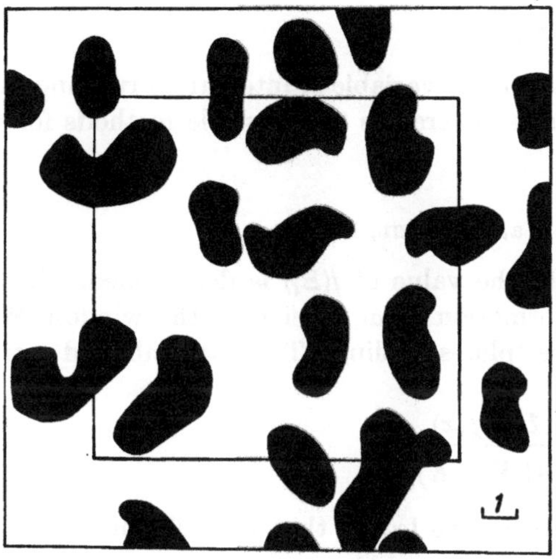

Fig. 6.4. An area pattern in a sampling window to be considered together with Example 6.2. The sampling window W is the inner square; the outer square is used for 'naive' determination of N_A and N_A^+ (neglecting edge-effects).

Image analyzer measurements through the window give

$$A(\Xi \cap W) = 41.2$$

$$L(\Xi \cap W) = 97.3$$

using the units of length of Figure 6.4. The connectivity number of $\Xi \cap W$ is 14. Thus (6.4.8) yields the estimates

$$A_A = 0.37$$

$$L_A = 0.74$$

$$N_A = 0.08.$$

All individual grains have connectivity number one. So the value N_A can be interpreted as estimating λ the mean number of grains per unit area. This number can also be estimated by the naive approach of counting all the lower left boundary points of grains falling wholly or partly in W, which comes to 10, and so leads to an estimate of N_A as 0.09. The specific convexity number N_A^+ can be estimated as 0.11, by counting the lower positive tangent points with respect to the bottom line of Figure 6.4. However, both these measurement procedures utilize information exterior to the measurement window W. Thus they provide a comparison with the estimate of 0.08 above, which was obtained by using only information from within the window.

Determination of mean values for convex grains

Let M be the distribution of a typical grain: so M is a probability measure on the space of compact convex sets. Given a measurable non-negative functional f, the problem is to estimate

$$M_f = \int f(K)\, M(\mathrm{d}K)$$

where here as before K is a dummy variable of integration running through the space of compact convex sets. There are two possible methods for the determination of M_f.

Plus-sampling (see Miles, 1974a, and Weil, 1982a)

For all grains Ξ_i hitting W the value of $f(\Xi_i)$ is determined. This of course means that in general information from outside of the window W must be gathered: hence the phrase 'plus-sampling'. Then an unbiased estimator for λM_f is $\hat{M}_f$,

$$\hat{M}_f = \sum_{[x,K]\in\Psi} f(K)\, \frac{\mathbf{1}_{W\oplus\check{K}}(x)}{A(W\oplus\check{K})}. \tag{6.4.8}$$

The sum is only over those grains $[x, K]$ that actually hit W as described above, for otherwise $\mathbf{1}_{W\oplus\check{K}}(x) = 0$.

The unbiased nature of M_f follows by an application of the Campbell theorem for marked point processes to Ψ.

Integral geometry gives formulae that simplify the evaluation of the area $A(W\oplus\check{K})$.

It is clear that in many cases plus-sampling will be unrealistic, since only the interior of the window W will be available for observation. In such cases the minus-sampling technique suggested by Miles (1974a) can be used.

Minus sampling

Only the grains lying completely within W are considered. The quantity $\overline{M}$ is an unbiased estimator for λM_f and is analogous to $\hat{M}_f$ of (6.4.8).

$$\overline{M}_f = \sum_{[x,K]\in\Psi} f(K)\, \frac{\mathbf{1}_{W\ominus\check{K}}(x)}{A(W\ominus\check{K})}.$$

Again, the sum is restricted since $\mathbf{1}_{W\ominus\check{K}}(x) = 0$ unless $x + K \subset W$. So the sum is only over those grains fully contained in W. Matheron (1978) and Weil (1982a) discussed means of determination of $A(W\ominus K)$.

7. Random measures

7.1 Fundamentals

7.1.1 Introduction

In this chapter we discuss random measures on $\mathbb{R}^d$ that are locally finite: special random set functions taking finite values on bounded sets. In this chapter only *non-negative* set-functions are considered; however, a general theory can be developed for signed random measures which follows the non-negative theory closely. Signed random measures are of interest in the study of curvatures, for example (see ZÄHLE, 1984a, c, 1985).

This chapter only gives an impression of the theory of random measures. It is a short sketch of the theory in so far as it is relevant to stochastic geometry. A more detailed account can be found in KALLENBERG (1976, 1983b).

Random measures are mostly auxiliary entities in stochastic geometry, giving additional means of describing random structures. Random closed sets, fibre processes, and surface processes all have various random measures associated to them. This allows the use of the notation and methods from measure and integration theory.

The following two examples illustrate the concept of a random measure. First, consider a collection Ξ of random fibres in space. For a Borel set B let $\Phi(B)$ be the total length of the intersections of all the fibres with B. Then $\Phi(B)$ is random so Φ is a random set function. For every sample realization of Ξ the function Φ has the properties of a measure. Therefore Φ is indeed a *random* measure.

For the second example, consider a random distribution of mass in space (perhaps the distribution of ore in a geological deposit). If $\Phi(B)$ denotes the total mass in the Borel set B then again Φ is a random measure.

Further (more abstract) examples of random measures are given in 7.3.

7.1.2 Definitions

Only random measures on Euclidean spaces are discussed here. The discussion can be extended easily to more general spaces.

Let M be the set of all locally-finite measures ψ on $[\mathbb{R}^d, \mathcal{B}^d]$. Let $\mathcal{M}$ denote the smallest σ-algebra making all functions $\psi(B)$ of ψ measurable (B runs through the Borel sets: $B \in \mathcal{B}^d$).

A *random measure* Ψ on $\mathbb{R}^d$ is a random element of $[\mathsf{M}, \mathcal{M}]$. So Ψ is a measurable mapping of a probability space $[\Omega, \mathcal{A}, \mathsf{P}]$ into the measurable space $[\mathsf{M}, \mathcal{M}]$. The *distribution* of Ψ is the measure $P = \mathsf{P}\Psi^{-1}$ given by

$$P(Y) = \mathsf{P}(\Psi \in Y) \quad \text{for } Y \text{ in } \mathcal{M}.$$

For $\psi \in \mathsf{M}$ let $\psi_x = T_{-x}\psi$ be the *translated* measure given by

$$\psi_x(B) = (T_{-x}\psi)(B) = \psi(B - x) \quad \text{for } B \text{ a Borel set.}$$

Then a random measure Ψ and its distribution P are said to be *stationary* if Ψ_x has a distribution not depending on x in $\mathbb{R}^d$.

$$P(Y_x) = P(Y) \quad \text{for } x \in \mathbb{R}^d, \quad Y \in \mathcal{M}.$$

Here $Y_x = \{\psi_x : \psi \in Y\}$.

The measure Λ on $[\mathbb{R}^d, \mathcal{B}^d]$ defined by

$$\Lambda(B) = \mathsf{E}\Psi(B) \quad \text{for } B \text{ a Borel set}$$

is the *intensity measure* of Ψ. This measure is not necessarily locally finite. If Ψ is stationary then Λ is a multiple of Lebesgue measure ν_d and for some λ in $[0, \infty]$

$$\Lambda = \lambda\nu_d.$$

The constant λ is then called the *intensity* of Ψ. In the following it is assumed that λ is in $(0, \infty)$.

There are important *operations* that transform given random measures into others, similar to the operations on point processes given in 5.1. By way of example, superposition and compounding are described here.

Suppose Ψ_1 and Ψ_2 are two random measures defined on the same probability space. Then Ψ defined by

$$\Psi(B) = \Psi_1(B) + \Psi_2(B) \quad \text{for all Borel sets } B$$

is again a random measure, the *superposition* of Ψ_1 and Ψ_2.

If Ψ is a random measure and $\zeta = (\zeta_x : x \in \mathbb{R}^d)$ is a non-negative random field of Ψ then the *ζ-compound* of Ψ (or measure *derived* from Ψ by ζ) is defined by

$$\zeta \cdot \Psi(B) = \int_B \zeta_x \Psi(\mathrm{d}x) \quad \text{for } B \text{ a Borel set.}$$

Some properties of such derived measures are studied by KARR (1978).

7.1.3 Palm distributions

Just as in the case of point processes, the definition of the Palm distribution for random measures is based on Campbell's theorem. Consider a random measure Ψ with distribution P. Its *Campbell measure* $\mathcal{C}$ is given by

$$\begin{aligned}\mathcal{C}(B \times Y) &= \mathsf{E}(\Psi(B); \Psi \in Y) \\ &= \int \psi(B)\, \mathbf{1}_Y(\psi)\, P(\mathrm{d}\psi) \quad \text{for Borel } B \text{ and for } Y \text{ in } \mathcal{M}.\end{aligned}$$

Again in analogy with the case of point processes, if Λ is a σ-finite measure on $\mathbb{R}^d$ then there is a family (P_x) of distributions on $(\mathsf{M}, \mathcal{M})$ such that

$$\mathscr{C}(B \times Y) = \int_B P_x(Y)\, \Lambda(\mathrm{d}x).$$

The P_x are called *Palm distributions*. If Ψ is stationary then it is possible to write

$$P_x(Y) = P_o(Y_{-x}) \quad \text{for } Y \text{ in } \mathcal{M} \text{ and for } Y_x = \{\psi_x : \psi \in Y\} \text{ as above.}$$

So in the stationary case it is enough to consider P_o, the Palm distribution with respect to the origin.

In that case

$$\lambda P_0(Y) = \int_{\mathsf{M}} \int_{\mathbb{R}^d} g(x)\, 1_Y(\psi_{-x})\, \psi(\mathrm{d}x)\, P(\mathrm{d}\psi) \quad \text{for every } g \text{ with } \int g(x)\, \mathrm{d}x = 1.$$

Mecke (1967) proved the following for the Palm distribution P_o of a stationary random measure Ψ.

Theorem 7.1. (a) *For all non-negative measurable functions f on $\mathbb{R}^d \times \mathsf{M}$,*

$$\lambda \int\int f(x, \psi)\, \mathrm{d}x\, P_o(\mathrm{d}\psi) = \int\int f(x, \psi_{-x})\, \psi(\mathrm{d}x)\, P(\mathrm{d}\psi). \tag{7.1.1}$$

(b) *For all non-negative measurable functions u on M, such that u is zero if evaluated at the null measure,*

$$\int u(\psi)\, P(\mathrm{d}\psi) = \lambda \int\int h(x, \psi_x)\, u(\psi_x)\, \mathrm{d}x\, P_o(\mathrm{d}\psi) \tag{7.1.2}$$

where h is an arbitrary non-negative measurable function on $\mathbb{R}^d \times \mathsf{M}$ with

$$\int h(x, \psi)\, \psi(\mathrm{d}x) = 1 \quad \textit{for all } \psi \textit{ except the null measure.}$$

(c) *For all f as in* (a)

$$\int\int f(-x, \psi_{-x})\, \psi(\mathrm{d}x)\, P_o(\mathrm{d}\psi) = \int\int f(x, \psi)\, \psi(\mathrm{d}x)\, P_o(\mathrm{d}\psi). \tag{7.1.3}$$

Remarks. Equation (7.1.1) corresponds to the equation (4.4.3). There is only one distribution P_o satisfying (7.1.1) for any given stationary P.

Formula (7.1.2) allows the determination of P for given P_o. So it is the basis for the so-called *inversion formulae*.

A symmetry property for P_o is given by (7.1.3). The Palm distributions are characterized by this property. If a probability measure P_0 on $[\mathsf{M}, \mathcal{M}]$ satisfies (7.1.3) then it is the Palm distribution of a stationary random measure.

7.1.4 Weighted random measures

Weighted random measures are to ordinary random measures as marked point processes are to ordinary point processes. Let W be the set of possible *weights*, with σ-algebra $\mathscr{W}$. Let M_W be the set of all measures φ on $[\mathbb{R}^d \times \mathsf{W}, \mathscr{B}^d \otimes \mathscr{W}]$ with $\varphi(B \times \mathsf{W}) < \infty$ for all bounded Borel sets B. Let $\mathcal{M}_\mathscr{W}$ be the smallest σ-algebra on M_W making all mappings $\varphi \to \varphi(B \times U)$ measurable for B a bounded Borel set and U in $\mathscr{W}$.

A weighted random measure on $\mathbb{R}^d \times \mathsf{W}$ is then a random variable Ψ with values in $[\mathsf{M}_\mathsf{W}, \mathcal{M}_\mathsf{W}]$.

Translations T_x of weighted random measures are defined to be shifts that don't alter the weights:

$$T_x A = A - x = \{(a, b) : (a - x, b) \in A\} \quad \text{for } A \in \mathcal{B}^d \otimes \mathcal{W}$$

If $\varphi \in \mathsf{M}_\mathsf{W}$ then $\varphi_x = T_{-x}\varphi$ is the translated weighted random measure:

$$\varphi_x(A) = T_{-x}\varphi(A) = \varphi(A - x).$$

Using this notion of translation, a stationary weighted random measure can be defined.

It is possible to define an intensity λ (in fact the intensity of the non-weighted random measure Ψ where $\Psi(B) = \Phi(B \times \mathsf{W})$), also moment measures (see STOYAN and OHSER, 1984), Palm distributions, etc. As with the mark distribution for a marked point process, so one can define a weight distribution W for a weighted random measure by

$$W(U) = \mathsf{E}\Phi([0, 1]^d \times U)/\lambda \quad \text{for } U \in \mathcal{W}.$$

There is a Campbell theorem which is analogous to (4.2.4):

$$\mathsf{E} \int f(x, w)\, \Phi\big(d(x, w)\big) = \lambda \int\int f(x, w)\, W(\mathrm{d}w)\, \mathrm{d}x.$$

Examples of weighted random measures occur in Section 9. These concern fibre and surface processes. The random measures that arise are the length or surface measures; and weights are the directions of tangents or normals or curvatures.

7.2 Moment measures and related characteristics

7.2.1 The Laplace functional

Every random measure Ψ on $\mathbb{R}^d$ is connected with a functional L_Ψ defined on the set U of all bounded non-negative measurable functions u of bounded support. The functional is given by

$$L_\Psi(u) = \mathsf{E} \exp\Big(-\int u(x)\, \Psi(\mathrm{d}x)\Big).$$

Note that $1 \geq L_\Psi(u) > 0$. The functional is called the *Laplace functional*. It depends only on the distribution P so strictly speaking the more cumbersome notation L_P should be used:

$$L_P(u) = L_\Psi(u) = \int \exp\Big(-\int u(x)\, \psi(\mathrm{d}x)\Big)\, P(\mathrm{d}\psi).$$

In the theory of random measures the Laplace functional plays a role similar to that of the Laplace transform in the theory of random variables.

A distribution on $(\mathsf{M}, \mathcal{M})$ is uniquely determined by its Laplace functional. If Ψ is a random measure and $B_1, \ldots, B_n$ are bounded Borel sets then the Laplace transform of the random vector $(\Psi(B_1), \ldots, \Psi(B_n))$ can be determined from knowledge of L_Ψ:

$$\mathsf{E} \exp\left(-s_1\Psi(B_1) - \cdots - s_n\Psi(B_n)\right) = L_\Psi(s_1 \mathbf{1}_{B_1} + \cdots + s_n \mathbf{1}_{B_n})$$

for $s_1, \ldots, s_n \geqq 0$.

If Ψ_1 and Ψ_2 are independent then the Laplace functional L_Ψ of their superposition $\Psi = \Psi_1 + \Psi_2$ is given by

$$L_\Psi = L_{\Psi_1} L_{\Psi_2}.$$

7.2.2 Moment measures

Consider as before a random measure Ψ on $\mathbb{R}^d$ with distribution P. The nth *moment measure* $\mu^{(n)}$ on $\mathcal{B}^{dn}$ is given by

$$\begin{aligned}
&\int_{\mathbb{R}^{nd}} f(x_1, \ldots, x_n)\, \mu^{(n)}(d(x_1, \ldots, x_n)) \\
&= \mathsf{E} \int \cdots \int f(x_1, \ldots, x_n)\, \Psi(\mathrm{d}x_1) \cdots \Psi(\mathrm{d}x_n) \\
&= \int \left(\int \cdots \int f(x_1, \ldots, x_n)\, \psi(\mathrm{d}x_1) \cdots \psi(\mathrm{d}x_n)\right) P(\mathrm{d}\psi)
\end{aligned}$$

for non-negative measurable f. In particular

$$\begin{aligned}
\mu^{(n)}(B_1 \times \cdots \times B_n) &= \mathsf{E}\Psi(B_1) \cdots \Psi(B_n) \\
&= \int \psi(B_1) \cdots \psi(B_n)\, P(\mathrm{d}\psi)
\end{aligned}$$

and $\mu^{(n)}(B^n) = \mathsf{E}(\Psi(B))^n$. Thus, the first moment measure coincides with the intensity measure.

The role of the moment measures is similar to that of moments of random variables. Only those moment measures which are locally-finite are of practical interest. They can be obtained from the Laplace functional by differentiation:

$$\mu^{(n)}(B_1 \times \cdots \times B_n) = (-1)^n \lim_{s_1, \ldots, s_n \downarrow 0} \frac{\partial}{\partial s_1} \cdots \frac{\partial}{\partial s_n} L_{P_\Psi}(s_1 \mathbf{1}_{B_1} + \cdots s_n \mathbf{1}_{B_n}),$$

for $B_1, \ldots, B_n$ bounded Borel sets.

Conversely, under certain conditions the Laplace functional can be determined by the moment measures. Suppose all moment measures $\mu^{(n)}$ of a distribution P are locally-finite, and the right-hand side below converges. Then for u in U with $0 \leq u(x)$ for all x in $\mathbb{R}^d$.

$$L_P(u) = 1 + \sum_{n=1}^{\infty} \frac{(-1)^n}{n!} \int u(x_1) \cdots u(x_n)\, \mu^{(n)}(d(x_1 \cdots x_n))$$

Zessin (1983) gives conditions to ensure uniqueness of a random measure when the moment measures are given. All this is in close analogy to the *moment problem* for random variables.

As in the case of point processes (4.5) it is possible to *reduce* the moment measures.

Cross-correlation measures and reduced cross-correlation measures can be defined for any two random measures that are based on the same probability space. They can be used to study correlations between two different random structures such as a point process and a fibre process. STOYAN and OHSER (1982, 1984) studied examples of this, for instance when the point process was derived from tree locations and the fibre process from the courses of rivers.

7.3 Examples of random measures

7.3.1 Random measures constructed from point processes

Every point process on $\mathbb{R}^d$ can be interpreted as a random integer-valued measure, since Φ,

$$\Phi(B) = \text{number of points in } B, B \text{ Borel}$$

is a random measure. It is clearly nonnegative and σ-additive. The realizations of such a random measure are (with probability one) so-called *counting measures*; integer-valued locally-finite measures on $[\mathbb{R}^d, \mathcal{B}^d]$. Consequently every point process corresponds to a random counting measure. If the notion of a point process is generalized to allow for multiple points then the correspondence is one-to-one.

The Laplace functional of Ψ_Φ is connected with the generating functional of Φ

$$L_{\Psi_\Phi}(u) = G(\mathrm{e}^{-u}) \quad \text{if } u \text{ is in } \boldsymbol{U}.$$

For $\Phi = \{[x_n; m_n]\}$, a marked point process in $\mathbb{R}^d$ with real-valued non-negative marks, consider the associated set-function

$$\Psi_M(B) = \sum_{[x,m]\in\Phi} \mathbf{1}_B(x) \cdot m$$

the sum of marks of all points in B where B is Borel. Then Ψ_M is a random measure, the *mark-sum measure* (STOYAN, 1984a). For instance x_n might be the position of the nth tree in a forest, m_n the stem-volume of the nth tree; and then $\Psi_M(B)$ would be the total stem-volume in the area B. The second-order properties of Ψ_M are studied in STOYAN (1984a).

7.3.2 Random measures constructed from random fields

If $\{\zeta_x : x \in \mathbb{R}^d\}$ is a non-negative random field on $\mathbb{R}^d$ then

$$\zeta \cdot \nu_d(B) = \int_B \zeta_x \, \mathrm{d}x \quad \text{for Borel } B$$

defines a set function (tacitly assuming ζ to be integrable over all Borel sets B). In fact $\zeta \cdot \nu_d$ is a random measure on $[\mathbb{R}^d, \mathcal{B}^d]$, which is almost surely diffuse. For ω in Ω, the realization $\zeta \cdot \nu_d(\omega)$ of $\zeta \cdot \nu_d$ has the density $\{\zeta_x(\omega) : x \in \mathbb{R}^d\}$ with respect to ν_d.

7.3.3 Completely random measures

Let Ψ be the random counting measure corresponding to a Poisson process on $\mathbb{R}^d$. Then for every family of pairwise disjoint bounded Borel sets, $B_1, \ldots, B_n$, the random variables $\Psi(B_1), \ldots, \Psi(B_n)$ are independent. It is natural to ask, which other random measures have this property. Such random measures are called *completely random*.

It is clear that the mark-sum measure of an independently marked Poisson process is completely random. More generally, suppose that H is a measure on the system of Borel sets of $\mathbb{R}^d \times (0, \infty)$, with

$$\int_{B\times(0,\infty)} (1 - e^{-a})\, H(d(x, a)) < \infty$$

if B is any bounded Borel set. Let Φ_H be the random counting measure corresponding to a Poisson process on $\mathbb{R}^d \times (0, \infty)$ with intensity measure H (if H has an atomic part then Φ can have multiple points). The measure Ψ_H that is given by

$$\Psi_H(B) = \int_{B\times(0,\infty)} a\Phi_H(d(x, a)) \quad \text{for Borel } B$$

is also completely random. The general structure of completely random measures was studied by KINGMAN (1967).

The Laplace functional of Ψ_H is

$$L_{\Psi_H}(u) = \exp\left(-\int (1 - \exp(-au(x)))\, H(d(x, a)\right) \quad \text{for } u \text{ in } U.$$

Its intensity measure Λ_{Ψ_H} is given by

$$\Lambda_{\Psi_H}(B) = \int_{B\times(0,\infty)} aH(d(x, a)) \quad \text{for } B \text{ a bounded Borel set.}$$

Some examples of higher moment measures are given by

$$\begin{aligned}
\mu^{(2)}(B_1 \times B_2) &= h_1(B_1)\, h_1(B_2) + h_2(B_1 \cap B_2)\\
\mu^{(3)}(B_1 \times B_2 \times B_3) &= h_1(B_1)\, h_1(B_2)\, h_1(B_3) + h_1(B_1)\, h_2(B_2 \cap B_3)\\
&\quad + h_1(B_2)\, h_2(B_3 \cap B_1) + h_1(B_3)\, h_2(B_1 \cap B_2)\\
&\quad + h_3(B_1 \cap B_2 \cap B_3) \quad \text{for } B_1, B_2, B_3 \text{ Borel sets.}
\end{aligned}$$

Here

$$h_n(B) = \int_{B\times(0,\infty)} a^n H(d(x, a)) \quad \text{for } B \text{ a Borel set.}$$

The nth-moment measure is locally-finite if and only if h_n is locally finite.

7.3.4 Random measures accompanying random closed sets — Minkowski and curvature measures

The volume measure V_Ξ

Given a random closed set Ξ, its volume measure is defined by

$$V_\Xi(B) = \nu_d(\Xi \cap B) \quad \text{for } B \text{ Borel.}$$

It is always well defined; see MATHERON (1975). If Ξ is stationary and of volume fraction p then V_Ξ is likewise stationary and its intensity is p. If $p > 0$ then it is useful to consider $C(r)$, and it can be shown that $C(r)$ is the second product density of V_Ξ.

The surface measure S_Ξ

This is the boundary measure for the random set Ξ;

$$S_\Xi(B) = \nu_{d-1}(\partial\Xi \cap B) \quad \text{for } B \text{ Borel.}$$

It is well defined only if the boundary $\partial\Xi$ has certain smoothness or convexity properties. An example is if Ξ is a grain model (a union of a locally finite family of convex bodies as described in 6.4). An analogue can be defined for more general sets by means of the notion of Hausdorff measure, as described in ZÄHLE (1982, 1983a) and in 9.6.

If Ξ is stationary then so is S_Ξ. Likewise S_Ξ inherits isotropy from Ξ. The intensity of S_Ξ is denoted by $S_{V^{(d)}}$ and is called the *specific surface area* of Ξ. Formula (6.3.2) connects it to $C'(0)$. In the planar and spatial cases special symbols are used:

$$S_V = S_{V^{(3)}},$$

$$L_A = S_{V^{(2)}}.$$

Minkowski or curvature measures

In $\mathbb{R}^d$ there are further random measures $W_{k,\Xi}$ connected with a grain model Ξ, where k runs through $0, \ldots, d$. The first two have already been described:

$$W_{0,\Xi} = V_\Xi,$$

$$W_{1,\Xi} = S_\Xi.$$

For all $k > 0$ the measures $W_{k,\Xi}$ are concentrated on $\partial\Xi$. They will be defined here as (positive) random Radon measures for grain models, following MATHERON (1975). SCHNEIDER (1979a, b, 1980), WEIL (1983, 1984), and ZÄHLE (1984a, c, 1986) have defined similar, but signed, measures connected with W_k in the same way that the $W_{k,\Xi}$ are connected with W_k^+. Of course, for $k = 0$ and $k = 1$ the positive and signed measures coincide. ZÄHLE (1984a, c, 1986) used the geometric measure theory of FEDERER (1959) to define generalizations of these measures for locally-finite unions of compatible sets of positive reach. The intensities of the signed measures (actually, densities of the corresponding generalized Minkowski functionals) are denoted by w_k, with $w_0 = V_V = p$, $w_1 = S_V{}^d/d$ etc. The particular case of $k = d$ is discussed briefly at the end of the chapter.

Let $\Pi_\Xi(x)$ be the set of projections of x in $\mathbb{R}^d$ onto Ξ, i.e. the set of points $x' \in \Xi$ for which there exists an open set O with $x' \in O$ and

$$\|x - y\| > \|x - x'\|$$

for any $y \in O \cap \Xi$ with $y \neq x'$ (see MATHERON, 1975, page 119). This set is almost surely locally-finite; and with probability one $\Pi_\Xi(x) \cap b\operatorname{int}(x, r)$ is finite for $0 < r < \infty$.

If we set

$$n(\Xi, B, r; x) = \sum_{z \in \Pi_\Xi(x) \cap b^{\mathrm{int}}(x,r)} \mathbf{1}_B(z)$$

for $r \geqq 0$, B a Borel set, then the number defined is the random number of distinct points in $\Pi_\Xi(x) \cap b^{\mathrm{int}}(x, r) \cap B$. The integral

$$\int n(\Xi, B, r; x)\, \mathrm{d}x$$

is finite for B a bounded Borel set.

MATHERON (1975) showed that there exist Radon measures $W_{k,\Xi}$ yielding the expansion

$$\int n(\Xi, B, r; x)\, \mathrm{d}x = \sum_{k=0}^{d} \binom{d}{k} r^k W_{k,\Xi}(B)$$

when B is Borel. This is how the (positive) *Minkowski measures* $W_{k,\Xi}$ of Ξ can be defined.

If Ξ is stationary and isotropic then the intensities w_k^+ of $W_{k,\Xi}$ are called the *densities* of the (positive) Minkoswki *functionals with respect to* Ξ. Note that it is possible in general for w_k^+ to be infinite.

When Ξ belongs to the convex ring $\mathcal{R}$ then $W_{k,\Xi}(\mathbb{R}^d) = W_k^+(\Xi)$, where W_k^+ denotes the positive extension to $\mathcal{R}$ of the kth Minkowski functional.

The quantity w_d^+/ω_d can be interpreted as a *specific convexity number* as in MATHERON (1975, p. 143); his arguments for the Boolean model apply equally to general grain models. The measurement procedure to estimate w_d^+/ω_d for the case $d = 2$ (when the notation $w_2^+/\omega_2 = N_A^+$ is used) uses the fact that N_A^+ is the intensity of the point process of lower positive tangent points (see Figure 7.1). If N_A^+ is to be estimated then all these points that fall in a given window W of observation must be counted. Their number $n^+(W)$ yields the estimator

$$\hat{N}_A^+ = n^+(W)/\nu_2(W).$$

It is useful to measure these intensities for various orientations (instead of using only the upwards direction) and then to average. BADDELEY (1984)

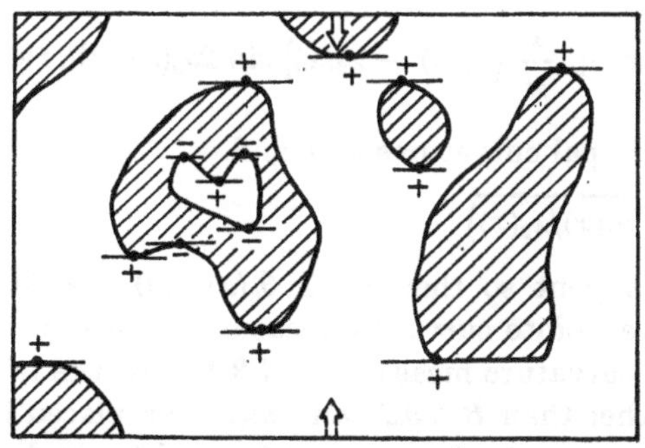

Fig. 7.1. Tangent count with respect to the bottom line. There are six lower positive tangent points and one lower negative tangent point; there are four upper positive tangent points and three upper negative tangent points.

considered the accuracy of the tangent count method, which is of course sensitive to image noise.

The quantity $N_A^- = N_A^+ - N_A$ has a similar interpretation. It is the intensity of lower negative tangent points (see Figure 7.1). In the spatial case the symbol N_V^+ is used for w_3^+/ω_3. The concept of N_A^+, N_A^- and the tangent count is due to HAAS, MATHERON, and SERRA (1967), DE HOFF (1967), and CAHN (1967).

In conclusion we consider the measure $W_{2,\Xi}$ in the spatial case, which has an interesting interpretation in terms of curvature. Let $\partial\Xi^+$ be the subset of the boundary $\partial\Xi$ in which all curvatures are non-negative. Since Ξ is a grain model, $\partial\Xi \setminus \partial\Xi^+$ must consist of curves generated by the intersections of the surfaces of the constituent convex bodies (see Figure 7.2).

Thus, if $m(x)$ is the mean crrvature at x,

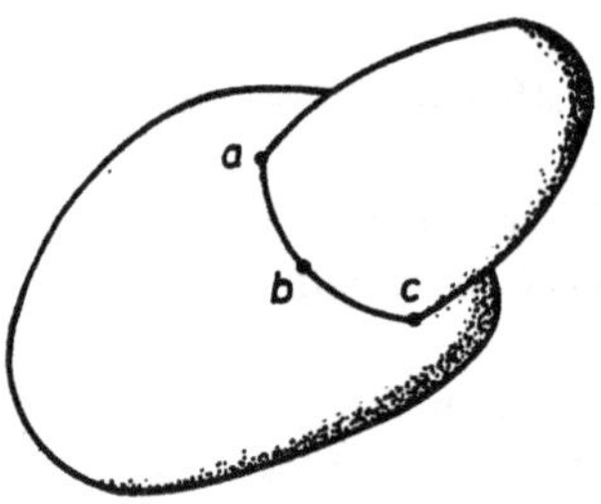

Fig. 7.2. Intersection curve (—a — b — c —) of two convex bodies

$$W_{2,\Xi}(B) = \int_{B \cap \partial\Xi^+} m(x)\, dx \tag{7.3.1}$$

for B a Borel set. Hence, $W_{2,\Xi}$ is called the *positive mean curvature measure*: its intensity (if it exists) is denoted by M_V^+ and is called the *specific positive mean curvature*.

A further (positive) curvature measure is related to the *absolute curvature* described in BADDELEY and AVERBACK (1983). The absolute curvature $m^a(x)$ of a surface point x is given by

$$m^a(x) = \begin{cases} |m(x)| & \text{if } \varkappa_1(x),\ \varkappa_2(x) \text{ have the same sign} \\ \left(\frac{4}{\pi}\,\theta_0(x) - 1\right) m(x) + \frac{1}{\pi}\left(\varkappa_1(x) - \varkappa_2(x)\right) \sin 2\theta_0(x) & \text{otherwise} \end{cases}$$

where $\varkappa_1(x)$, $\varkappa_2(x)$ are the principal curvatures at x and

$$\theta_0(x) = \arctan \sqrt{\left(-\varkappa_1(x)/\varkappa_2(x)\right)}$$

chosing the branch of arc-tangent so that $\pi/2 \geqq \theta_0(x) \geqq 0$; see BADDELEY (1980). The definition of the absolute curvature measure is entirely analogous to that of the positive mean curvature measure at (7.3.1), but the integration is performed over $B \cap \partial\Xi$ rather than $B \cap \partial\Xi^+$ and $m(x)$ is replaced by $m^a(x)$. The intensity of the absolute curvature measure (if it exists) is denoted by K_V. Note that in this case the intersection curves provide positive contributions to the integral.

Important signed random measures are defined for grain models. They are related to the mean curvature number and the connectivity number. For $d = 3$ the *mean curvature measure* μ_m^{Ξ} is defined by a formula analogous to (7.3.1) but the integral is performed over all of $B \cap \partial\Xi$ rather than just $B \cap \partial\Xi^+$. The intersection curves here give negative contributions. In case of stationarity the intensity of μ_m^{Ξ} is denoted by M_V.

The intensity of the random measure connected with the *connectivity number* is denoted by N_A ($d = 2$), and N_V' ($d = 3$). It is related to the intensity w_d by $w_2 = \omega_2 N_A$ ($d = 2$) and $w_3 = \omega_3 N_V$ ($d = 3$). If the grain model Ξ is a stationary planar union of non-overlapping compact sets, then N_A can be expressed in terms of an alternative point process representing Ξ. Note that the non-overlapping grains need not be the original grains, but may be unions of these or *secondary* grains. Let λ_s be the mean number of (secondary) grains per unit area (so that λ_s is the intensity of a point process representing these grains) and let $\bar{\chi}$ be the mean connectivity number of the typical (secondary) grain. Then

$$N_A = \lambda_s \bar{\chi} .$$

Analogous formulae hold for the intensities of the other (Minkowski) measures considered above.

In case $d = 2$ a further signed measure can be defined: the *planar curvature measure* μ_C^{Ξ}. It is defined by a formula similar to (7.3.1) but the integral is performed over $B \cap \partial\Xi$ and $m(x)$ is replaced by $c(x)$ the curvature of the boundary at x. The corresponding intensity is denoted by C_A. It can be shown (see for example WEIBEL, 1980) that

$$C_A = 2\pi N_A . \tag{7.3.2}$$

In the particular case of non-overlapping grains mentioned above this relation is clear since

$$C_A = \lambda_s \bar{c} \quad \text{and} \quad \bar{c} = 2\pi\bar{\chi}$$

where $\bar{c}$ denotes the mean total curvature of the typical grain. When using the tangent count method for N_A it is possible to determine C_A.

8. Random processes of geometrical objects

8.1 Introduction

This chapter introduces the theory of random patterns of objects with special geometrical structure. Two examples are discussed in detail to demonstrate the concepts and methods involved.

The first example is of random planar patterns of lines. Such random patterns can be treated directly as random sets; Chapter 9 describes this approach. However, they can also be considered as point processes whose constituent 'points' lie not in Euclidean space but in the space of lines in the plane. This space can be parametrized as a cylinder in $\mathbb{R}^3$. Thus the theory of planar line processes becomes a special chapter in the general theory of point processes. Moreover, the geometric structure leads to remarkable constraints on the possible variety of regular planar line processes. In Sections 8.2 to 8.4 this geometric structure is explored; Section 8.5 describes generalizations to higher dimensions.

The second example is of processes of shapes of triangles, in which the triangles arise from patterns of planar points. The triangle shapes are themselves points in a representation space which can be parametrized as a sphere in $\mathbb{R}^3$. The statistics of the process of triangle shapes depends on the underlying point process, though in simple cases important conclusions can be drawn by means of geometrical arguments. Because the representation space is a sphere the point process of triangle shapes can be displayed graphically, and this is of importance in data analysis. Section 8.6 introduces the concept of the shape of a triangle. Sections 8.7 and 8.8 describe the geometry of the corresponding representation space. Section 8.9 describes statistical applications and section 8.10 briefly discusses further developments.

8.2 Planar line processes

Line processes are the simplest examples of *fibre processes* (which are discussed in Chapter 9) and so their treatment serves to introduce fibre process theory. They also provide a clear example of the use of representation space in stochastic geometry. Practical applications generally concern Poisson line processes, which generate special tessellations and thus random sets in the plane (as described in Chapter 10) and also provide useful models for random line probe sampling in stereology.

A representation space for directed lines in the plane

A *directed line* is a line with an arrow placed upon it to indicate a preferred direction along the line. The family of all undirected lines in the plane is denoted by $F(2, 1)$, and that of all directed lines by $F^*(2, 1)$. To each member of $F(2, 1)$ there correspond two elements of $F^*(2, 1)$; the two families are thus linked. In the following, we concentrate on the case of directed lines, as the representation is simpler to describe. Directed lines in the plane can be put into one to one correspondence with the set of points on a cylinder in $\mathbb{R}^3$.

A convenient set of coordinates for a directed planar line l is based on its perpendicular distance from the origin and the angle that it makes with the x_1-axis. The first co-ordinate p is the *signed* perpendicular distance of l from o; the sign is positive if o lies to the left of l and otherwise negative. The second coordinate α is the angle between l and the x_1-axis, measured in an anticlockwise direction. Thus $p \in \mathbb{R}$ and $\alpha \in (0, 2\pi]$. The co-ordinatization is illustrated by Figure 8.1. The one to one correspondence relates $F^*(2, 1)$ to the cylinder

$$C^* = \{(\cos\alpha, \sin\alpha, p) : p \in \mathbb{R}, \alpha \in (0, 2\pi]\} \text{ in } \mathbb{R}^3.$$

To each *undirected line* in $F(2, 1)$ correspond a pair of directed lines and thus a pair of points in C^*. These pairs of points are reflections of each other in the origin. We select as representative of the undirected line that point which lies in the half-cylinder

$$C = \{(\cos\alpha, \sin\alpha, p) : p \in \mathbb{R}, \alpha \in (0, \pi]\}.$$

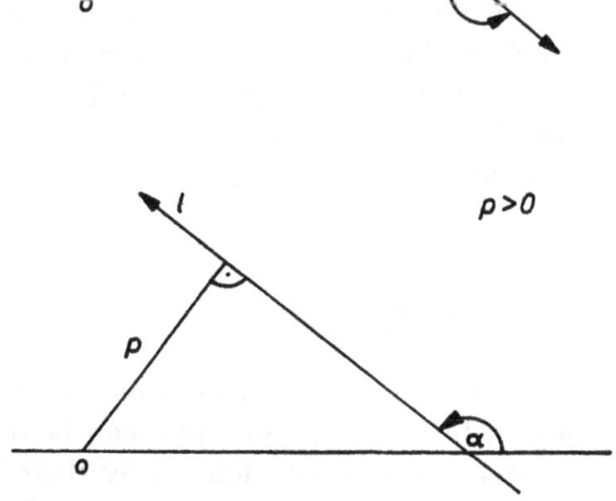

Fig. 8.1. Construction of the representation space for directed lines in the plane. The two diagrams illustrate the cases of negative and positive signed perpendicular distance p.

Note that as the angle of the undirected line passes through o so the representative point jumps from one edge of the half-cylinder to the other, and also the value of the p-coordinate is multiplied by -1.

Symmetries of representation spaces

The representations described above depend on the particular choice of origin o and x_1-axis for the plane. The stochastic geometry of line processes must take such arbitrary choices into account as they correspond to important symmetries in the representations.

The *motion-symmetry group* is generated by those transformations of $F^*(2, 1)$ (or $F(2, 1)$) induced by plane translations and rotations. The *translation-symmetry group* is composed of those transformations induced by plane translations alone. Processes and measures invariant under these groups are described as being respectively *motion-symmetric* or *translation-symmetric*. Up to a constant multiple there can be only one measure on C^* invariant under the motion-symmetry group, and measures invariant under the translation-symmetry group must have a special form. Line processes arising in stochastic geometry typically have properties that are invariant under at least the translation-symmetry group.

Discussion of the symmetries entails study of the transformations induced on the representation cylinder C^* by translations and rotations of the plane. From the construction of C^* it follows that rotations about the origin of the plane induce rotations of C^*. The effect of translations is seen by considering a translation through distance s parallel to the x_1-axis. Directions, and thus α, are unchanged but the perpendicular distance p of a line l from the origin is changed to $p + s \cos \alpha$. Thus a translation induces a *shear* on the cylinder C^*. More generally the corresponding transformation $T_{(s,\gamma)}$ to a translation through distance s in direction γ acts on C^* by

$$T_{(s,\gamma)}(p, \alpha) = \big(p + s \cos(\alpha - \gamma), \alpha\big). \tag{8.2.1}$$

To see this note that when $\mathfrak{r}$ is a rotation about the origin through γ then $\mathfrak{r}T(s, \gamma)\, \mathfrak{r}^{-1}$ is a translation parallel to the x_1-axis. So the symmetries of the Euclidean motions correspond to rotations of the cylinder and *vertical shears* along the cylinder's axis. Note that rigid symmetries of C^* do not arise from the Euclidean symmetries of the plane underlying $F^*(2, 1)$.

8.3 Measures on representation spaces

Line processes give rise to, and are controlled by, measures on the representation space C^*. For example Poisson directed-line processes can be defined as Poisson processes on C^*; so their distributions are defined by their intensity measures on C^*. Invariance properties holding for a line process lead to invariance properties for the corresponding measures, so the study of invariant measures is a necessary preliminary to the study of invariant line processes.

Measures invariant under the stationary symmetry group induced by translations must be of a specific form. If invariant under the motion symmetry group they are unique up to a multiplicative factor.

Theorem 8.1. (a) *Suppose μ is a locally finite measure on C^* invariant under the stationary-symmetry group (plane-translation-invariant). Then μ is of the form*

$$\mu(\mathrm{d}p\,\mathrm{d}\alpha) = \mathrm{d}p \cdot \varkappa(\mathrm{d}\alpha) \tag{8.3.1}$$

for some finite measure $\varkappa$ on the circle $(0, 2\pi]$.

(b) *Suppose μ is a locally finite measure on C^* invariant under the motion-symmetry group. Then μ is a constant multiple of Lebesgue measure on C^**

$$\mu(\mathrm{d}p\,\mathrm{d}\alpha) = s\,\mathrm{d}p\,\mathrm{d}\alpha \tag{8.3.2}$$

for some constant s with $0 \leqq s < \infty$. If $s = 1$ then μ is the standard invariant measure on C^*

This theorem can be proved by using differential forms (see Santaló, 1976). The encyclopaedia article by Crofton (1885) gives a heuristic argument that can be made precise as follows.

Let $T_{(s,\gamma)}$ be the translation through a distance s in a direction γ. For given α, β in $(0, 2\pi]$ consider the measure defined on intervals of $\mathbb{R}$ by

$$[a, b] \to \mu([a, b] \times [\alpha, \beta]).$$

By translation-invariance of μ

$$\mu([a, b] \times [\alpha, \beta]) = \mu\big(T_{(s,\gamma)}([a, b] \times [\alpha, \beta])\big)$$

and by an approximation argument it follows that

$$\begin{aligned}\mu([a, b] \times [\alpha, \beta]) &= \int_\alpha^\beta \mu\big(T_{(s,\gamma)}([a, b] \times \mathrm{d}\gamma)\big)\\ &= \int_\alpha^\beta \mu([s + a, s + b] \times \mathrm{d}\gamma)\\ &= \mu([s + a, s + b] \times [\alpha, \beta]).\end{aligned}$$

So μ is invariant under vertical shifts of the cylinder C^*.

The formula (8.3.1) follows from general results on the uniqueness of translation-invariant measures on $\mathbb{R}$, while (8.3.2) follows from the implications of rotational invariance (which of course carries over directly to C^*).

All possible invariant measures for $F(2, 1)$ can be derived from those for $F^*(2, 1)$. Formula (8.3.1) remains valid except that $\varkappa$ is now a measure on $(0, \pi]$. In formula (8.3.2) the factor becomes $2s$.

Geometric arguments can also be used to derive formulae for the invariant measure of hitting sets. The ***hitting set*** of a compact subset K is defined as

$$[K] = \{l \in F^*(2, 1) : l \cap K \text{ is not empty}\}.$$

Theorem 8.2. *If μ is a motion-invariant measure on C^* and K is compact, convex, and of piecewise-smooth boundary then*

$$\mu([K]) = s \cdot \nu_1(\partial K). \tag{8.3.3}$$

Here s is the constant factor in (8.3.2).

Theorem 8.2 can be established as follows. If S is a line segment then invariance considerations show that $\mu([S])$ is proportional to the length $\nu_1(S)$ of S, and does not depend on the orientation nor on the location of S. If K is a convex polygon of sides $S_1, \ldots, S_n$ then any l in $[K]$ intersects exactly two of the S_i, unless l penetrates K by a vertex. Thus, using indicator functions,

$$2\,\mathbf{1}_{[K]} \equiv \sum_i \mathbf{1}_{[S_i]}$$

where $\equiv$ denotes equality save on a set that is the finite union of hitting sets of points. Points $\boldsymbol{p}$ are segments of zero length and so $\mu([\boldsymbol{p}]) = 0$. Thus

$$2\mu([K]) = \sum_i \mu([S_i])$$

which is proportional to $\nu_1(\partial K)$ the length of the boundary of K. By an approximation argument the proportionality of $\mu([K])$ to $\nu_1(\partial K)$ holds good for all convex compact K with piecewise smooth boundary.

The above formula can be used to define boundary length for general convex compact K.

If K is the unit disc $b(o, 1)$ then

$$\mu\big([b(o, 1)]\big) = 2\pi c$$

for some constant of proportionality c. On the other hand, if $\mu\big(d(r, \alpha)\big) = s\, \mathrm{d}p\, \mathrm{d}\alpha$ then

$$\mu\big([b(o, 1)]\big) = \mu\big((-1, 1) \times (0, 2\pi]\big) = 2\pi s.$$

So c equals s.

This result can be generalized to higher dimensional cases. One can also consider the evaluation of invariant μ on the hitting set of a finite union of line segments (the *Sylvester problem*) and on general hitting sets of finite unions of convex compact sets. Ambartzumian (1982a) describes the recent full solution of this problem and generalizations.

8.4 Line processes as point processes on the representation space

8.4.1 General theory

A *line process* is a random collection of lines in the plane which is *locally-finite*; only finitely many lines hit each compact planar set. It is convenient to consider directed line processes: results transfer easily to the undirected case. A directed line process can be formally defined by considering it as a random subset of the representation space C^*. The process is locally-finite exactly when the representative random subset is a locally finite subset, hence a point process, on C^*. Such point processes are particular cases of point processes on $\mathbb{R}^2$ because, as suggested by the parametrisation (p, α) of C^*, the cylinder can be cut and embedded as the subset $\mathbb{R} \times (0, 2\pi]$ of $\mathbb{R}^2$. So the theory of line processes is a special chapter in the theory of planar point processes as described in Chapters 2, 4, and 5.

Stationary-symmetric and motion-symmetric line processes

The definition of stationary-symmetric and motion-symmetric line processes is entirely analogous to those of stationary and motion-invariant point processes. The difference lies in the replacement of the translation and motion symmetry groups by the stationary-symmetry and motion-symmetry groups. Thus the line process $\Phi = \{l_1, l_2, \ldots\}$ is *stationary-symmetric* if $\Phi_T = \{Tl_1, Tl_2, \ldots\}$ has the same distribution (as a line process) for every translation T, and this is to say that in the C^* representation the point process

$$\Big\{\big(p(l_1) + s \cdot \sin\big(\alpha(l_1) + \gamma\big), \alpha(l_1)\big), \big(p(l_2) + s \cdot \sin\big(\alpha(l_2) + \gamma\big), \alpha(l_2)\big), \ldots\Big\}$$

has the same point process distribution as

$$\Big\{\big(p(l_1), \alpha(l_1)\big), \big(p(l_2), \alpha(l_2)\big), \ldots\Big\}$$

for each s in $\mathbb{R}$ and γ in $(0, 2\pi]$.

Motion-symmetric line processes have in addition the property that

$$\Big\{\big(p(l_1), \alpha(l_1) + \gamma\big), \big(p(l_2), \alpha(l_2) + \gamma\big), \ldots\Big\}$$

has the same point process distribution as

$$\Big\{\big(p(l_1), \alpha(l_1)\big), \big(p(l_2), \alpha(l_2)\big), \ldots\Big\}$$

for each γ in $(0, 2\pi]$, where the additions of angles $\alpha(l_k) + \gamma$ are interpreted modulo 2π.

The intensity measure

A directed line process Φ when regarded as a point process on C^* yields an intensity measure Λ on C^*:

$$\Lambda(A) = \mathsf{E}\,\#\{l \in \Phi \cap A\} \tag{8.4.1}$$

for each Borel subset A of C^*.

If Φ is stationary-symmetric then so is Λ, and if Λ is locally finite then formula (8.3.1) can be applied

$$\Lambda\big(d(p, \alpha)\big) = L_A \, \mathrm{d}p \cdot \mathsf{R}(\mathrm{d}\alpha). \tag{8.4.2}$$

Here L_A is a constant, the averaged intensity of Φ, and $\boldsymbol{R}$ is a probability measure on $(0, 2\pi]$, the *rose of direction* of Φ. The constant L_A is discussed further under the next heading. The rose of direction can be interpreted as the distribution of direction of a 'typical' line of Φ.

In the case of motion-symmetry the formula (8.3.2) can be applied

$$\Lambda(A) = L_A \, \mathrm{d}p \cdot \mathrm{d}\alpha/(2\pi) \tag{8.4.3}$$

The constant $L_A/(2\pi)$ is then the intensity of the representing point process with respect to the standard invariant measure $\mathrm{d}p \cdot \mathrm{d}\alpha$. In the motion-symmetric case L_A is called the intensity of the line process Φ.

Undirected line processes satisfy similar formulae, but $\mathrm{d}p \cdot \mathrm{d}\alpha/(2\pi)$ in (8.4.3) is replaced by $\mathrm{d}p \cdot \mathrm{d}\alpha/\pi$, and the rose of direction $\boldsymbol{R}$ becomes a distribution on $(0, \pi]$.

The averaged intensity L_A

If Φ is stationary-symmetric then invariance arguments can be applied to

$$\Lambda^*(B) = \mathsf{E} \sum_{l \in \Phi} \nu_1(l \cap B) \tag{8.4.4}$$

for planar Borel sets B. Since Λ^* is a translation-invariant measure on $\mathbb{R}^2$ there is a constant c such that

$$\Lambda^*(B) = c\nu_2(B) \tag{8.4.5}$$

for planar Borel sets B. However, $\Lambda^*(B)$ can also be calculated using the cylinder representation of Φ. If B is chosen to be $B = b(o, 1)$ the unit disc then

$$\begin{aligned}\Lambda^*\big(b(o, 1)\big) &= \mathsf{E} \sum_{l \in \Phi} \nu_1\big(l \cap b(o, 1)\big) \\ &= \mathsf{E} \sum_{(p,\alpha)\in\Phi:|p|\leqq 1} 2\sqrt{(1 - p^2)} \\ &= 2 \int_{|p|\leqq 1} \sqrt{(1 - p^2)}\, \Lambda\big(d(p, \alpha)\big) \\ &= 2L_A \int_{-1}^{1} \sqrt{(1 - p^2)}\, \mathrm{d}p = L_A\pi .\end{aligned}$$

Since (8.4.5) yields $\Lambda^*\big(b(o, 1)\big) = c\pi$ the averaged intensity L_A is also the intensity of the random length measure

$$B \to \sum_{l \in \Phi} \nu_1(l \cap B) .$$

This point is returned to in Section 9.2, where line processes are considered as random measures in a prelude to the discussion of more general fibre processes.

Moment measures

Since line processes can be regarded as point processes on the representing space the definition of second- and higher-order moment measures is straightforward. We shall not specifically discuss these concepts here, although they play an important part in the general theory of line processes.

8.4.2 Poisson line processes

A Poisson line process is the line process produced by a Poisson process on C^*. Consequently, it is completely characterized by its intensity measure Λ. Thus stationary-symmetric Poisson line processes are characterized by the averaged intensity L_A and the rose of directions $\boldsymbol{R}$. Moreover stationary-symmetric Poisson line processes are motion-symmetric precisely when the rose of direction is the uniform probability measure on $(0, 2\pi]$.

Poisson line processes provide examples of stationary-symmetric line processes which are free of parallel lines, since a Poisson point process on $\boldsymbol{C}^*$ with intensity $L_A \cdot \mathrm{d}p \cdot \boldsymbol{R}(\mathrm{d}\alpha)$ can have no more than one point per vertical line when the rose of directions $\boldsymbol{R}$ is diffuse. All this remains true for the case of undirected lines except that $\boldsymbol{R}$ is then a distribution on $(0, \pi]$.

Example 8.1. *The motion-symmetric Poisson line process.*

Let Φ be a motion-symmetric Poisson line process of intensity L_A. Then the number of lines of Φ hitting a convex compact set K is of Poisson distribution of mean

$$(L_A/\pi) \cdot \nu_1(\partial K).$$

The Poisson distribution is established by noting that the number of lines hitting K is precisely the number of representing points lying in a certain subset of $\boldsymbol{C}^*$. The value for the mean is derived by applying Theorem 8.2 to the intensity measure of Φ, which is $\Lambda(d(p, \alpha)) = L_A \cdot \mathrm{d}p \cdot \mathrm{d}\alpha/(2\pi)$.

If $K = b(o, r)$ then the number of lines of Φ hitting K is Poisson of mean $2rL_A$.

A simulation of a Poisson line process is illustrated in Figure 8.2.

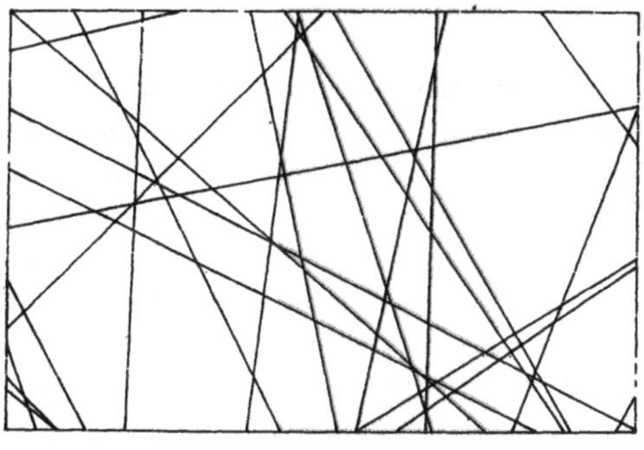

Fig. 8.2. A simulation of a Poisson line process

8.4.3 Cox line processes

Cox line processes are line processes produced by Cox processes on $\boldsymbol{C}^*$. They are characterized by their driving random measures Λ as in Chapter 5. Conditioned on the realization of the random measure Λ the process Φ is Poisson of intensity measure Λ. The driving random measures of stationary-symmetric Cox line processes are themselves stationary-symmetric. Thus a complete characterization of stationary-symmetric Cox line processes depends on the analysis of stationary-symmetric random measures on $\boldsymbol{C}^*$. It is a remarkable fact that if a stationary-symmetric random measure Λ on $\boldsymbol{C}^*$ has locally finite second-moment measure, and does not charge any vertical line

$$\alpha = \text{constant}$$

then it must factorize into the form

$$\Lambda\big(d(p, \alpha)\big) = \mathrm{d}p\, \Lambda^\Theta(\mathrm{d}\alpha) \tag{8.4.6}$$

where Λ^Θ is a random measure on $(0, 2\pi]$. This fact was first observed by DAVIDSON (1974b) for stationary-symmetric random measures with second-order moment density. The density condition was lifted by KRICKEBERG (1974) using disintegration of the second-order moment measure.

This factorization implies that under suitable regularity conditions stationary-symmetric Cox line processes must be mixtures of stationary-symmetric Poisson line processes in which the averaged intensity L_A and the rose of directions $\boldsymbol{R}$ are randomized. Moreover, motion-symmetric Cox line processes must be mixtures of motion-symmetric Poisson line processes in which the only randomization is of the intensity L_A. In brief, stationarity and motion-invariance force Cox line processes to be *mixed Poisson* line processes.

The central position occupied by mixed Poisson processes is also indicated by a class of results concerning intersections of the lines of a Cox line process. These intersections form a point process which is stationary if the line process is stationary-symmetric. The intensity of the intersection process (suitably normalized) is maximized by the isotropic mixed Poisson processes see DAVIDSON (1974c).

8.4.4 Davidson's conjecture

The problem posed by DAVIDSON (1974a) concerned the existence of motion-symmetric line processes of locally finite second-moment measure that neither were Cox nor contained parallel lines. He conjectured that there were no such line processes. This was later established (subject to regularity conditions) by KALLENBERG (1976, 1980, 1981), KRICKEBERG (1974), and PAPANGELOU (1974, 1976), in the more general case of stationary-symmetric line processes satisfying technical regularity conditions. Geometrical properties of C^* place strong restrictions on the characteristics of stationary-symmetric line processes. These restrictions imply the Cox property under suitable regularity conditions. However KALLENBERG (1977) has shown that there do exist counterexamples to Davidson's conjecture.

Example 8.2. *A line process with no parallel lines that is not Cox.*

The description of this line process is best given using not the cylindrical parametrization but an alternative coordinatization. Each line l in the plane is parametrized by a pair (a, b) where

$a(l) = \cot(\alpha(l))$ using $\alpha(l)$ as in the cylindrical representation, and

$b(l)$ is the signed distance from the origin of the intercept of l with the x_1-axis.

So the representation space is a plane. This co-ordinatization breaks down if l is parallel to the x_1-axis. However, in our example that eventuality is of probability zero.

Translations $T_{(s,\gamma)}$ of $\mathbb{R}^2$ correspond to linear shears of the representation plane:

$$T_{(s,\gamma)}(a, b) = (a, b - s(a \sin\gamma - \cos\gamma))$$

and so stationary-symmetries will send a linear lattice on the representation plane into another linear lattice. Thus the corresponding class of patterns of lines (patterns that correspond under the planar parametrization to lattices of points) will be left invariant under the group of translations. Kallenberg constructed a probability measure on the space of lattices which is invariant under the action of the group of linear shears and hence is translation-symmetric. In a letter published in KALLENBERG (1977) Kingman pointed out that the existence of this probability measure follows from number-theoretic calculations of SIEGEL (1945). This example suffers the imperfection that a lattice on (a, b)-space corresponds to a line pattern that is not locally finite (there are infinitely many lines of small angle γ hitting each compact set). However, this imperfection is removed by a thinning depending only on the angular parameter, for example by deleting all points of the lattice lying outside a region in (a, b)-space bounded by two lines of the form

$$a = \text{constant}.$$

The resulting process has a finite second-order moment measure.

A simulation of the resulting lattice process is displayed in Figure 8.3. It is clearly not a Cox process; the lattice structure means that there is positive probability of observing more than two lines meeting at some intersections. This failure to have a

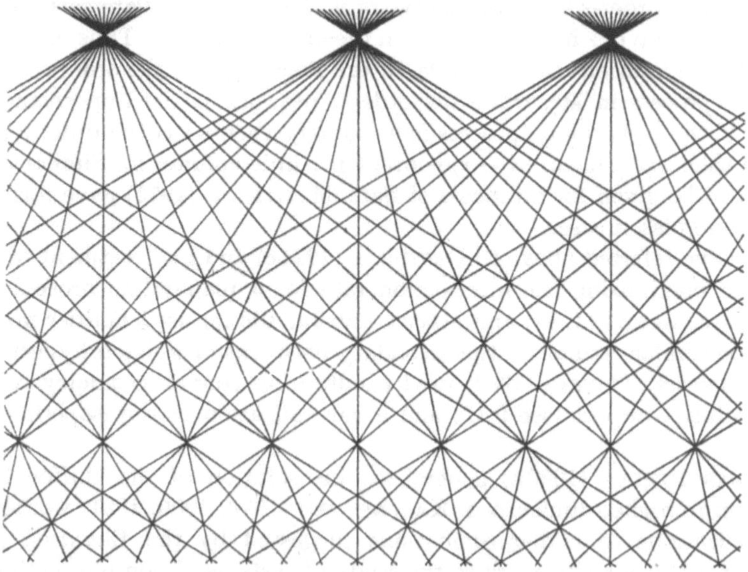

Fig. 8.3. A simulation of a Kallenberg lattice process. This is clearly not a Poisson line process: many intersections involve more than two lines and moreover the pattern is very regular

bundle-free property follows immediately from the observation that lines meeting in a common point in the (a, b)-representation correspond to points that are collinear. One might hope that all stationary-symmetric line processes that are both free of parallel lines and bundle-free must also be Cox. KALLENBERG (personal communication) pointed out that this cannot be so; if his example is modified by independent parallel translations of the lines then the bundles are resolved but the modification can be shown not to be Cox.

The Kallenberg lattice-type line process is completely defined by a fundamental cell of the lattice in the (a, b)-space representation. MECKE (1979b) constructs the process using only five real-valued random variables.

8.5 Generalizations

Some authors have studied random collections of lines, hyperplanes, and more generally flats, in $\mathbb{R}^d$. The approach described above generalizes directly. Fundamental work has been done by KALLENBERG (1976, 1978, 1980, 1983a), KRICKEBERG (1972b, 1974, 1975), MATHERON (1975), MILES (1974b), and PAPANGELOU (1974, 1976). Statistical problems for Cox processes of hyperplanes and flats have been considered by KRICKEBERG (1975), FELLOUS and GRANARA (1981) and FELLOUS, GRANARA, and KRICKEBERG (1978). KALLENBERG (1983a) and AMBARTZUMIAN (1984) considered generalizations of the Davidson problem. RIPLEY (1976b, c) discussed general issues in the theory of representation spaces.

A process of hyperplanes in $\mathbb{R}^d$ generates *secondary* networks of intersections of flats. The order-r secondary network is the process of flats each of which is the intersection of r hyperplanes from the original process. Thus under suitable regularity conditions (for example, in the Poisson case) the order-d network is a point process in $\mathbb{R}^d$, while the order-1 network is the hyperplane process itself. If the hyperplane process is stationary then so is the order-r network, which then has a mean $(d - r)$-content per unit volume, n_r. Similar constructions can be given for a process of flats. MATHERON (1975) gave formulae for n_r in the Poisson case.

We have referred above to DAVIDSON (1974c) and his treatment of the problem of maximizing the intersection rate for a planar line process. JANSON and KALLENBERG (1981) and MECKE (1983a, 1984c) used Fourier methods to produce similar results for the secondary networks of Poisson hyperplane and flat processes. THOMAS (1984) used results from convex geometry concerning the Steiner compact (see 9.12) and a generalization of the isoperimetric inequality.

AMBARTZUMIAN (1982a) generalized his solution of the Buffon–Sylvester problem to higher dimensions. KALLENBERG (1976, 1980, 1981) worked on the generalizations to higher dimensions suggested by the Davidson problem. Kallenberg applied his work to the study of systems of non-interacting particles.

Poisson and Cox line processes can be used as simple mathematical models of random systems of very long fibres of weak curvature, or of trajectories of moving particles. They do not describe the global behaviour of the real structures correctly, but may be of value as a local description. Formulae for the point process of self-intersection points and for the induced polygons (see Section 10.5) are of interest. KALLENBERG (1983a) discussed the approximation of segment processes by Cox line processes.

8.6 The statistical theory of shape

Intitial statistical analysis of point patterns in $\mathbb{R}^2$ can be summarized under four headings: those of *location*, *scale*, *orientation*, and *interactions*. Questions about location are superseded if the investigator assumes the pattern to be derived via a window from a larger point pattern that is statistically stationary.

Questions about scale lead to discussion of the *intensity* of the pattern. Orientation is connected to the matter of whether or not the pattern is *isotropic*. The remaining issue of inter-point interactions is bound up with the interdependence of the points of the pattern, and so cannot be expressed as simply as the other three. However some interactions only involve measurements that are *invariant* under changes of location, scale, and orientation. For example, one might be concerned with the triples of points arising from the pattern, and in particular with the statistical significance of the observation of a high number of nearly collinear triads of points of the pattern. It is reasonable to suppose such a matter concerns the angles of these triangles, and not at all their location, scale, or orientation. Such measurements and concerns fall into the domain of the *statistics of shape*.

In the study of line processes the constituent lines are associated with points on a representation space. A similar representation is fundamental to the theory of shape: the shapes of sequences of labelled points are associated with points on the *shape-space* that is appropriate for the problem. The specific case which will illustrate our discussion is that of triads of labelled planar points. To each triad of labelled pattern points is associated a shape-point in the representation space. This is done in such a manner that triads correspond to the same shape point if and only if they share the same shape. In general the shape-space of k labelled points in d-dimensional space is denoted by $\sum_d^k$. This can have a complicated topology and geometry especially in dimensions $d \geqq 3$. We confine ourselves to the well-behaved planar setting and in particular to the case $k = 3$ of planar triads, which has found the most application so far. Discussion of the geometry of $\sum_2^3$ is a necessary prerequisite for the statistical theory of shape.

Applications of the statistical theory of shape typically investigate apparent departures from pure randomness in a point pattern. Thus Broadbent (1980) discussed the problem of whether the locations of neolithic sites were chosen so as to define straight lines, while the problem of whether quasar observations define significantly too many alignments has been considered for example by Edmunds and George (1981) and Zuiderwijk (1982). In both cases, the emphasis is on data analysis rather than stochastic modelling; one is concerned not so much to explain structure as to evaluate the evidence for it being really there at all.

To assess the significance of various observed configurations it is helpful to investigate the distributions of shape that arise from simple stochastic models for point patterns. These distributions provide a framework for the discussion of shapes induced from planar point patterns. A more powerful approach to practical problems uses simulation-based techniques.

However, an important contribution to be made by the theory of shape concerns the presentation of data, without any reference to probability. Once the geometry of the shape-space has been established then the various shapes derived from the original point pattern can be displayed as points on the shape space $\sum_2^3$. Examination of this new point pattern, in a parametrization suggested by the geometry of the shape-space, allows one to investigate aspects of the data-set that would not otherwise be evident to the eye.

8.7 The shape-space of triads of labelled planar points

Consider the triad formed by three labelled planar points x_1, x_2, x_3 selected from a point pattern in the plane. This triad defines a *shape* which is invariant when the triad is translated to a different location, rescaled, or turned to a different orientation (although the shape may change if the three points are relabelled). Thus measurements of shape correspond to *invariant statistics* under the symmetry group generated by translations, dilations, and rotations. The shape of the triad itself is an example of a *maximal invariant statistic* or *maximal invariant* (see for example, Cox and Hinckley, 1974, pp. 159ff, or Small, 1983) and statistics that are maximal invariant subject to a given symmetry group are all essentially equivalent. So a maximal invariant statistic provides a representation of shape. The maximal invariant statistic should be chosen in as natural a way as possible, avoiding arbitrary choices of co-ordinates. This makes it easier to interpret the statistic in geometrical terms.

8.7.1 Construction of the shape-space

In the sequel we identify the plane with the complex plane $\mathbb{C}$. So the symmetries of the plane correspond to the basic operations of complex addition and multiplication. This produces economy of description.

Each triad (x_1, x_2, x_3) lies in $\mathbb{C}^3$. In the language of group theory the shape of the triad is its *orbit* under the symmetry group generated by (a) addition of the same complex number z to each point x_i (this corresponds to translation); (b) multiplication of each point x_i by the same non-zero complex number w (this corresponds to combinations of rotation and dilation). This orbit is the set

$$\{(wx_1 + z, wx_2 + z, wx_3 + z) : z \in \mathbb{C}, w \in \mathbb{C} \setminus \{o\}\}$$

of triads produced from (x_1, x_2, x_3) by all possible translations, rotations, and dilations.

Triads of totally coincident points are degenerate in shape and are excluded from further consideration. A maximal invariant statistic or shape for the non-degenerate triad (x_1, x_2, x_3) is its orbit in $\mathbb{C}^3 \setminus \{(z, z, z) : z \in \mathbb{C}\}$ under the symmetry group. However, a more explicit description is required for practical purposes. In fact, we need a good *parametrization* of the orbit space.

Any triad (x_1, x_2, x_3) can be moved by a translation so that $(x_1 + x_2 + x_3)/3 = 0$. Each possible nondegenerate triad (x_1, x_2, x_3) is associated to a unique centred triad. The effect on the corresponding point of $\mathbb{C}^3$ is to move it to a subspace of $\mathbb{C}^3 \setminus \{(z, z, z) : z \in \mathbb{C}\}$ which is a copy of $\mathbb{C}^2 \setminus \{o\}$, and comprises the family of all *centred triads*. This subspace of centred triads can be parametrized by the pair of complex coordinates

$$\left.\begin{aligned} u &= (x_2 - x_1)/\sqrt{2} \\ v &= (2x_3 - x_2 - x_1)/\sqrt{6} \end{aligned}\right\} \qquad (8.7.1)$$

this parametrization being convenient because it uses orthogonal co-ordinates. By this means the effect of translation symmetries is removed.

Suppose a triad (x_1, x_2, x_3) is subjected to a dilation and rotation corresponding to multiplication by a non-zero complex number w. Then its centred representative is dilated and rotated by the same transformation. Two triads have the same shape precisely when their centred triads can be transformed into each other in this way. It follows that the shape of a nondegenerate triad (x_1, x_2, x_3) is represented by the set

$$\Pi(x_1, x_2, x_3) = \{(wu, wv) : w \in \mathbb{C} \setminus \{o\}\}.$$

This set $\Pi(x_1, x_2, x_3)$ is a complex plane (minus its origin) embedded in $\mathbb{C}^2 \setminus \{o\}$. Every such complex plane can arise from a suitable (x_1, x_2, x_3), so the shape-space for triads of planar points is the space of complex planes in the two-complex-dimensional space $\mathbb{C}^2$, the *projective complex plane*. It is denoted by $\mathbb{C}P^1$ and is topologically and smoothly equivalent to the two-sphere $S^2 = \partial b(o, 1)$.

The construction above provides a proof of a special case of a geometrical theorem that is fundamental to the theory of shape:

Theorem 8.3 (D. G. Kendall (1984)). *The shape-space of k-ads of planar points is given by*

$$\Sigma_2^k = \mathbb{C}P^{k-2}. \tag{8.7.2}$$

Here $\mathbb{C}P^{k-2}$ is the space of complex planes passing through the origin in $\mathbb{C}^{k-1}$. In the case $k = 3$ of planar triads we have further

$$\Sigma_2^3 = S^2.$$

Having fixed the metric for shapes by a natural choice (see Section 8.7.2), these relations are found to be *isometries* when the scale of the representing space is appropriately chosen. Thus for example Σ_3^2 turns out to be isometric with the sphere $S^2(1/2) = \partial b(o, 1/2)$ of curvature 4.

8.7.2 Geometry of the shape-space

The shape-space Σ_2^3 carries a geometrical structure that is of great relevance to the statistical theory. D. G. Kendall (1984) points out that the *procrustean metric* on the configuration space $\mathbb{C}^3$ makes Σ_2^3 into $S^2(1/2)$ (the metric sphere of radius 1/2). We concentrate here on the more mundane task of identifying the correspondence between regions of Σ_2^3 and classes of shapes of triads.

The nub of the matter is to fix on a good co-ordinatization of $\Sigma_2^3 = S^2(1/2)$. The natural correspondence with $\mathbb{C}P^1$ given above, together with the orthogonal parametrization of the space of centred triads at (8.7.1), provides a suitable candidate. If (x_1, x_2, x_3) is a triad of planar points with centred triad parametrized by (u, v) then the centred triad is made up of points

$$2v/\sqrt{6}, \qquad (\sqrt{3}\,u - v)/\sqrt{6}, \qquad (-\sqrt{3}\,u - v)/\sqrt{6}.$$

The projective representation is the composition of two maps: the *projective co-ordinate map* ζ of $\mathbb{C}^2$ to the extended complex plane $\mathbb{C} \cup \{\infty\}$

$$\zeta(u, v) = \begin{cases} v/u & \text{if } u \neq 0 \\ \infty & \text{if } u = 0 \end{cases} \tag{8.7.3}$$

and the *stereographic* projection q of $\mathbb{C} \cup \{\infty\}$ to Σ_2^3

$$\begin{aligned} q(\zeta) &= \left(\frac{1}{2}, 2 \operatorname{arc\,tan} |\zeta|, \arg(\zeta)\right) \\ q(\infty) &= \left(\frac{1}{2}, \pi, 0\right) \end{aligned} \tag{8.7.4}$$

where the result is expressed in spherical polar co-ordinates and arctangent takes the range $[-\pi/2, \pi/2]$. Here $\arg(\zeta)$ is the angle made by the segment from o to ζ with the positive real axis. Note that the definition of q uses the fact that the radius of Σ_2^3 is $1/2$.

The two transformations q and ζ allow the representation of the shape of a centred triad as a point of the sphere $S^2(1/2)$ by the *spherical co-ordinate*

$$q_o \zeta = \left(\frac{1}{2}, \theta, \psi\right) \quad \text{where } \theta \in [0, \pi] \text{ and } \psi \in (0, 2\pi]$$

and as a point on the extended plane $\mathbb{C} \cup \{\infty\}$ by the *projective co-ordinate* ζ. (The spherical coordinates refer to the point $(r \sin\varphi \sin\theta, r \cos\varphi \sin\theta, r \cos\theta$.) It is convenient to hold both representations in mind. The spherical representation allows graphical representation, while the projective representation is more suitable for computations.

We note the useful relation

$$\zeta = q^{-1}\left(\frac{1}{2}, \theta, \psi\right) = e^{i\psi} \tan(\theta/2) \tag{8.7.5}$$

and the fact that for finite ζ the projective co-ordinate ζ represents the shape of the triad

$$\left(\zeta, (\sqrt{3} - \zeta)/2, -(\sqrt{3} + \zeta)/2\right).$$

These constructions can be used to find the locations of specific geometric shapes on Σ_2^3. Table 8.1 summarizes the results for important shapes.

The most important result in applications is summarized in the first entry in the table. It can be established by noting that x_1, x_2, x_3 are collinear precisely when they can be rotated and translated so as to lie simultaneously on the real axis of the complex plane. This occurs precisely when the ratio ζ above is real. Thus the shape lies on $\mathbb{R} \cup \{\infty\}$ (when expressed in projective coordinates on $\mathbb{C} \cup \{\infty\}$), and on the great circle of collinearities (when considered as an element of Σ_2^3)

$$\left\{\left(\frac{1}{2}, \theta, 0\right), \left(\frac{1}{2}, \theta, \pi\right) : 0 \leqq \theta \leqq \pi\right\}$$

(expressed here in spherical coordinates on $S^2(1/2)$).

Table 8.1

Class of shapes of triads	Corresponding class of points on the shape-space
Colinear triads	The great circle of collinearities $(1/2, \theta, 0)$ and $(1/2, \theta, \pi)$ for $0 \leqq \theta \leqq \pi$
Equilateral triads	$(1/2, \pi/2, \pm\pi/2)$
Isoceles triads	Three great circles, all passing through $(1/2, \pi/2, \pm\pi/2)$, one each through $(1/2, 0, 0)$, $(1/2, 2\pi/3, 0)$, $(1/2, 2\pi/3, \pi)$
One of the angles of triad fixed	Arcs of lesser circles with endpoints at a pair pair of the three points $(1/2, \pi/3, 0)$, $(1/2, \pi/3, \pi)$,. $(1/2, \pi, 0)$
The triad forming a right angle at some vertex	Lesser circles cutting the great circle of collinearities orthogonally at two of the three points $(1/2, \pi/3, 0)$, $(1/2, \pi/3, \pi)$, $(1/2, \pi, 0)$

The shapes of all triads yielding an angle α at $A = x_1$ can be produced by referring to the triad produced above from the projective co-ordinate ζ. Basic complex analysis reveals that for such shapes

$$\arg \frac{\sqrt{3}\,\zeta - 1}{\sqrt{3}\,\zeta + 1} = \pm\alpha \tag{8.7.6}$$

and hence such shapes (except for those for which x_1 coincides with x_2 or x_3) make up the set

$$\{\zeta : \zeta = 3^{-1/2} \cdot (1 - \varrho\, e^{\pm i\alpha})/(1 + \varrho\, e^{\pm i\alpha}) \quad \text{for } \varrho \in (0, \infty)\}.$$

This is the union of two arcs of circles on $\mathbb{C}$ with endpoints $1/\sqrt{3}$ and $-1/\sqrt{3}$, and projects into the union of two arcs of circles on $S^2(1/2)$ as indicated in the table. The symmetries induced by permuting the vertices of the triad produce formulae giving the shapes with fixed angles at the other vertices. These symmetries are discussed further in the next section.

8.8 The spherical blackboard

While the sphere $\sum_2^3$ can be visualized easily it is not convenient for graphical representation of point patterns of shapes. In many applications no real distinction is to be drawn between shapes differing only in the labelling of their vertices. This allows for selection of a convenient representative sub-region of $\sum_2^3$.

Permutations in labelling of the vertices induce symmetries generated by the two operations of (a) interchanging x_1 and x_2, and (b) cyclically permuting the vertices as in

$$(x_1, x_2, x_3) \to (x_2, x_3, x_1).$$

By considering the effect of symmetry (a) on a centred triad and the effect of symmetry (b) on

$$\left(1, (\sqrt{3} - \zeta)/2), -(\sqrt{3} + \zeta)/2\right)$$

the symmetries can be identified as reflections and rotations on the sphere $\sum_2^3$. Symmetry (a) reflects shape-points through the axis formed by the shapes $[x_2 = x_3]$ and $[x_1$ is the midpoint of the segment from x_2 to $x_3]$. Symmetry (b) rotates $\sum_2^3$ through an angle of $2\pi/3$ about the axis formed by the equilateral shapes.

Because of these symmetries, it is enough to consider the shapes as lying in one of six equivalent *lunes*. One example is the lune bounded by the great circles of triangles isoceles at x_1 or x_2 and containing collinear shapes for which x_2 lies between x_1 and x_3. The upper half of this lune is shaded in Figure 8.4.

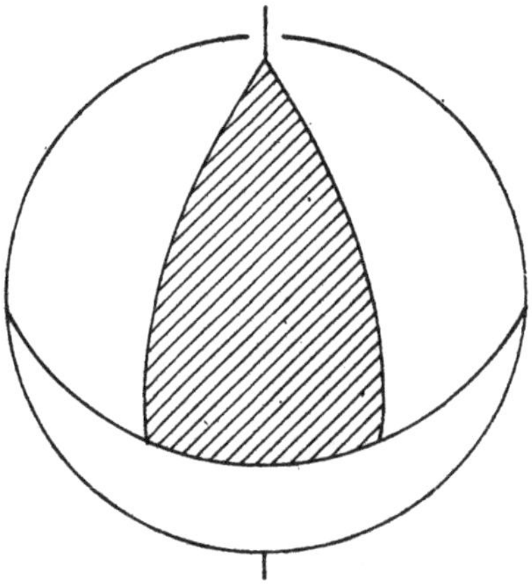

Fig. 8.4. The spherical lune used to construct the spherical blackboard

The presentation of data is considerably improved by this step since the whole lune can be displayed at one time. Moreover, no data need be lost: shapes falling in other lunes can be relabelled so as to fall into the chosen region.

It is often appropriate to consider in addition that symmetry induced by reflection of the triad. This corresponds to complex conjugation of the projective co-ordinate, and thus to reflection through the plane defined by collinear shape points in the spherical representation. If this symmetry is taken into account then $\sum_2^3$ can be divided into twelve equal half-lunes. This step effectively doubles the intensity of the point process of shapes, which is a further aid to interpretation. Thus the shape of a triad can be studied as a point on the spherical triangle that is the chosen half-lune.

The half-lune can be flattened out by application of a projection of the sphere onto a suitably chosen cylinder. The chosen half-lune projects onto a curvilinear triangle on the surface of the cylinder, which can be rolled out into a flat curvilinear triangle. This flat curvilinear triangle is called the *spherical blackboard*. The importance of this construction lies in it being equi-areal: regions of equal area on the half-lune project into regions of equal area on the spherical blackboard. Thus a point pattern of shapes resulting from data can be presented on a flat surface without any distortion of apparent intensity.

The form of the spherical blackboard is illustrated by Figure 8.5, which also displays small triangle shapes located at the corresponding points on the blackboard.

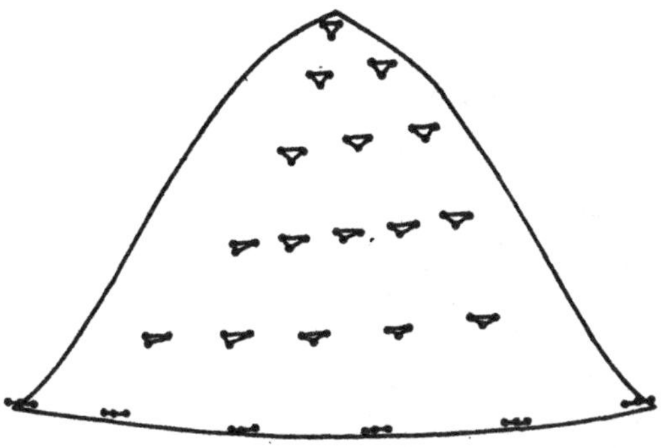

Fig. 8.5. Shapes on the spherical blackboard. Each small triangle is centred on the point of the blackboard representing the shape of that triangle

Example 8.3. *Irish ore bodies.*

Russell (1973) discussed a dataset of locations of ore bodies in Ireland. From geological arguments one expects a tendency for the locations to be aligned. The locations are displayed in Figure 8.6a, together with locations of some ore bodies that were discovered subsequent to Russell's paper. There are only 18 locations. So this dataset is too small to allow a satisfactory investigation of the possible alignments. However, it serves to demonstrate the use of the spherical blackboard.

Figure 8.6b uses the spherical blackboard to display the 816 shapes of triads formed by the dataset. The observed intensity of the point process of shapes increases towards the collinear edge of the blackboard (that edge that corresponds to the great circle of collinear shapes on Σ_2^3). Interpretation of this increase requires the probabilistic theory that is presented in the next section.

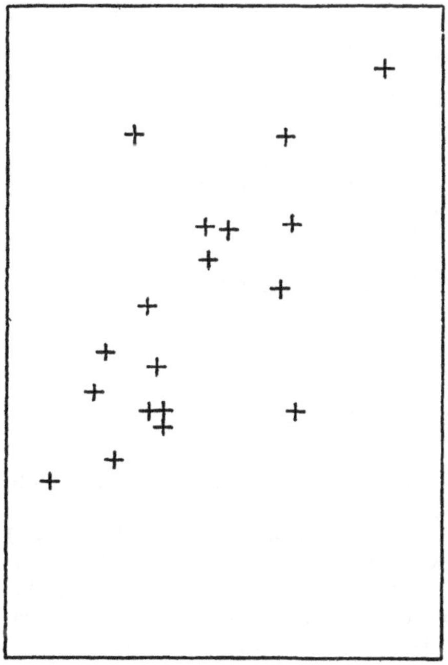

Fig. 8.6 a) (Fig. 8.6 (b) and (c) on next page)

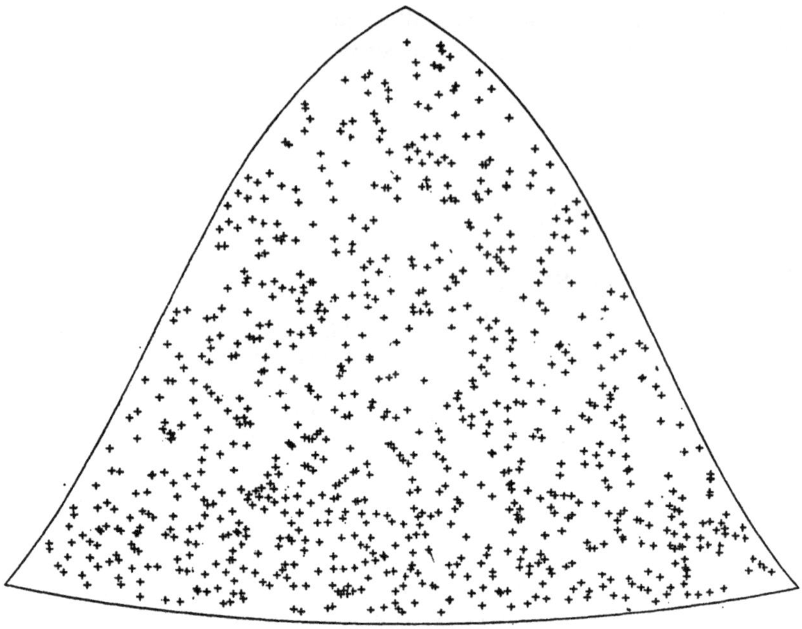

b)

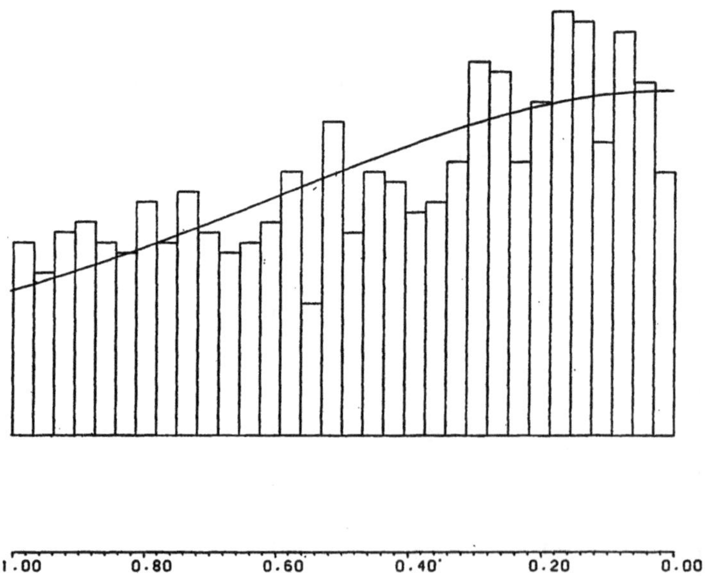

c)

Fig. 8.6. Irish ore bodies. (a) The relative locations of the 18 ore bodies; (b) The 816 shape points corresponding to the triangles formed by the orebody locations. The outline of the spherical triangle has been slightly enlarged so as not to interfere with the display of the shape points; (c) A histogram of the Z-values of the shape points, together with a curve representing the expected frequencies of Z-values from a set of 18 Gaussian points with variance-covariance matrix matched to that of the ore body locations.

8.9 Probability theory on the shape-space of triangles

The shape-density

It is important to know what sort of behaviour is to be expected of the point process of shapes on the spherical blackboard. Thus it is helpful to determine the distribution of the shape of (x_1, x_2, x_3) when the x_i's are independent with common distribution of a particular form. The density of the distribution of shape with respect to uniform measure on the sphere is called the ***shape-density***.

The simplest case is when the common distribution is Gaussian. If the Gaussian distribution is symmetric on the plane $\mathbb{C}$ then the ensuing shape-density is uniform on the sphere Σ_2^3. This is immediate from general properties of the symmetric Gaussian distribution and the orthogonal coordinatization employed in the construction of the shape-manifold. The joint distribution of two angles of the triangle can be computed explicitly. In particular the distribution and density of the maximum-angle of the triad can be found; a graph of the density is given in Figure 8.7.

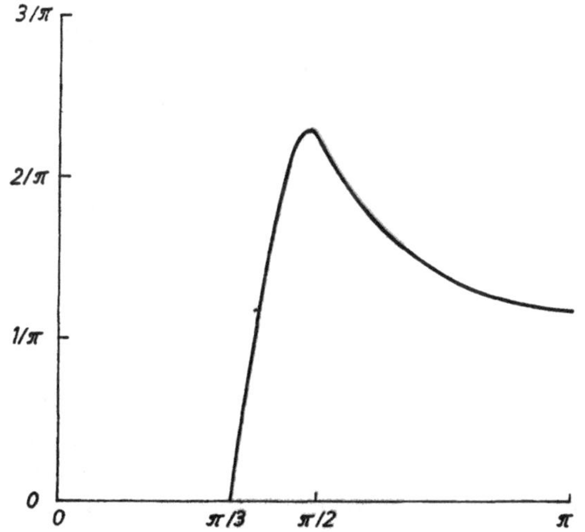

Fig. 8.7. The maximum angle density of a Gaussian triangle (the Gaussian distribution having circular symmetry). Notice the flatness of the density near π

If the common distribution is Gaussian but without circular symmetry then the shape-density is no longer uniform, but depends on the distance in Σ_2^3 of the shape from the great circle of collinear shapes. It also depends on the *Broadbent factor* $f = (s + s^{-1})/2$, where

$$\begin{bmatrix} s & 0 \\ 0 & s^{-1} \end{bmatrix}$$

is the variance-covariance matrix for the common Gaussian distribution. D. G.

KENDALL (1984) obtained an exact formula for the shape-density as

$$\frac{f/2}{[1 + (f^2 - 1) z^2]^{3/2}} \tag{8.9.1}$$

where $-1 \leq z \leq 1$. Here $z = \sin \theta \cdot \sin \psi$ is proportional to the distance of the shape point $(1/2, \theta, \psi)$ from the plane defined by the great circle of collinear shapes. The density of the maximum-angle of the random shape is graphed for varying s in KENDALL and KENDALL (1980).

Example 8.3 (continued) *Irish ore bodies.*

Figure 8.6c gives a histogram of the values of $|z|$ for the shapes formed by the data points. Superimposed on the histogram is a curve graphing the expected number of shapes with $|z|$-value falling in the appropriate cell; this curve is derived from (8.9.1) using for Broadbent factor f an estimate provided via the empirical variance-covariance matrix of the data. It should be borne in mind that the histogram is based on only 18 planar points, so that a formal analysis is inappropriate. However, the comparison of theory with practice gives no grounds to suppose an excess of aligned shapes over that to be expected for such an f-value.

An alternative analysis can be based on the approach used in Example 8.4.

Shape-densities have been computed either theoretically or by simulation for other cases; see D. G. KENDALL (1984), D. G. KENDALL (1985), Mardia and Edwards as reported by MARDIA (1980), and SMALL (1982).

One can also consider triads formed by sampling from a point process. In this case, care has to be taken to define the sampling procedure. A connection can be made with the theory of marked point processes. D. G. KENDALL (1983) recast work of MILES (1970) in the setting of shape theory, the triads in this case being formed from a Poisson–Delaunay tessellation.

The shape-constant

Geometrical arguments lend weight to the use of the z-statistic in analysis of collinearity; it is closely related to the procrustean distance between two shapes. However, an angular criterion for collinearity is in some ways more easily interpreted. Consider the set of all triads formed by a collection of n independent identically distributed random points. An elementary angular criterion is the proportion of triads whose maximum-angles are near to π. This value can be calculated once the maximum-angle distribution is known. From explicit calculation of such densities it is apparent that in many cases the density is nearly flat for high values of the maximum-angle: the graph for the symmetric Gaussian case in Figure 8.7 is typical. If $N(\varepsilon)$ is the number of triads of maximum-angle closer than ε to π then

$$\mathsf{E}N(\varepsilon) \simeq \binom{n}{3} \lambda_c \varepsilon + o(\varepsilon) \quad \text{for small } \varepsilon \tag{8.9.2}$$

where λ_c is the value of the maximum-angle density at π radians.

The value of the maximum-angle density at π is called the *collinearity-constant* for the given parent distribution. For a given distribution let $\lambda_c(s)$ be the collinearity-constant produced from a shear-transform of the distribution. Broadbent (1980) and Kendall and Kendall (1980) showed that the *Broadbent factor* $f = \lambda_c(s)/\lambda_c(1)$ is given by

$$f = (s + s^{-1})/2$$

for a wide class of symmetric distributions.

Table 8.2 displays collinearity constants for the symmetric Gaussian, uniform disc, and uniform square distributions. In each case the Broadbent factor can be employed and so collinearity constants may be obtained for general Gaussian, uniform ellipse, and uniform rectangle distributions.

Table 8.2

Distribution	Collinearity constant
Symmetric Gaussian	$3/\pi - 1/\sqrt{3} = 0.37758$
Square	$1/3 = 0.33333$
Circular disc	$1/\pi = 0.31831$

The observed number of flat triads

The statistical problem that motivated the theory of shape is the *collinearity problem*: how to assess whether observed values of $N(\varepsilon)$ are significantly high for suitably small ε.

The calculations of collinearity constants above allow one to compare $N(\varepsilon)$ to its expected value under simple stochastic models. It is also possible to calculate the asymptotic variance of $N(\varepsilon)$: Kendall and Kendall (1980) give the results for uniform rectangle and uniform disc distributions while W. S. Kendall (1979) deals with the Gaussian case. So if the tolerance ε is fixed in advance then crude significance tests are possible. These can be justified by an appeal to results proved by Hoeffding (1948) on the asymptotic normality of U-statistics.

Example 8.4. *Standing stones at Lands End.* Broadbent (1980), Kendall and Kendall (1980).

Figure 8.8 displays the locations of 52 standing stones from the Lands End region of Britain. It has been suggested that the standing stones were deliberately positioned by their constructors so as to be aligned on straight lines. It is reasonable to expect that if this were so then there would be an excessive number of triads whose maximum angles lay within 40 minutes, say, of 180 degrees. The threshold of 40 minutes is selected arbitrarily but is at least plausible on physical grounds.

In fact out of 22,100 triads there are 108 triads whose largest angle is within 40 minutes of 180 degrees. This compares with the expected values tabulated in Table 8.3 for three simple stochastic models (circular Gaussian, uniform square, and uniform disc), and for the models obtained by stretching these distributions by a factor of $s = 1.661$ (which is a ratio obtained by principal component analysis of the data points). Standard deviations are also tabulated: they have been calculated using formulae taken from the sources referred to above.

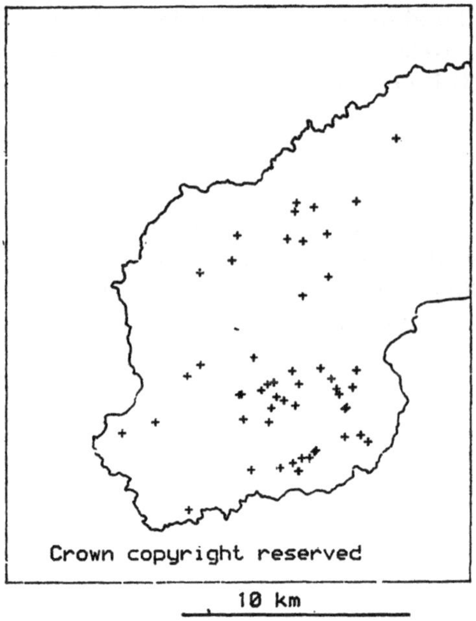

Fig. 8.8. The locations of 52 standing stones in the Lands End region, U. K. The outline of the coastline in this figure is taken from the 1 : 50000 (2nd series) Ordnance Survey Landranger map number 203. *Crown copyright reserved.*

Table 8.3. The expected number of triads whose largest angle is within 40 minutes of 180 degrees, when 52 points are sampled at random from the corresponding distributions. Where the value $s = 1.661$ is given, it refers to the ratio of stretch used.

Distribution	Expected value	Standard deviation
Circular Gaussian	97.09	10.90
Elliptic Gaussian ($s = 1.661$)	109.86	14.12
Uniform square	85.72	10.62
Uniform rectangle ($s = 1.661$)	96.99	12.17
Uniform disc	81.85	10.28
Uniform ellipse ($s = 1.661$)	92.62	12.36

The expected numbers of tetrads and pentads which are flat at tolerance of 40 minutes, when 52 points are sampled at random from the corresponding distributions. Where the value $s = 1.661$ is given, it refers to the ratio of stretch used.

Distribution	Expected numbers of:	
	tetrads	pentads
Circular Gaussian	6.90	0.46
Elliptic Gaussian ($s = 1.661$)	9.81	0.88
Uniform square	5.16	0.21
Uniform rectangle ($s = 1.661$)	7.33	0.40
Uniform disc	4.13	0.17
Uniform ellipse ($s = 1.661$)	5.86	0.33

It is clear from the table that the number of flat triads is not very different from that might be obtained from any reasonable simple stochastic model, once allowances have been made for the elongated nature of the data.

The analysis given in Example 8.3 has also been applied to this dataset (see D. G. KENDALL, 1984).

There are two main deficiencies in the analysis given above. Firstly, it is necessary to select the threshold value ε. Secondly, it may be unreasonable to assume that the point pattern is produced by random sampling from a planar distribution. The first point can be met by making use of results of SILVERMAN and BROWN (1978). These imply that $N(\varepsilon)$ behaves like the counting measure of a Poisson process, at least for large n and for small ε. One can then impose the hypotheses of a *change-point problem*, supposing that $N(\varepsilon)$ is the counting measure of a Poisson process of rate that is constant save possibly for a change from high to low level as ε increases beyond some unknown threshold value. The second point is answered by using simulation to perturb the data points, thus effectively using a stochastic model that is based on the broad features of the data set itself. A more sophisticated analysis based on these features is to be found in KENDALL and KENDALL (1980).

8.10 Further developments

The alignment problem has been studied from the point of view of Palm measures; see HANISCH (1983a) and HANISCH and STOYAN (1984), who assumed that the point pattern is a sample of a stationary point process.

D. G. KENDALL (1984) proved a general case of Theorem 8.3, identifying the shape-manifold $\sum_2^k$. It is possible to find the asymptotic probability of a k-ad being flat (KENDALL and KENDALL, 1980; SMALL, 1982; W. S. KENDALL, 1987) and the calculations in the references were used to produce the values in the second part of Table 8.3.

Example 8.4 (continued).

There are 10 tetrads and 1 pentad which are flat at tolerance of 40 minutes. Here a k-ad is said to be flat at ε (in radians) if two of its points are such that the remaining $k-2$ points all form angles on these two points that are within ε of π radians. Comparison with Table 8.3 shows that these figures are not surprising on the basis of elliptic Gaussian or uniform rectangular variation, though the number of tetrads might be regarded as surprising on the basis of random variation derived from the uniform distribution in an ellipse.

SILVERMAN (BROADBENT, 1980: discussion) pointed out that the Poisson limit arguments of SILVERMAN and BROWN (1978) apply when the expected number of k-ads is small. The pentad is flat at 22.66 minutes, and if one assumes a distribution uniform within an ellipse (with $s = 1.661$) the significance probability is 0.95. This is marginally significant, but its significance is reduced by noting the generosity of the distributional assumptions.

For work on collinearities of k-ads in $\mathbb{R}^2$ and on higher-dimensional shapes (see D. G. KENDALL, 1984; D. G. KENDALL, 1979; and SMALL, 1983).

The analogous problem when alignments are replaced by circles has been addressed by BRADLEY and SMALL (1985).

9. Fibre and surface processes

9.1 Introduction

This chapter is concerned with the study of systems of fibres, and of systems of (fragments of) surfaces, distributed at random on the plane or in space. Interest is focused first on general results holding in the cases of stationarity or of motion invariance. Particular models are also considered. In this introductory section

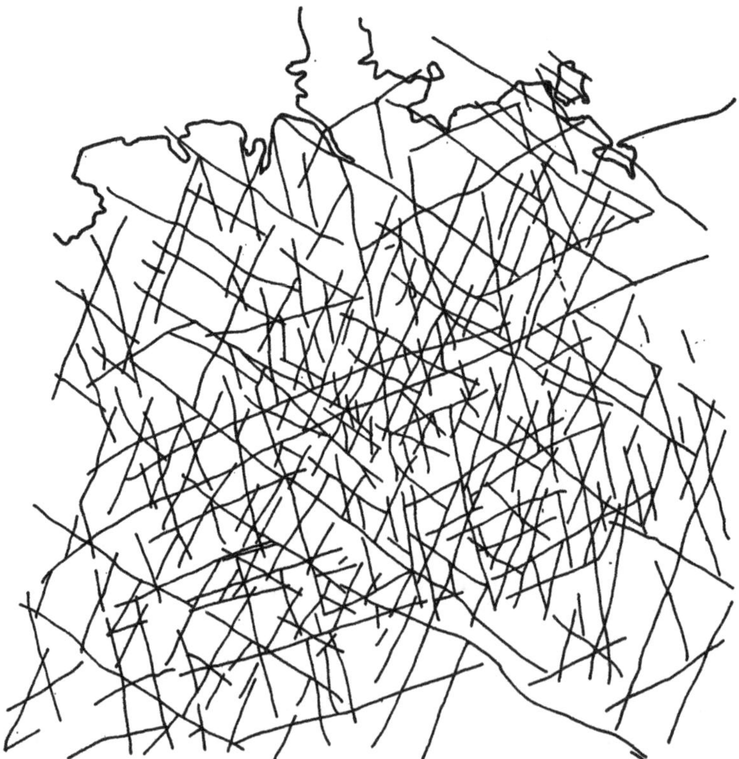

Fig. 9.1. Distribution of linear fault zones in Central Europe (from WEBER, 1977). The system of curves can be interpreted as a sample of a planar fibre process. A line process could also serve at least as a rough approximation. The structure is clearly not isotropic. There are four dominating directions (ENE–WSW, NNE–SSW, NW–SE, ESE–WNW). The apparent lower intensity of fault zones in the northern region of the figure is due to incomplete information about that region.

some illustrations and simple examples are discussed in order to motivate basic definitions.

The theory of fibre and surface processes is not yet tidily linked up with the theories of random sets and of point processes. Thus this chapter presents the preliminary stages of a theoretical development, rather than a polished final product. From a practical point of view the importance of the concepts of fibre and surface processes makes the inclusion of this chapter well worthwhile.

As elsewhere in this book, in the case of stationarity one basic characteristic is *intensity*, describing in some sense the average quantity of the system present in a region. Both fibre and surface processes also associate directions (tangents or normals) to their constituent points and this leads to a further important characteristic: the *rose of directions*.

Figures 9.1 and 9.2 display two realizations of irregular systems of curves in the plane, to be considered as samples of planar fibre processes. Figure 9.3 displays a spatial system of fibres such as might arise in fibrated material, or as systems of blood vessels or nerves. The intensity here is the mean total fibre length per unit area or volume. The rose of direction summarizes the spatial distribution of the collection of tangents.

Planar fibre systems can arise as sections of spatial surface systems such as the system of black graphite lamella given in Figure 3.6. If the lamella are taken to be of negligible thickness then the system can be represented as a random collection of fragments of surfaces. The intensity is then the mean surface

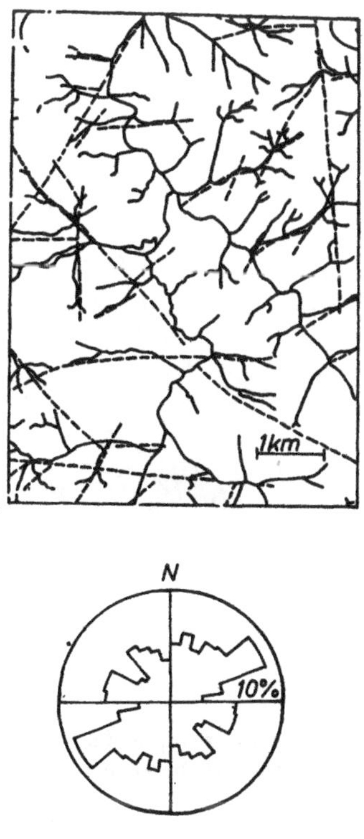

Fig. 9.2. A part of the drainage network in the Eastern Erzgebirge (G. D. R.) and the corresponding rose of direction (derived from a larger region). The main flow direction is SE–NW but the rose has its maximum at SW–NE. This is because of the large number of tributary rivers. The broken lines are geological fault lines.

area per unit volume while the rose of direction is found by considering the collection of normals.

Statistical investigation begins with estimation of the intensity and rose of direction. A further stage considers the suitability of various stochastic models

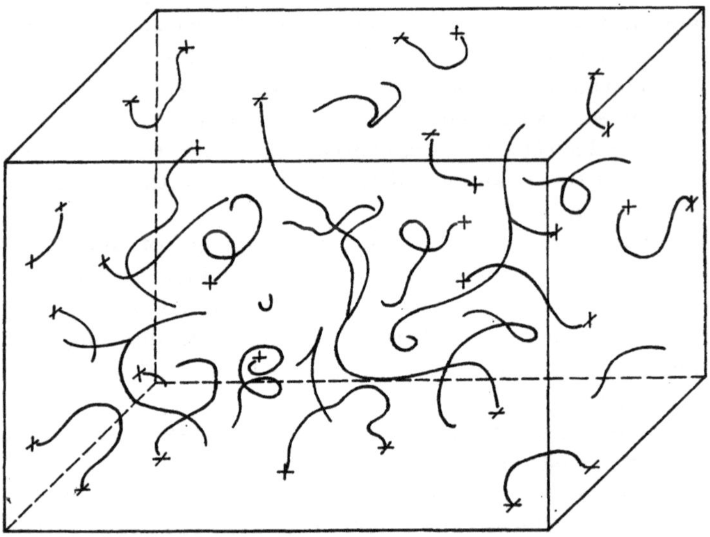

Fig. 9.3. A sample of a spatial fibre process

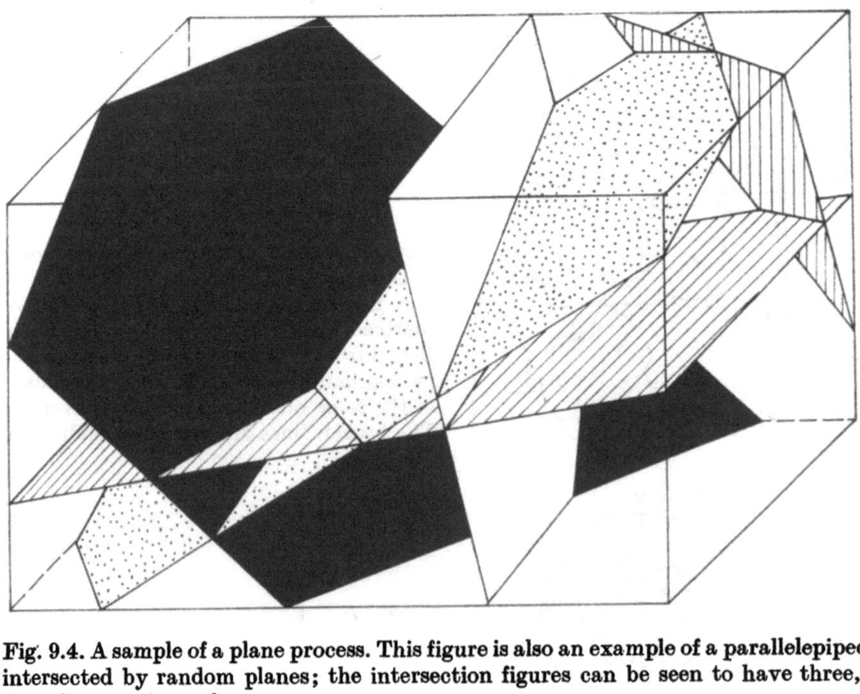

Fig. 9.4. A sample of a plane process. This figure is also an example of a parallelepiped intersected by random planes; the intersection figures can be seen to have three, four, five, or six angles

for the sample. On occasion such a description can shed light on the manner in which the process originated.

Stochastic models of these systems are called *fibre processes* and *surface processes* (as in the term 'point process' no dependency on time is implied here). Some examples are given below.

Figure 8.2 resulted from the simulation of a Poisson line process. This is an example of the class of *line processes* discussed in Chapter 8 and 9.2; fibre processes in which each fibre can be taken to be an entire line. There is a considerable theory concerning such processes, discussed in Chapter 8. Higher dimensional analogues are the *plane* or *flat processes* in $\mathbb{R}^3$, and more generally *hyperplane* and *(hyper)flat* processes in $\mathbb{R}^d$. Figure 9.4 results from a simulation of a plane process in $\mathbb{R}^3$. All these structures are best understood as point processes in suitable representing spaces, as discussed in Chapter 8.

Random tessellations of space, as described in Chapter 10, yield fibre and surface processes from their boundary lines and planes.

Fibre and surface processes also arise from Boolean models; in particular some surface processes arise as the unions of boundaries of grains of Boolean models in $\mathbb{R}^3$. More direct examples of Boolean fibre and surface processes are given below.

Example 9.1. *Boolean fibre and surface processes.* (See also Example 3.2 and Figure 3.8).

(a) *A planar Boolean fibre process.* Boolean models are defined by the intensity of the underlying Poisson process and the distribution of the typical primary grain Ξ_0. In this example Ξ_0 is defined by ξ a uniformly distributed point on the unit circle and l a positive random variable; the random set Ξ_0 is the line segment between o and $l\xi$. By construction Ξ_0 is isotropic. The corresponding Boolean model forms a fibre process in the plane. A simulation of this model is illustrated in Figure 3.8.

(b) *A Boolean surface process.* If the typical primary grain Ξ_0 is a random isotropic two-dimensional disc in $\mathbb{R}^3$, of random radius R, then the corresponding Boolean model can be interpreted as a surface process in $\mathbb{R}^3$.

The surface process of Example 9.1(b) has been proposed as a model for discontinuities in engineering geology (such as joints in rocks or fissures in soil masses). Planar sections are visible on slopes or on outcrops and so stereological methods can be used to determine the statistical characteristics of the underlying process. See Koschitzki (1980), Molek, Pohlmann, Reuter, and Stoyan (1981), and Baecher (1983).

To conclude this section we note that we confine ourselves to processes of *smooth* fibres and surfaces. Zähle (1982, 1983b) has considered fibre and surface processes (and higher-dimensional generalizations) under smoothness assumptions that have been considerably relaxed. The relatively stringent assumptions of Sections 9.3 and 9.5 are replaced in Zähle's work by the requirement that the sets involved are *Hausdorff-rectifiable* closed sets, and Hausdorff measure is used following Federer (1969). This more general theory is developed to give stereological formulae generalizing (9.3.7) and (9.5.12) and to define direction distributions analogous to the roses of directions mentioned above and defined in Sections 9.3 and 9.5. Moreover, the formulae for intersections of fibre and surface processes can be generalized.

ZÄHLE (1982, 1983b) discusses a further avenue of generalization: the case of fibre and surface processes on homogeneous Riemannian spaces. The most immediately applicable case of this is of fibre processes on the sphere.

9.2 Undirected line processes regarded as fibre processes

Before a discussion of the general class of fibre processes (which is to be found in Section 9.3) we consider the particular case of line processes as discussed in Chapter 8. Line processes were regarded there as point processes on a representation parameter space. Here they are reinterpreted as fibre processes in order to introduce a corresponding random measure and associated concepts that are used in the general theory of fibre processes.

Suppose that Φ is a random system of undirected lines. It can be represented as a point process on a half-cylinder, the constituent points of which correspond to the lines of the system Φ (see Chapter 8). It also can be considered as a random closed subset of $\mathbb{R}^2$, the union of the constituent lines. The theory of random closed sets as expounded in 6.1 then applies; the family of (locally finite) line systems is a measurable subset of the family of closed sets (when the latter is endowed with the hit-or-miss σ-algebra) and the natural map provided by this procedure of taking the union is itself measurable. Thus the *distribution of the line process* can be considered as a probability measure on the space of closed sets, as well as a measure on the half-cylinder. From this relationship to the theory of random closed sets, the definitions of stationarity and motion-invariance are immediate.

If an undirected line process Φ is considered as a set $\{l_1, l_2, \ldots\}$ of lines in $\mathbb{R}^2$ then it induces a random measure on $\mathbb{R}^2$ also denoted by Φ

$$\Phi(B) = \sum_{l \in \Phi} \nu_1(l \cap B) \quad \text{for } B \text{ a planar Borel set.} \tag{9.2.1}$$

This ambiguous notation follows the example set in the treatment of point processes (it should not cause confusion since the meaning of Φ will always be clear from the context). From the theory of random measures the corresponding *intensity* L_A of Φ is defined when Φ is stationary (for then Φ is also stationary as a random measure):

$$L_A \nu_2(B) = \mathsf{E}\Phi(B) \quad \text{for any planar Borel set } B. \tag{9.2.2}$$

If Φ is motion-invariant then the intensity measure of Φ as a point process on its representation space is determined up to a multiplicative constant; see equations (8.3.3) and (8.4.3) which show that L_A as defined by (9.2.2) is proportional to the intensity of Φ as a point process on the representation space.

If the direction $\alpha(l)$ of each line in Φ is taken into account then a more refined random measure can be introduced. The direction of an (undirected) line lies in $(0, \pi]$ so one can define

$$\Psi(B \times L) = \sum_{l \in \Phi, \alpha(l) \in L} \nu_1(B \cap l) \tag{9.2.3}$$

where B is a planar Borel set and L lies in Π the family of Borel subsets of $(0, \pi]$. For general Φ one can define the intensity measure Λ of the random measure Ψ. In the stationary case invariance arguments lead to a factorization

$$\Lambda = L_A \nu_2 \times \boldsymbol{R} \tag{9.2.4}$$

where $\boldsymbol{R}$ is a probability measure on $(0, \pi]$, the *rose of direction*. This can be interpreted as follows: $\boldsymbol{R}(L)$ is the probability that a 'typical' line l of Φ has direction in L. Compare the measure given by (8.3.2) which is appropriate for a *directed* line process.

Further invariance arguments show that if Φ is also isotropic then so is Λ, and $\boldsymbol{R}$ must then be the uniform distribution on $(0, \pi]$

$$\boldsymbol{R}(L) = \frac{1}{\pi} \nu_1(L) \quad \text{for } L \text{ in } \Pi.$$

The invariance arguments follow those of Theorem 8.1.

9.3 Planar fibre processes

9.3.1 Fundamentals

As mentioned above planar fibre processes model random collections of curves in the plane $\mathbb{R}^2$. Figures 9.1 and 9.2 give examples of systems of curves or fibres which could be interpreted as samples of fibre processes. The discussion below is based on the papers of Mecke and Stoyan (1980a) and Mecke and Nagel (1980). Earlier references on fibre processes are Ambartzumian (1973, 1977), Ambartzumian and Oganian (1975), and Cowan (1979). In these papers the fibres are usually segments and the processes are usually motion-invariant. The following treatment is more general.

Fibres and fibre systems

A fibre is a sufficiently smooth simple curve in the plane, of finite length. Thus a fibre γ is a subset of R^2 which is the image of a curve $\gamma = (\gamma_1, \gamma_2)$ such that

(i) $\gamma : [0, 1] \to \mathbb{R}^2$ is once continuously differentiable;

(ii) $|\gamma'(t)|^2 = |\gamma_1'(t)|^2 + |\gamma_2'(t)|^2 > 0$ for all t;

(iii) the mapping γ is one-to-one, so that a fibre does not intersect itself.

As in Section 9.2 so here there is a deliberate ambiguity; γ also stands for the measure

$$\gamma(B) = \nu_1(\gamma \cap B) = \int_0^1 \boldsymbol{1}_B(\gamma(t)) \sqrt{(\gamma_1'(t)^2 + \gamma_2'(t)^2)}\, dt$$

for planar Borel sets B. Note that the *measure* γ depends only on the curve $\gamma = \gamma([0, 1])$ and not on the precise representing map γ.

The definition of a fibre system is straightforward given this definition of a fibre. A *fibre system* φ is a closed subset of $\mathbb{R}^2$ which can be represented as an (at most countable) union of fibres $\gamma^{(i)}$ with the property that any compact set is intersected by only a finite number of the fibres, and such that distinct fibres have only endpoints in common

$$\gamma^{(i)}\big((0,1)\big) \cap \gamma^{(j)}\big((0,1)\big) \quad \text{is empty if } i \neq j.$$

The length measure corresponding to the fibre system φ is then defined by

$$\varphi(B) = \sum_{\gamma^{(i)} \in \varphi} \gamma^{(i)}(B).$$

Despite the condition on endpoints φ, actually may consist of intersecting curves, since fibres can end at intersection points. All the definition requires is that such curves can be dissected into a locally finite collection of fibres as above (of course, this representation of φ as a union of fibres is not unique). This prevents fibre systems from having locally dense accumulations of self-intersection points. The locally-finite and smoothness conditions ensure that the measure φ is locally-finite; it depends only on the closed set that is the union of all the fibres. In particular line processes can be interpreted as fibre processes.

The family of all planar fibre systems is denoted by $\mathbf{S}$ and it is endowed with a σ-algebra $\mathcal{S}$ generated by sets of the form

$$\{\varphi \in \mathbf{S} : \varphi(B) < x\}$$

for planar Borel sets B and real numbers x. The theory of fibre processes is thus a special section in the theory of random measures, and from this point of view $\mathcal{S}$ can be related to $\mathcal{M}$ as described in Section 7.1.2, or $\mathcal{N}$ as in Section 4.1. It can be shown that $\mathcal{S}$ is the same as $\mathcal{F} \cap \mathbf{S}$, where $\mathcal{F}$ is Matheron's σ-algebra for the family of all closed subsets of $\mathbb{R}^2$.

Fibre processes

A (planar) *fibre process* Φ is a random variable with values in $[\mathbf{S}, \mathcal{S}]$, that is to say, a measurable mapping from an underlying probability space $[\Omega, \mathcal{A}, \mathsf{P}]$ to $[\mathbf{S}, \mathcal{S}]$. The same symbol Φ is also used to denote the corresponding *random length measure*. The *distribution* of the fibre process is the measure P generated on $[\mathbf{S}, \mathcal{S}]$ by Φ.

Stationarity and isotropy

The fibre process Φ is said to be *stationary* if it has the same distribution as the translated fibre process Φ_x. Thus

$$P(Y) = P(Y_x) \quad \text{for all } Y \text{ in } \mathbf{S}, \text{ all } x \text{ in } \mathbb{R}^2.$$

Likewise, it is *isotropic* if the distribution remains invariant under rotations about the origin (these definitions are special cases of the general definitions for random measures).

Intensity and the rose of Direction

The *intensity measure* Λ of a fibre process is given by

$$\Lambda(B) = \mathsf{E}\Phi(B) = \mathsf{E}\left(\sum_{\gamma\in\Phi} \nu_1(\gamma \cap B)\right)$$

for B a planar Borel set. If Φ is stationary then invariance considerations show that

$$\Lambda = L_A\nu_2$$

for a constant $L_A (0 \leq L_A \leq \infty)$, the *intensity* of Φ. The notation here is that of stereology as described in Table 11.1; L_A stands for 'mean line length per unit area'. In fact L_A is precisely the intensity of Φ regarded as a stationary random measure. In the following L_A is assumed to be positive and finite.

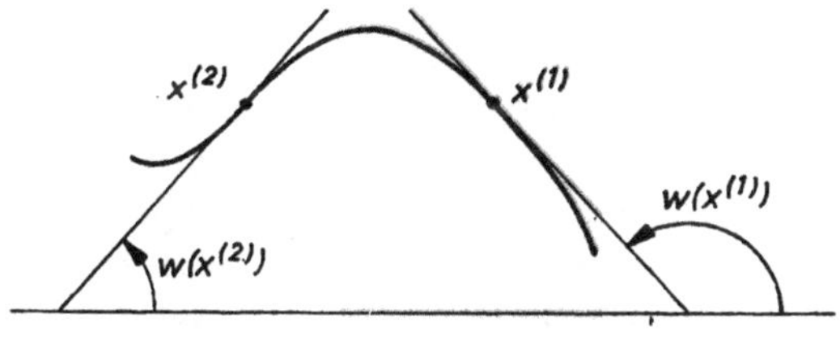

Fig. 9.5. Definition of the tangent direction $w(x)$

The rose of directions arises from consideration of fibre tangents. Let $w(x)$ denote the direction of the fibre tangent at x, presuming that precisely one fibre of Φ passes through x. Then $w(x)$ is some number between 0 and π; see Figure 9.5 (in the case when more than one fibre passes through x or an endpoint occurs at x let $w(x)$ equal π; such points form a subset of ν_1-measure zero by the conditions on fibre systems). A weighted random measure Ψ can be defined on $\mathcal{B}^2 \times \Pi$ by

$$\Psi(B \times L) = \int_B \mathbf{1}_L(w(x))\, \Phi(\mathrm{d}x)$$

for B in $\mathcal{B}^2$ and L in Π. Thus $\Psi(B \times L)$ is the total length of all fibre pieces of Φ in B that have tangent directions lying in L.

If Λ_Ψ is the intensity measure of Ψ, so that

$$\Lambda_\Psi(B \times L) = \mathsf{E}\Psi(B \times L)$$

and if Φ is stationary then invariance considerations dictate that Λ_Ψ can be written as

$$\Lambda_\Psi(B \times L) = L_A\nu_2(B)\, \boldsymbol{R}(L)$$

where $\boldsymbol{R}$ is a distribution on Π which is called the *rose of direction* of Φ. Reference to (7.1.4) shows that this is a weight distribution. In analogy to similar constructions in the theory of point processes $\boldsymbol{R}$ can be interpreted as the distribution of the direction of the tangent to a fibre at a 'typical' point. Figure 9.2 includes an example of an empirical rose of directions.

This analogy carries further: a *Campbell theorem* holds for stationary fibre processes

$$\mathsf{E} \int f(x, w(x))\, \Phi(\mathrm{d}x) = \iint f(x, w(x))\, \varphi(\mathrm{d}x)\, P(\mathrm{d}\varphi)$$
$$= L_A \iint f(x, \alpha)\, \boldsymbol{R}(\mathrm{d}\alpha)\, \mathrm{d}x. \qquad (9.3.1)$$

By definition both L_A and $\boldsymbol{R}$ are invariant under translation of the underlying fibre process. Furthermore L_A does not vary under rotations of the fibre process (though $\boldsymbol{R}$ does of course change). If $\mathfrak{r}$ is a rotation about o through an angle ϱ then the rotated fibre process has rose of direction $\boldsymbol{R}^0\,\mathfrak{r}$ where

$$\boldsymbol{R}^0\,\mathfrak{r}(L) = \boldsymbol{R}(\mathfrak{r}^{-1}(L)).$$

Finite measures on $(0, \pi]$ can be represented in terms of their *Steiner compacts* S, first used by MATHERON (1975) for this purpose. These form a particular class of symmetric compact convex sets defined by corresponding measures, the class of *zonoids* described by SCHNEIDER and WEIL (1983). For a stationary fibre process Φ one can produce an associated Steiner compact S as follows. The modified support function r_S of S is given by

$$r_S(\beta) = \frac{1}{2} L_A \int\limits_{(0,\pi]} |\cos(\alpha - \beta)|\, \boldsymbol{R}^{\perp}(\mathrm{d}\alpha). \qquad (9.3.2)$$

Here $\boldsymbol{R}^{\perp}$ is the rose of direction obtained by rotating Φ through a right angle. It may be thought of as the *rose of normal directions* of Φ. This concept aids the presentation of the important formula (9.3.11). The notion of a modified support function for a symmetric convex set is described in 1.6.

If the rose of direction $\boldsymbol{R}^{\perp}$ is uniformly distributed on $(0, \pi]$ then the resulting S is the ball

$$b(o, L_A/\pi).$$

The Steiner compact is used in the study of intersections of fibre processes with lines; see Section 9.3.2. Two further relationships with stationary fibre processes are:

(1) if S is the Steiner compact of Φ then

$$L_A = \frac{1}{2} U(S)$$

(2) if Φ_1, Φ_2 are independent stationary fibre processes with Steiner compacta S_1 and S_2 then the Steiner compact of the union $\Phi_1 \cup \Phi_2$ is

$$S = S_1 \oplus S_2.$$

The second moment measure and the reduced second moment measure

Moment measures for fibre processes can be defined as for general random measures. In particular the *second moment measure* $\mu^{(2)}$ of Φ is given by

$$\mu^{(2)}(B_1 \times B_2) = \mathsf{E}\Phi(B_1)\,\Phi(B_2)$$

for B_1 and B_2 planar Borel sets.

In the stationary case, this measure can be decomposed just as for point processes in Section 4.5. Given $\mu^{(2)}$ for a stationary fibre process a measure $\mathcal{K}$ can be defined on the plane by

$$\mu^{(2)}(B_1 \times B_2) = L_A^2 \iint \mathbf{1}_{B_1}(x)\,\mathbf{1}_{B_2}(x+h)\,\mathrm{d}x\,\mathcal{K}(\mathrm{d}h)$$

for B_1, B_2 planar Borel sets.

This $\mathcal{K}$ is the *reduced second moment measure* of Φ. Its interpretation follows the point process case; if x is a 'typical' fibre point of Φ then $L_A\mathcal{K}(B)$ is the mean total length of fibre pieces of Φ in B_x. This is further discussed in STOYAN (1981).

In analogy with the use of $\mathcal{K}$ in 4.5, and in parallel to the rose of directions $\boldsymbol{R}$, the reduced second moment measure can be used as a directional characterisation of a fibre process. In particular (4.5.2) holds for fibre processes.

In the case of stationarity and isotropy it is enough to consider the *reduced second moment function* K

$$K(r) = \mathcal{K}\big(b(o, r)\big) \quad \text{for } r \geqq 0.$$

If K is differentiable then, just as in the point process case, one can consider the *pair correlation function* g

$$g(r) = \frac{1}{2\pi r}\,\frac{\mathrm{d}K(r)}{\mathrm{d}r} \quad \text{for } r \geqq 0.$$

Figure 9.6 shows the pair correlation function for two simple fibre process models. As r tends to infinity so they tend to 1, but for $r \downarrow 0$ they diverge to infinity. This divergence is typical for fibre processes. It is explained by noting that even in the circles about a 'typical' fibre point there must be pieces of at least one fibre. Thus $\mathrm{d}K/\mathrm{d}r$ does not become small for small r. The corresponding quantity for point processes uses the *factorial* second moment measure, deleting the 'typical' point itself (Note, however, that even in the point process case there are examples with pair correlation function infinite at zero. The point process illustrated in Figure 5.5 is an example). A similar approach for fibre processes (using for example $K_0(r) = K(r) - 2r/L_A$) does not in general dispose of the infinity at zero.

Example 9.2. *Basic characteristics of two simple fibre processes.*

(a) Let Φ be a motion-invariant Poisson line process of intensity L_A. The reduced second-moment function K is given by

$$L_A K(r) = 2r + L_A \pi r^2 \quad \text{for } r \geqq 0. \tag{9.3.3}$$

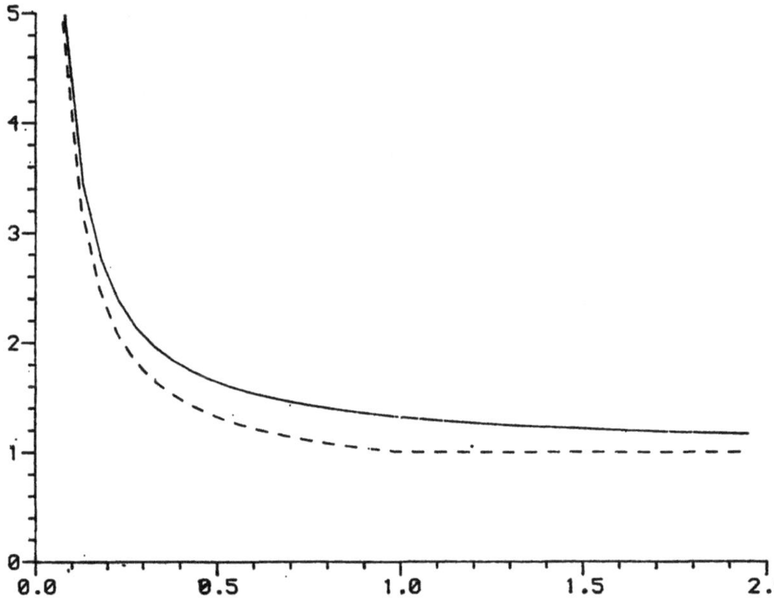

Fig. 9.6. Pair correlation function of a Poisson line process with unit intensity and of a Boolean segment process with segments of constant unit length and unit intensity. Poisson line process p. c. f., ———; Boolean segment process p. c. f., – – – –

The pair correlation function is therefore

$$g(r) = 1 + 1/(L_A \pi r).$$

To make a heuristic argument for (9.3.3) note that $L_A K(r)$ has two components: one is the contribution from the line containing the 'typical' point and the other is formed from the remainder of the process. Clearly the first component is $2r$. By Slivnyak's theorem the remainder of the process has distribution equal to that of the original process and thus yields a contribution $L_A \nu_2(b(o, r))$.

(b) Let Φ be a Boolean segment process as in Example 9.1. If the length l of the 'typical' segment has distribution L and mean $\bar{l}$ then the intensity of Φ is $L_A = \lambda \bar{l}$, where λ is the intensity of the germ process. The reduced second moment function K is given by

$$L_A K(r) = L_A \pi r^2 + \lambda \left(\int_0^r x^2 \, \mathrm{d}L(x) + \int_r^\infty (2xr - r^2) \, \mathrm{d}L(x) \right)$$

for $r \geqq 0$, as in STOYAN (1983). Again the Slivnyak theorem explains this; this time the second term arises from the segment on which the 'typical' point lies. The distribution of the length of the 'typical' segment is length-biased and the 'typical' point lies uniformly at random on this segment.

In Figure 9.6 graphs of g are given for both examples. HANISCH (1984d) has calculated g for a series of more complicated fibre processes.

9.3.2 Intersections of fibre processes

Important methods of measurement for fibre processes use enumeration of the intersections of the process with test systems of lines, circles, or other curves. Therefore it is useful to derive formulae for the random point patterns that arise from such intersections and more generally from the intersections of fibre processes with other fibre processes.

Intersections with lines

Let Φ be a stationary fibre process with distribution P, intensity L_A, and rose of direction $\boldsymbol{R}$. We consider the intersection of Φ with a fixed line e. To be definite e is taken to be the x_1-axis.

The process $\Phi \cap e$ is a point process on e and is marked by the angles of intersection. This marked point process $\Psi = \{[y_n; w(y_n)]\}$ was first considered by DAVIDSON (1974c), and later discussed in AMBARTZUMIAN (1974b, 1977, 1982), and MECKE and STOYAN (1982a). The characteristics L_A and $\boldsymbol{R}$ can be recovered from its distribution. Furthermore (under assumptions of isotropy and non-linearity of fibres for Φ), the reduced second moment function K can be determined from knowledge of the distribution of Ψ.

Let P_L be the intensity of Ψ and H the mark distribution on $(0, \pi]$. The mark distribution is the distribution of the intersection angle at a 'typical' intersection point of $\Psi = \Phi \cap e$.

The following formula connects the distributions of Φ and Ψ

$$P_L \int\int h(z, \alpha) \, H(\mathrm{d}\alpha) \, \mathrm{d}z = L_A \int\int h(z, \alpha) \sin \alpha \boldsymbol{R}(\mathrm{d}\alpha) \, \mathrm{d}z \tag{9.3.4}$$

where h is a non-negative and measurable function on $\mathbb{R} \times (0, \pi]$. Formula (9.3.4) can be proved as follows. Simple geometry shows that

$$\int m(x, w(x)) \sin w(x) \, \varphi(\mathrm{d}x) = \int \sum m(x_1, x_2, w(x_1, x_2)) \, \mathrm{d}x_2$$

where the sum is over x_1 such that $x = (x_1, x_2) \in \varphi$. This holds for every fibre system φ and every non-negative measurable function m on $\mathbb{R}^2 \times (0, \pi]$. Here $w(x)$ is the direction of the fibre tangent of φ at x.

If g is a non-negative measurable function with $\int g(z) \, \mathrm{d}z = 1$ then the Campbell theorem (9.3.1) can be employed on Φ using

$$f(x, \alpha) = h(x_1, \alpha) \, g(x_2) \sin \alpha$$

$$\mathsf{E} \int h(x_1, w(x)) \, g(x_2) \sin w(x) \, \Phi(\mathrm{d}x) = L_A \int h(x_1, \alpha) \, g(x_2) \sin \alpha \, \boldsymbol{R}(\mathrm{d}\alpha) \, \mathrm{d}x .$$

The right-hand side equals

$$L_A \int h(z, \alpha) \sin \alpha \, \boldsymbol{R}(\mathrm{d}\alpha) \, \mathrm{d}z .$$

By the simple geometrical formula given above and by stationarity the left-hand side equals

$$\iint \sum_{x_1:(x_1,x_2)\in\varphi} h\big(x_1, w(x)\big)\, g(x_2)\, \mathrm{d}x_2 P(\mathrm{d}\varphi)$$

$$= \iint \sum_{x_1:(x_1,x_2)\in\varphi+(0,x_2)} h\big(x_1, w(x_1, 0)\big)\, P(\mathrm{d}\varphi)\, g(x_2)\, \mathrm{d}x_2$$

$$= \int \sum_{x_1:(x_1,0)\in\varphi} h\big(x_1, w(x_1, 0)\big)\, P(\mathrm{d}\varphi)$$

$$= \mathsf{E} \sum_{[y;w(y)]\in\Psi} h\big(y, w(y)\big).$$

The proof is completed by applying the Campbell theorem for Ψ to this last formula.

From formula (9.3.4) one can derive some important formulae, relating characteristics of Φ to those of Ψ. For example, the intensity of Ψ is related to the intensity and rose of directions of Φ by

$$P_L = L_A \int_{(0,\pi]} \sin \alpha \boldsymbol{R}(\mathrm{d}\alpha) \tag{9.3.5}$$

as can be seen by putting $h(z, \alpha) = \mathbf{1}_{[0,1]}(z)$.

If $h(z, \alpha) = \mathbf{1}_{[0,1]}(z)\, \mathbf{1}_{(0,\beta]}(\alpha)$ for some β in $(0, \pi]$ then a more refined relation is obtained that uses the mark distribution H

$$P_L H\big((0, \beta]\big) = L_A \int_{(0,\beta]} \sin \alpha \boldsymbol{R}(\mathrm{d}\alpha). \tag{9.3.6}$$

It follows that the distribution function F_H for H is given by

$$F_H(\beta) = H\big((0, \beta]\big) = \frac{\int_{(0,\beta]} \sin \alpha \boldsymbol{R}(\mathrm{d}\alpha)}{\int_{(0,\pi]} \sin \alpha \boldsymbol{R}(\mathrm{d}\alpha)}$$

for $0 \leqq \beta \leqq \pi$.

When $\boldsymbol{R}$ is the uniform distribution on $(0, \pi]$, as in the case of isotropy, then

$$P_L = 2L_A/\pi \tag{9.3.7}$$

and

$$F_H(\beta) = \frac{1}{2}\,(1 - \cos \beta) \quad \text{for } 0 \leqq \beta \leqq \pi \tag{9.3.8}$$

with corresponding density function

$$f_H(\beta) = \frac{1}{2} \sin \beta \quad \text{for } 0 \leqq \beta \leqq \pi.$$

This density also arises in the case of intersection with curves; see Morton (1966).

If $\boldsymbol{R}(\{\pi\}) = 0$ (so that a typical point on a fibre has fibre tangent almost surely not parallel to the sampling line e) then the characteristics of Φ can be expressed in terms of those of Ψ

$$\boldsymbol{R}((0, \beta]) = \frac{P_L}{L_A} \int\limits_{(0,\beta]} (\sin \alpha)^{-1} H(\mathrm{d}\alpha) \tag{9.3.9}$$

$$L_A = P_L \int\limits_{(0,\pi]} (\sin \alpha)^{-1} H(\mathrm{d}\alpha). \tag{9.3.10}$$

A further formula for $\boldsymbol{R}$ involves the so-called *rose of intersections* $P_L(\cdot)$. The quantity $P_L(\beta)$ is the intensity of intersections of Φ with a sampling line at angle β to e. Up to a factor this is the same as the modified support function of the Steiner compact S related to $L_A\boldsymbol{R}^{\perp}$, as described above in Section 9.3.1. Consider the point process of intersections of Φ with a straight line forming an angle β with the reference axis (the x_1-axis) e. The fibre system rotated by $\mathfrak{r}$ has rose of directions $\boldsymbol{R}\circ\mathfrak{r}$ as given in Section 9.3.1. Thus if $\mathfrak{r}$ rotates through an angle $-\beta$ sending the intersection line to e then formula (9.3.5) can be applied to the rotated fibre process. This shows that the point process of intersections has intensity

$$P_L(\beta) = L_A \int\limits_{(0,\pi]} |\sin (\alpha - \beta)| \, \boldsymbol{R}(\mathrm{d}\alpha). \tag{9.3.11}$$

This formula can be found in Hilliard (1962).

Formula (9.3.2) together with a change of variables shows that

$$P_L(\beta) = 2r_S(\beta) \quad \text{for } \beta \text{ in } (0, \pi].$$

If $\boldsymbol{R}$ has a continuous density $f_{\boldsymbol{R}}$ then formula (9.3.11) can be differentiated to show

$$\frac{\mathrm{d}^2}{\mathrm{d}\beta^2} P_L(\beta) + P_L(\beta) = 2L_A f_{\boldsymbol{R}}(\beta) \tag{9.3.12}$$

for β in $(0, \pi]$. Mecke (1981a) generalizes this formula for the case when $\boldsymbol{R}$ has no density. In principle the formula allows for estimation of $f_{\boldsymbol{R}}$ if the rose of intersections $P_L(\cdot)$ is calculated empirically. In practise one must deal with the serious problems of smoothing the numerical differentiation of $P_L(\beta)$.

As a simple example of an anisotropic fibre process one can consider the 'pressed' fibre process discussed in Stoyan, Mecke, and Pohlmann (1980). This is produced from an isotropic fibre process by the 'pressing' mapping

$$p_c : (x_1, x_2) \to (x_1, c \cdot x_2) \quad \text{for some } c \text{ in } (0, 1).$$

Characteristics of the pressed fibre process can be easily expressed in terms of the original isotropic fibre process: for example the intensity of the pressed fibre process can be estimated to good accuracy and independently of c by the quantity

$$(\pi/2)\, P_L(\alpha) \quad \text{when } \alpha = 40°.$$

Figure 9.7 illustrates this. See Stoyan, Mecke, and Pohlmann (1980) for details.

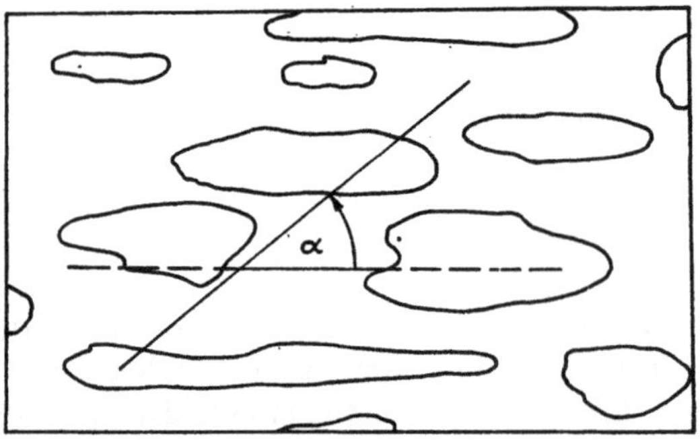

Fig. 9.7. A sample of a pressed fibre process together with a 40°-measurement line

For an isotropic fibre process Φ the reduced second moment function $K(r)$ can be determined in terms of the marked intersection point process Ψ (if additional marks are used). If $[y; w(y)]$ belongs to Ψ then the additional mark on y required is

$$s_r(y) = L(y) + 2\pi \sum_{\substack{(x_1,0)\in\Phi,\\ y\le x_1\le y+r}} \frac{x_1 - y}{\sin \alpha(x_1, 0)}$$

where $\alpha(x_1, 0)$ is the intersection angle between the fibre at $(x_1, 0)$ and the x_1-axis (so $0 < \alpha(x_1, 0) \leqq \pi$), and $L(y)$ is the total length of all straight line pieces of fibre lying on lines radial to $(y, 0)$ within the ball $b((y, 0), r)$.

Note that if Φ has no straight line pieces of fibre then $s_r(y)$ is determined entirely by the information contained within the marked point process Ψ. However the additional information represented by the lengths $L(y)$ is required if there are straight pieces of fibre in Φ.

Consider the marked point process

$$\{[y; w(y), s_r(y)] : [y; w(y)] \in \Psi\}$$

and let J be the joint mark distribution of the mark pair $(w(y), s_r(y))$. Then

$$L_A K(r) = \frac{2}{\pi} \int \frac{s}{\sin \alpha} J(\mathrm{d}(\alpha, s)) \tag{9.3.13}$$

as shown by Stoyan (1981).

Ambartzumian (1981) studied the particular case of segment processes and gave formulae for the density of K. The properties of intersection of fibres processes having segments, triangles, and circles as fibres are studied in Oganian (1978) and Sukiasan (1978, 1980). Some results can be found in the following paragraphs.

Intersections with fibre systems

Suppose that Φ is a stationary fibre process of intensity L_A and rose of direction $\boldsymbol{R}$ and ψ is a (non-random) fibre system with finite total length l. Mecke (1981a) and Ohser (1981) studied the intersection process $\Phi \cap \psi$. This intersection will be a point process with only a finite number of points (because

ψ has finite total length). The fibre system ψ controls the basic characteristics of $\Phi \cap \psi$ by means of two functional parameters:

(a) the angle distribution η_ψ, a probability measure on $(0, \pi]$ given by

$$\eta_\psi(A) = \nu_1\{x \in \psi : w_\psi(x) \in A\}/l$$

(here $w_\psi(x)$ is the ψ-fibre tangent angle at x);

(b) the total length $\varrho_\psi(\beta)$ (counted by multiplicity) of the projection of ψ in the direction β, given by

$$\varrho_\psi(\beta) = \int_{l_\beta{}^\perp} \# \{\psi \cap (l_\beta - y)\} \, dy$$

for $0 < \beta \leqq \pi$. Here l_β is a line of direction β and $l_\beta{}^\perp$ is a line perpendicular to l_β.

Here are examples of these two characteristics in two simple cases:

- if ψ is a circle of radius R then η_ψ is uniform on $(0, \pi]$ and $\varrho_\psi(\beta) = l/\pi$ (note $l = 2\pi R$)
- if ψ is the union of two segments of equal lengths $l/2$ which make angles α and $\alpha + \pi/2$ on the x_1-axis respectively then η_ψ is the obvious two-point distribution on $\{\alpha, \alpha + \pi/2\}$ and

$$\varrho_\psi(\beta) = l(|\sin(\beta - \alpha)| + |\cos(\beta - \alpha)|)/2.$$

The mean number of points $\overline{N}_\psi$ in $\Phi \cap \psi$ satisfies

$$\overline{N}_\psi = L_A l \int |\sin(\alpha - \beta)| \, \boldsymbol{R}(d\alpha) \, \eta_\psi(d\beta). \tag{9.3.14}$$

If ψ is a circle of radius R then

$$\overline{N}_\psi = 4RL_A. \tag{9.3.15}$$

If $\overline{N}_\psi(A)$ denotes the mean number of points in $\Phi \cap \psi$ for which the Φ-fibre direction is in A then

$$H_\psi(A) = \overline{N}_\psi(A)/\overline{N}_\psi$$

defines a distribution on $(0, \pi]$ which can be interpreted as the angle distribution for the intersections. It satisfies

$$\overline{N}_\psi H_\psi(A) = L_A \int_A \varrho_\psi(\alpha) \, \boldsymbol{R}(d\alpha) \tag{9.3.16}$$

and if ψ is a circle then

$$H_\psi = \boldsymbol{R}. \tag{9.3.17}$$

Intersection of two fibre processes

Let Φ_1, Φ_2 be two independent stationary fibre processes of characteristics $L_{A,i}$, $\boldsymbol{R}_i$ for $i = 1, 2$. The intersection $\Phi_1 \cap \Phi_2$ is a stationary point process;

denote its intensity by λ. Then

$$\lambda = L_{A,1}L_{A,2} \int\limits_{(0,\pi]} \int\limits_{(0,\pi]} |\sin(\alpha_1 - \alpha_2)| \, \boldsymbol{R}_1(\mathrm{d}\alpha_1) \, \boldsymbol{R}_2(\mathrm{d}\alpha_2). \tag{9.3.18}$$

If either of $\boldsymbol{R}_1$ or $\boldsymbol{R}_2$ is actually uniform on $(0, \pi]$ (and this will occur for example if either of the Φ_i is isotropic) then (9.3.18) simplifies to

$$\lambda = (2/\pi) \, L_{A,1}L_{A,2}. \tag{9.3.19}$$

A formula for the reduced second moment measure of $\Phi_1 \cap \Phi_2$ is given in HANISCH (1985).

9.3.3 Basic statistical methods for planar fibre processes

In general, estimation of basic characteristics of fibre processes follows the same path as the case of point processes. However, there is an important difference in the process of model fitting as there is no fibre process model holding a central position analogous to that of the Poisson point process. There seem to be only a few references on statistics for particular fibre process *models*. FELLOUS, GRANARA, and KRICKEBERG (1978) write on statistics for line processes. LASLETT (1982) studied the estimation of segment length distribution for segment processes.

In this section estimation procedures are discussed for the basic characteristics L_A, $\boldsymbol{R}$, and K. In every case the data arise as a sample of the fibre process Φ observed in a compact window W. It can be useful to approximate the observed image by systems of segments; see SCHEIDEGGER (1979), STOYAN and STOYAN (1984). In an analysis without the aid of an image analyzer this can be helpful, as it simplifies length and angle measurements. More generally it may serve as a useful smoothing procedure when the fibres are irregular.

Estimation of L_A

The total length $\Phi(W)$ of all fibre pieces in W can be measured easily if an image analyzer is employed, or if the fibres are all segments or straight lines. An unbiased estimator for L_A is given by

$$\hat{L}_A = \Phi(W)/\nu_2(W). \tag{9.3.20}$$

Otherwise, intersections with sampling lines or circles can be used and application made of formulae (9.3.7) and (9.3.15). If Φ is isotropic an unbiased estimator is given by

$$L_A{}^i = \frac{\pi \# \{T \cap \Phi\}}{2\nu_1(T)} \tag{9.3.21}$$

where T is a test system of segments (such as an array of parallel segments) in W. If Φ is not isotropic then one can use a test system C composed of n circles of fixed radius R. An unbiased estimator for L_A is

$$L_A{}^0 = \frac{\#\{C \cap \Phi\}}{4\pi R} \tag{9.3.22}$$

These estimators are demonstrated in Example 9.4.

Estimation of the rose of directions $\boldsymbol{R}$

In this section two methods are presented for the estimation of $\boldsymbol{R}$. One of these is based on (9.3.12) and the enumeration of intersection points; the other measures fibre directions at test points. A third method not described here involves approximation of the fibres by segments, and determination of the rose of directions from the end-point coordinates of the various segments.

The first method uses formula (9.3.12) to connect the rose of intersections $P_L(\cdot)$ with the density $f_{\boldsymbol{R}}$ of the rose of directions. The values of $P_L(\beta)$ for various β can be estimated by counting the mean numbers of intersection points made by Φ on lines forming the angle β with the x_1-axis. Then the distribution function for the rose of directions,

$$F_{\boldsymbol{R}}(\beta) = \int_0^\beta f_{\boldsymbol{R}}(\alpha)\,\mathrm{d}\alpha = \boldsymbol{R}\big((0, \beta]\big)$$

is estimated by

$$\hat{F}_{\boldsymbol{R}}(\beta) = \frac{1}{2L_A}\left(\frac{\mathrm{d}}{\mathrm{d}\beta}\hat{P}_L(\beta) + \int_0^\beta \hat{P}_L(\alpha)\,\mathrm{d}\alpha\right) \tag{9.3.23}$$

where $\mathrm{d}\hat{P}_L(\beta)/\mathrm{d}\beta$ is a numerical derivative of $\hat{P}_L(\cdot)$ at β. As has been noted this method is susceptible to measurement error and to estimation errors, and some kind of smoothing of $\hat{F}_{\boldsymbol{R}}$ is required. An alternative method first assumes a parametric form for $f_{\boldsymbol{R}}$ and then uses Fourier methods. Kanatani (1984) discussed this method and suggested that in most cases the first two terms of the Fourier series would be sufficient.

The second method considers $\Phi \cap T$ for T a subset of W which is a test system of circles or pieces of circles of radius R. If $w(x)$ is the direction of the fibre tangent at x then (Ohser, 1981)

$$\frac{1}{\#(\Phi \cap T)} \sum_{x \in \Phi \cap T} \mathbf{1}_{(\alpha_1, \alpha_2]}\big(w(x)\big)$$

is an unbiased estimator of $\boldsymbol{R}\big((\alpha_1, \alpha_2]\big)$. This is because the angles at the intersection points form a sample from a population of angles of distribution $\boldsymbol{R}$.

Estimation of the second reduced moment function K

The estimation of K is a rather more complicated matter. Various methods are sketched, the last of which is approximate. In the following Φ is assumed to be motion-invariant.

An unbiased edge-corrected estimator $\hat{k}_F(r)$ of $L_A{}^2K(r)$ is given by

$$\hat{k}_F(r) = \int_{W^{(r)}} k(x, r)\, \Phi\big(S(x, r, W^{(r)})\big)\, \Phi(\mathrm{d}x)/\nu_2(W^{(r)}) \tag{9.3.24}$$

where $S(x, r, A)$ is the union of sectors of the circle $b(x, r)$ lying completely in A (see Figure 9.8), and $W^{(r)}$ and $k(x, r)$ are as on p. 125. This follows the ideas

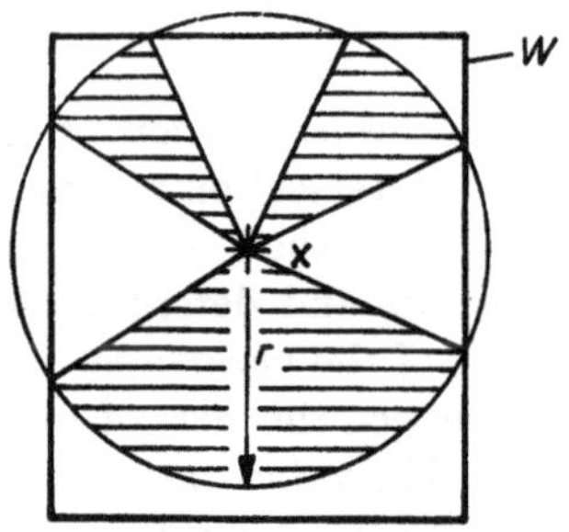

Fig. 9.8. The union $S(x, r, W)$ of the sectors is the hatched area.

used in estimating the point process K in Section 4.6. It may well be unrealistic in many cases, involving as it does an integration along fibres with respect to Φ.

A second method avoids this difficulty by considering $\Phi \cap T$ the intersection of Φ with a test system T. Since isotropy is assumed T can be a test system of parallel lines. Thus formula (9.6.3) provides the unbiased estimator $\bar{l}(r)$ of $L_AK(r)$

$$\bar{l}(r) = \frac{1}{\#\{\Phi \cap T\}} \sum_{x \in \Phi \cap T} k(x, r)\, \Phi\big(S(x, r, W)\big).$$

This is produced by weighting the fibre point x by the total length of all fibre pieces within the circle $b(x, r)$ and using formula (9.6.2) to show the estimator to be unbiased.

Hanisch (1985) suggests the use of point process methods. For example, consider a line process. On each line independently of the others a Poisson point process can be generated with intensity λ. The union of all the points so generated is a sample of a Cox process with driving measure $\lambda\Phi$. By (5.2.3) the second reduced moment function of the Cox process coincides with K for the line process and so estimation of one yields the other.

A further approach considers the intersection of the fibre process with a Poisson line process χ of intensity λ (χ being isotropic of course). This is a motion-invariant point process of intensity

$$\lambda_m = (2/\pi)\, \lambda L_A .$$

The 'factorial' moment measure of $\Phi \cap \chi$ is given by

$$\alpha_m^{(2)}(B_1 \times B_2) = \mathbf{E} \sum_{x_1, x_2 \in \Phi \cap \chi}^{+} \mathbf{1}_{B_1}(x_1)\, \mathbf{1}_{B_2}(x_2)$$

summing over pairs of points in $\Phi \cap \chi$ that arise from *different* lines of χ. Then $\alpha_m^{(2)}$ has a density $\varrho_m^{(2)}$ with respect to Lebesgue measure, and the pair-correlation function g_A of Φ satisfies

$$g_A(r) = \left(\frac{\pi}{2\lambda L_A}\right)^2 \varrho_m^{(2)}(r) \quad \text{for } r \geqq 0.$$

Since

$$L_A^2 K(r) = \int_{b(o,r)} \varrho^{(2)}(\|x\|)\,\mathrm{d}x$$

we have

$$\lambda_m^2 K(r) = \int_{b(o,r)} \varrho_m^{(2)}(\|x\|)\,\mathrm{d}x.$$

An edge-corrected unbiased estimator for $\lambda_m^2 K(r)$ is given by

$$\hat{\mathfrak{K}}_m(r) = \sum_{\substack{x,y \in \Phi \cap \chi \cap W \\ 0 < \|x-y\| \leq r}} \frac{k(x, y)}{\nu_2(W^{\|x-y\|})}$$

in analogy to (4.6.3). The summation here is again over pairs of points arising from different lines of χ. A variation of this method uses the intersection of Φ with a Boolean segment process.

The approximate method uses the analysis of the intersection of Φ with the x_1-axis (or with a system of straight lines). It is based on (9.3.13) and uses the simplifying assumptions that the fibres contain no straight line pieces and that the angles $\alpha(x_1, 0)$ in the weights $s_r(y)$ are independent (this second assumption cannot be true over short distances). Given these assumptions the mean on the right-hand side of (9.3.13) is approximately equal to

$$\pi^2 \int_0^r x \varrho_L^{(2)}(x)\,\mathrm{d}x / P_L$$

where $\varrho_L^{(2)}$ is the second-order product density of the (non-marked) point process of intersections of Φ with the x_1-axis. The product density $\varrho_L^{(2)}$ can be estimated by methods such as those proposed by Cox and Lewis (1966). If $\hat{\varrho}_L^{(2)}$ is such an estimator then $L_A K(r)$ can be estimated by

$$\hat{k}_a(r) = \pi^2 \int_0^r x \hat{\varrho}_L^{(2)}(x)\,\mathrm{d}x / \hat{P}_L$$

where $\hat{P}_L$ estimates P_L. The function g_A is approximately equal to the pair-correlation function g_L given by

$$g_L(r) = \varrho_L^{(2)}(r)/P_L^2.$$

Other approximate methods have been described and discussed in Hanisch, Klimanek, and Stoyan (1985), Hanisch, König, and Stoyan (1985), and Stoyan (1985c).

9.4 Spatial fibre processes

The definitions of spatial line and fibre processes follow the pattern of their planar counterparts, with occasional extra complexities; see MECKE and NAGEL (1980) and NAGEL (1983). Here only some brief remarks on stationary fibre processes are made.

As with planar fibre processes so in this case the *intensity* J_V of a stationary spatial fibre process Φ is the mean fibre length per unit volume, and

$$J_V \nu_3(B) = \mathsf{E}\Phi(B)$$

for B a spatial Borel set. The *rose of direction* $\boldsymbol{R}$ must now be defined as a measure on the space $\boldsymbol{L}$ of all straight lines through the origin. The corresponding σ-algebra $\mathcal{L}$ is precisely the system

$$\{A \subset \boldsymbol{L} : \bigcup \{l : l \in A\} \text{ is a spatial Borel set}\}.$$

The quantity $\boldsymbol{R}(A)$ is defined by

$$J_V \boldsymbol{R}(A) = \int_B \boldsymbol{1}_{\{l(x) \in A\}}(x)\, \Phi(\mathrm{d}x)$$

where $l(x)$ is the fibre tangent line through x.

If Φ is isotropic then $\boldsymbol{R}$ is the rotation-invariant, or uniform, probability distribution U_1 on $\boldsymbol{L}$. To describe this distribution more explicitly, and in order to aid descriptions in the remainder of this section, we introduce a coordinate system on $\boldsymbol{L}$. A line l in $\boldsymbol{L}$ makes an angle $\beta(l)$ with the x_3-axis, the *azimuthal angle*; and the projection of $l \cap \{x_3 > 0\}$ onto the $(x_3 = 0)$ plane makes an angle $\lambda(l)$ with the x_1-axis, the *geographical longitude*. These angles lie in the region

$$0 \leqq \beta(l) \leqq \pi/2$$

$$-\pi \leqq \lambda(l) \leqq \pi$$

(although if $\beta(l) = \pi/2$ then $(\beta(l), \lambda(l))$ and $(\beta(l), \pi - \lambda(l))$ refer to the same line). With these angles

$$U_1(\mathrm{d}(\beta, \lambda)) = \sin \beta \, \mathrm{d}\beta \, \mathrm{d}\lambda/(2\pi). \tag{9.4.1}$$

See Figure 9.9.

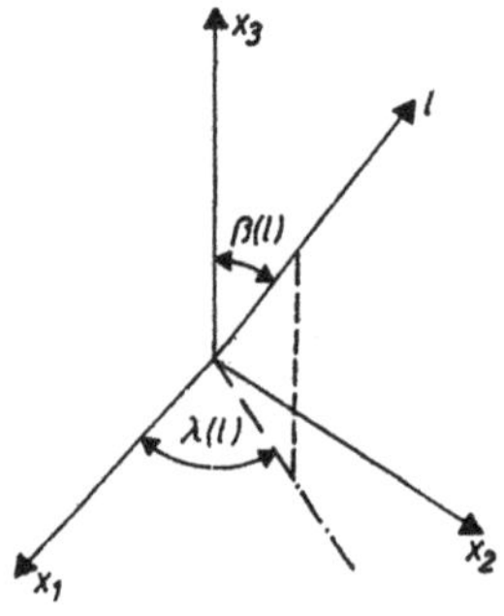

Fig. 9.9. A diagram to illustrate the system of spherical polar co-ordinates

The intersection of a spatial fibre process with a plane e (for simplicity we consider this plane e to be the plane $(x_3 = 0)$) generates a point process on the plane e. A point x of this point process can be marked with the line $l(x)$ in $\boldsymbol{L}$ which is the fibre tangent line through x. Thus the intersection generates a stationary marked point process

$$\Psi = \{[x; l(x)]\}.$$

If P_A is the intensity of Ψ, and Θ its mark distribution then by analogy with (9.3.4) these can be connected to the characteristics of the fibre process in a formula of Mecke and Nagel (1980)

$$P_A \iint h(x_1, x_2, l)\, \Theta(\mathrm{d}l)\, \mathrm{d}x_1\, \mathrm{d}x_2 = J_V \iint h(x_1, x_2, l) \cos \beta(l)\, \boldsymbol{R}(\mathrm{d}l)\, \mathrm{d}x_1\, \mathrm{d}x_2; \tag{9.4.2}$$

here h can be any non-negative measurable function on $\mathbb{R}^2 \times \boldsymbol{L}$.

If $\boldsymbol{R}$ is the uniform distribution U_1 on $\boldsymbol{L}$ then

$$\int \cos \beta(l)\, \boldsymbol{R}(\mathrm{d}l) = \frac{1}{2}$$

and so

$$J_V = 2P_A. \tag{9.4.3}$$

In this case the mark distribution H, the distribution of the fibre tangent at a 'typical' intersection point, has the form

$$H(\mathrm{d}l) = 2 \cos \beta(l)\, \boldsymbol{R}(\mathrm{d}l) = \sin 2\beta \, \frac{\mathrm{d}\beta\, \mathrm{d}\lambda}{2\pi}. \tag{9.4.4}$$

As in the planar case a *rose of intersections* $P_A(\cdot)$ can be defined. $P_A(l)$ is the intensity of the point process $\Phi \cap e_l$, where e_l is the plane perpendicular to l. Mecke and Nagel (1980) gave an integral formula connecting $\boldsymbol{R}$ and $P_A(\cdot)$, involving Legendre functions.

Further results on spatial fibre processes in the context of stereology can be found in Section 11.2.2, 11.6, and Nagel (1983), Zähle (1984b).

Laslett, Kendall, Gleadow, and Duddy (1982) used a spatial segment process to model the occurrence of fission particle tracks within apatite crystals. The objective was to measure the length distribution of the tracks, and in particular to gain information about whether the distribution was unimodal or not. Since most particle tracks are revealed by etching treatment only when exposed by a sheared plane of the crystal this is a complicated stereological problem. Numerical problems of the kind referred to in the section on planar fibre processes were very much in evidence. It was found to be more effective to sample only the infrequent complete tracks which were revealed because of an intersection with another exposed track.

Hanisch, Klimanek, and Stoyan (1985) used spatial fibre processes as models for systems of dislocation lines, see also Section 11.6.3 and Stoyan and Stoyan (1986).

9.5 Spatial surface processes

9.5.1 Plane processes in $\mathbb{R}^3$

Plane processes are used as stochastic models for systems of planes in space and can be regarded as special cases of spatial surface processes. Just as Section 9.2 used line processes to introduce fibre processes, so here the theory of plane processes is briefly reviewed as a prelude to the general theory of surface processes.

As in 9.2 the theory of random sets can be used to provide a measure-theoretic framework and to introduce the notions of *stationary* and *motion-invariant* plane processes. These notions, and the measure theory, tie up in a satisfactory manner with the representation space approach described in Chapter 8.

A plane process $\Phi = \{e_1, e_2, \ldots\}$ can also be described as a random measure

$$\Phi(B) = \sum_{e \in \Phi} \nu_2(e \cap B)$$

for B a spatial Borel set. If Φ is stationary then one can define the *intensity* S_V of Φ to be the mean plane area per unit volume. Moreover, a weighted random measure Ψ can be defined on $\mathbb{R}^3 \times \boldsymbol{L}$, where $\boldsymbol{L}$ is defined as in Section 9.4 to be the projective plane of lines through the origin; Ψ is defined by

$$\Psi(B \times A) = \sum \nu_2(B \cap e)$$

where the sum is over e in Φ with normal lying in A and where B is a spatial Borel set and A is a Borel subset of $\boldsymbol{L}$. If Φ is stationary then invariance arguments provide a factorization of the intensity measure Λ of Ψ as

$$\Lambda = S_V \nu_3 \times \boldsymbol{R} \tag{9.5.1}$$

where $\boldsymbol{R}$ is a distribution on $[\boldsymbol{L}, \mathcal{L}]$, the *rose of (normal) directions* of Φ. If Φ is isotropic then $\boldsymbol{R}$ is the uniform distribution U_1 on $[\boldsymbol{L}, \mathcal{L}]$. The rose of normal directions $\boldsymbol{R}$ can be interpreted as the distribution of the normal line to a 'typical' plane in Φ.

If Φ is isotropic then the mean number $\bar{n}_C$ of planes of Φ intersecting a compact convex set C is proportional to the mean average breadth $\bar{b}(C)$ of C

$$\bar{n}_C = S_V \bar{b}(C). \tag{9.5.2}$$

As in the case of planar fibres, it is useful to consider the *Steiner compact* S corresponding to the measure $S_V \boldsymbol{R}$ on $\boldsymbol{L}$. Note that in contrast to the Steiner compact for the fibre process, no rotation is required here. The Steiner compact is the symmetric compact convex set with modified support function r_S given by

$$r_S(l) = \frac{1}{2} S_V \int_{\boldsymbol{L}} |\langle l, l' \rangle| \, \boldsymbol{R}(\mathrm{d}l') \tag{9.5.3}$$

where $\langle l, l' \rangle = \cos \lambda$, for λ the angle between the lines l and l'.

If Φ is isotropic then its Steiner compact S is given by

$$S = b(o, S_V/4). \tag{9.5.4}$$

The Steiner compact is a useful concept in the study of intersections of Φ with planes and lines. Consider for example the intersection of Φ with a fixed line l in $\mathbf{L}$. Let $P_L(l)$ be the intensity of the point process $\Phi \cap l$, so that $P_L(\cdot)$ is the *rose of intersections* of Φ. Then

$$P_L(l) = 2r_S(l) \quad \text{for } l \in \mathbf{L} \tag{9.5.5}$$

where S is the Steiner compact of Φ.

If e is a fixed plane then $\Phi \cap e$ is a line process with Steiner compact S_e. It can be shown that

$$S_e = \mathfrak{p}_e S \tag{9.5.6}$$

where $\mathfrak{p}_e$ is the orthogonal projection onto e.

In the case of isotropy the formulae (9.5.5) and (9.5.6) give

$$P_L = S_V/2 \tag{9.5.7}$$

if P_L is the intensity of $\Phi \cap l$, and

$$L_A = (\pi/4)\, S_V \tag{9.5.8}$$

if L_A is the intensity of $\Phi \cap e$. Note that the left-hand sides do not depend on the choices of l and e.

9.5.2 General surface processes

Basic characteristics

The definition of the class of surface processes follows closely that for fibre processes; see Pohlmann, Mecke, and Stoyan (1981). A *surface system* is a set φ contained in $\mathbb{R}^3$ which can be represented as the boundary (that is, the surface) of a set A contained in $\mathbb{R}^3$, which in turn has the property

$$A \cap K \in \mathcal{R} \quad \text{for every } K \text{ in } \mathbf{K}, \text{ where } \mathcal{R} \text{ is the convex ring.}$$

Since A is locally a finite union of convex compact sets its surface φ has a well-defined area measure using ν_2. Thus the following defines a Radon measure in $\mathbb{R}^3$

$$\varphi(B) = \nu_2(\varphi \cap B)$$

for B a spatial Borel set.

The class of all surface systems is denoted by $\mathbf{S}$, and $\mathcal{S}$ is the σ-algebra of subsets of $\mathbf{S}$ generated by

$$\{\varphi \in \mathbf{S} : \varphi(B) < x\} \text{ for real } x \text{ and } B \text{ a spatial Borel set.}$$

A *surface process* Φ is a random variable defined on a probability space $[\Omega, \mathcal{A}, \mathsf{P}]$ and taking values in $[\mathbf{S}, \mathcal{S}]$. The probability measure generated by Φ on $[\mathbf{S}, \mathcal{S}]$

is called the *distribution* of Φ. The surface process Φ corresponds to a random measure, its random surface measure (see 7.3).

Stationarity and isotropy are defined as for random closed sets, or for random measures.

Stationary surface processes have characteristics of *intensity* S_V and *rose of normal directions* $\boldsymbol{R}$. These are defined in a manner entirely analogous to the definitions of the corresponding quantities for fibre processes, generalizing the definitions for plane processes given in Section 9.5.1. In particular $\boldsymbol{R}$ gives the distribution of the normal to a 'typical' surface point. Stationarity ensures that this normal is unique with probability one. These characteristics satisfy a *Campbell theorem* (which can be used to recover their definitions)

$$\mathsf{E} \int f(x, l(x))\, \Phi(\mathrm{d}x) = S_V \iint f(x, l)\, \boldsymbol{R}(\mathrm{d}l)\, \mathrm{d}x \tag{9.5.9}$$

for every non-negative measurable function f on $\mathbb{R}^3 \times \boldsymbol{L}$. Here $l(x)$ is the surface normal at x.

If Φ is isotropic then $\boldsymbol{R}$ is the uniform distribution U_1 on $\boldsymbol{L}$. Following Sections 9.3 and 9.5.1 the rose of normal directions $\boldsymbol{R}$ can be characterized by the associated Steiner compact.

Intersections by lines and planes

As in the case of fibre processes it is of interest to consider intersections of surface processes with lines and planes. Intersection with a line generates a point process on the line; intersection with a plane produces a fibre process on the plane. It suffices to consider plane intersections since linear intersections can be regarded as produced by first a planar intersection, and then a linear intersection of the ensuing planar fibre process. Linear intersections of planar fibre processes are dealt with in Section 9.3.2; see also formula (11.2.3).

Suppose that Φ is a stationary surface process and $\Psi = \Phi \cap e$ is the planar fibre process resulting from the intersection of Φ with the plane e. Here we take e to be the plane $(x_3 = 0)$. Furthermore, let Θ be the rose of tangent directions for Ψ. Then Θ is connected with $\boldsymbol{R}$ the rose of normal directions of Φ by

$$\begin{aligned} &L_A \iint h(x_1, x_2, \lambda)\, \Theta(\mathrm{d}\lambda)\, \mathrm{d}x_1\, \mathrm{d}x_2 \\ &= S_V \iint h(x_1, x_2, \lambda(l)) \sin \beta(l) \cdot \boldsymbol{R}(\mathrm{d}l)\, \mathrm{d}x_1\, \mathrm{d}x_2 \end{aligned} \tag{9.5.10}$$

for every non-negative measurable function h defined on $\mathbb{R}^2 \times (0, \pi]$. Here $\beta(l)$ is the azimuthal angle of l in $\boldsymbol{L}$, and $\lambda(l)$ is the angle of a line in e orthogonal to the projection of l onto e (so $0 < \lambda(l) \leqq \pi$).

If $h(x_1, x_2, \lambda) = \boldsymbol{1}_{[0,1]}(x_1)\, \boldsymbol{1}_{[0,1]}(x_2)$ then formula (9.5.10) gives

$$L_A = S_V \int \sin \beta(l)\, \boldsymbol{R}(\mathrm{d}l). \tag{9.5.11}$$

If $\boldsymbol{R}$ is the uniform distribution (for example, if Φ is isotropic) then by means of (9.4.1)

$$L_A = S_V \cdot \frac{1}{2\pi} \iint \sin^2 \beta(l)\, \mathrm{d}\beta(l)\, \mathrm{d}\lambda(l) = \frac{\pi S_V}{4}. \tag{9.5.12}$$

The formulae (9.5.5), (9.5.6) for Steiner compacta remain true for surface processes.

The definition of the *second-order moment measure* of a surface process follows that of a fibre process. If the surface process is stationary then its second order behaviour is summarized in a reduced second moment measure, and in the isotropic case a reduced second moment function K suffices. The interpretation of $S_V K(r)$ is that it is the mean area of all surface pieces within a ball of radius r centred on a 'typical' surface point.

Example 9.3. *Basic characterizations of two simple surface process models.*

(a) Let Φ be an isotropic Poisson plane process with intensity S_V. By isotropy the rose of normal directions is the uniform distribution on L. The intensity S_V is also the intensity of Φ interpreted as a surface process. By arguments as in Example 9.2 it can be shown that the reduced second moment function K satisfies

$$S_V K(r) = \pi r^2 + S_V(4/3)\,\pi r^3 \quad \text{for } r \geqq 0.$$

(b) Let Φ be a Boolean disc model as in Example 9.1, with constant disc radii R. Then

$$S_V = \lambda\pi R^2$$

where λ is the intensity of the germ process of disc centres. The reduced second moment function K has a more complicated form

$$S_V K(r) = (4/3)\,\pi r^3 S_V + F(r)$$

where

$$F(r) = \begin{cases} \pi R^2 & \text{if } r > 2R \\ 2/R^2 \int\limits_0^R x f(x)\,dx & \text{if } r \leqq 2R \end{cases}$$

and $f(x) = \nu_2(b(o, r) \cap b(\boldsymbol{x}, R))$ for $\|\boldsymbol{x}\| = x$, working on the plane.

The second term comes from the disc holding the 'typical' point while the first arises from contribution by other discs; this is reminiscent of Example 9.2.

9.6 Weighted fibre and surface processes

Just as for point processes so for fibre and surface processes it is useful to attach *weights* to the fibres and surfaces constituting the process. The weight might be a real number (for example, the thickness of the fibre at the point), or it might be more general; at any rate it gives further information about the structure being studied. The space of possible weights will be denoted here by W^*, and it must be endowed with a σ-algebra $\mathcal{W}^*$.

Figure 9.10 gives an example of a weighted fibre process. The fibre system is the system of cracks in a clay soil in plane section. The fibres are weighted by the width of the cracks (of course this is really a planar section of a weighted surface process!)

As in aspects of point process theory it can be useful to define weights for an unweighted fibre or surface process so that the weights describe distributional aspects of the process; see STOYAN (1984d). For example, the reduced second moment function $K(r)$ of a planar fibre process can be interpreted as follows:

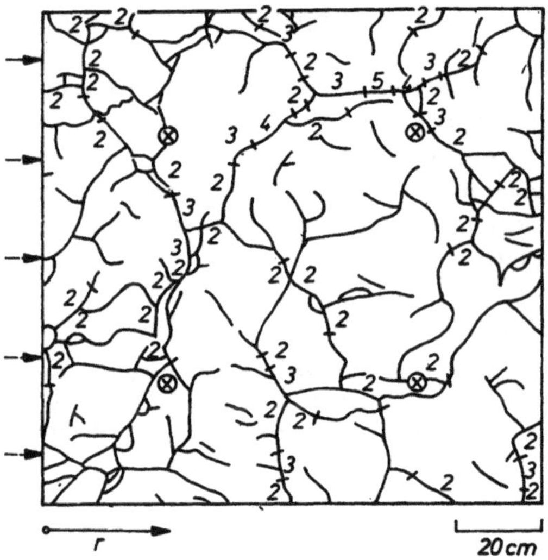

Fig. 9.10. Crack pattern in a clay soil, taken from a horizontal section at depth of 30 cm (DOLEŽAL, 1982). The numbers indicate the widths of the cracks in the section plane; unlabelled cracks all have width 1 mm.

$L_A K(r)$ is the mean of a weight distribution when the weight at a point x is the total length $S_r(x)$ of all fibre pieces within the circle $b(x, r)$.

MECKE and STOYAN (1980a), POHLMANN, MECKE, and STOYAN (1981), and STOYAN (1984d) give important formulae for weighted fibre and surface processes. In this section, a particular case is given brief consideration: the case of stationary planar weighted fibre processes. (We note in passing that stationarity and isotropy are defined just as in the case of marked point processes.)

The basic characteristics of a stationary planar weighted fibre process are its *intensity* L_A (simply the intensity of the fibre process when the weights are disregarded), and the *joint distribution* $\boldsymbol{J}$ *of tangent direction and weight*. This distribution $\boldsymbol{J}$ can be interpreted as the joint distribution of tangent direction and weight for a 'typical' fibre point. For A in Π and C in $\mathcal{W}^*$ the quantity $L_A \cdot \boldsymbol{J}(A \times C)$ is the mean length per unit area of all fibre fragments for which the constituent points have tangent direction in A and weight in C (its strict definition follows closely the definition of the rose of direction, which itself can be thought of as a weight distribution). Of course

$$\boldsymbol{J}(A \times \mathsf{W}^*) = \boldsymbol{R}(A) \quad \text{for } A \text{ in } \Pi .$$

The unconditional weight distribution $\boldsymbol{W}$ is given by

$$\boldsymbol{W}(C) = \boldsymbol{J}\big((0, \pi] \times C\big) \quad \text{for } C \text{ in } \mathcal{W}^* .$$

If the weighted fibre process is isotropic then $\boldsymbol{R}$ is uniform and $\boldsymbol{J}$ factorizes

$$\boldsymbol{J} = U_1 \times \boldsymbol{W}. \qquad (9.6.1)$$

A Campbell theorem holds.

Intersection of the stationary weighted fibre process with the x_1-axis yields a marked point process Ψ. The intersection points are marked by pairs of tangent directions and weights. Let P_L be the intensity of Ψ and $\boldsymbol{I}$ the joint distribution of a mark pair. MECKE and STOYAN (1980a) showed that for all non-negative measurable functions h on $\mathbb{R} \times (0, \pi] \times \mathsf{W}^*$

$$P_L \iint h(x, \lambda, w)\, \boldsymbol{I}(\mathrm{d}(\lambda, w))\, \mathrm{d}x = L_A \iint h(x, \lambda, w) \sin \lambda \boldsymbol{J}(\mathrm{d}(\lambda, w))\, \mathrm{d}x. \qquad (9.6.2)$$

In the case of isotropy

$$\boldsymbol{I}(\mathrm{d}(\lambda, w)) = \frac{1}{2} \sin \lambda\, \mathrm{d}\lambda\, \boldsymbol{W}(\mathrm{d}w). \qquad (9.6.3)$$

Thus in the isotropic case the mark distribution $\boldsymbol{I}$ also factorizes. One factor is the weight distribution $\boldsymbol{W}$ and the other the distribution given by (9.3.8).

Example 9.4. *Cracks in clay soil* (DOLEŽAL, 1982).

Figure 9.10 shows a pattern of cracks in soil in planar section. DOLEŽAL (1982) used stereological methods to draw inferences about the spatial crack structure. Here we consider the fibre system as a weighted fibre process with weights being the crack widths as shown in Figure 9.10.

Length measurements yield an estimate for L_A of $\hat{L}_A = 0.134\ \mathrm{cm}^{-1}$, using (9.3.20). Line and circle section methods were also used to estimate L_A. The arrows in Figure 9.10 define five line segments parallel to the bottom line of the window, and if these are used as test set and isotropy is assumed then formula (9.3.21) yields $L_A{}^l = 0.120\ \mathrm{cm}^{-1}$. The circle section method used four circles of radius $r = 30$ cm and centres at points marked in Figure 9.10 by $\otimes$. Formula (9.3.22) yields $L_A{}^\circ = 0.154\ \mathrm{cm}^{-1}$. This example may be instructive in illustrating actual errors achieved with these two latter methods.

Using both line and circle methods, weights were measured for the intersection points. By (9.6.3) the two families of weights so obtained can be considered as two samples drawn from a population with distribution $\boldsymbol{W}$, the weight distribution. The corresponding empirical distributions are tabulated in Table 9.1, together with the distributions of weights obtained by exhaustive length measurement.

Table 9.1. Empirical weights distributions for the crack-width weight distribution

Width (mm)	Exhaustive measurement	Line section method	Circle section method
1	0.786	0.783	0.770
2	0.163	0.174	0.203
3	0.039	0.043	0.027
4	0.009	0	0
5	0.004	0	0

9.7 Manifold processes in $\mathbb{R}^d$

Just as line and plane processes in $\mathbb{R}^2$ and $\mathbb{R}^3$ can be generalized to plane and flat processes in $\mathbb{R}^d$, so fibre and surface processes can be generalized to the *manifold processes* of MECKE (1981b). These are random systems of (fragments of) *manifolds* of fixed dimension embedded in some higher dimensional Euclidean space $\mathbb{R}^d$. Here a manifold is a generalization of a fibre or of a surface: it can be thought of as the solution set of some system of equations $f_1(x) = 0$, $\ldots, f_r(x) = 0$. Mecke studies the intersections of manifold processes with other manifold processes and also with fixed manifolds.

HILLIARD (1962) discussed intersections of fibre and surface processes in $\mathbb{R}^3$. In particular he derived a formula for the mean number P_V per unit volume of intersections of a surface process of intensity S_V with an independent fibre process of intensity J_V, in the case of motion invariance. The formula is

$$P_V = J_V S_V / 2 . \tag{9.7.1}$$

10. Random tessellations

10.1 Introduction and definitions

A *tessellation* or *mosaic* is a division of the plane into polygons, or of space into polyhedra. Figures 10.1, 10.2, and 10.7 show patterns that can be interpreted (to a greater or lesser degree) as the result of *tessellating* a plane. Various models are described in Section 10.2. Such geometrical patterns occur in many natural situations: for example, in crystalline structures, crack patterns, and foam structures.

We confine ourselves to the subject of random tessellations by *convex* polygons or polyhedra, although some of the results given are also true in the non-convex case. Precise definitions are given for the case of planar tessellations; the generalization to the spatial case (and the d-dimensional case) is obvious.

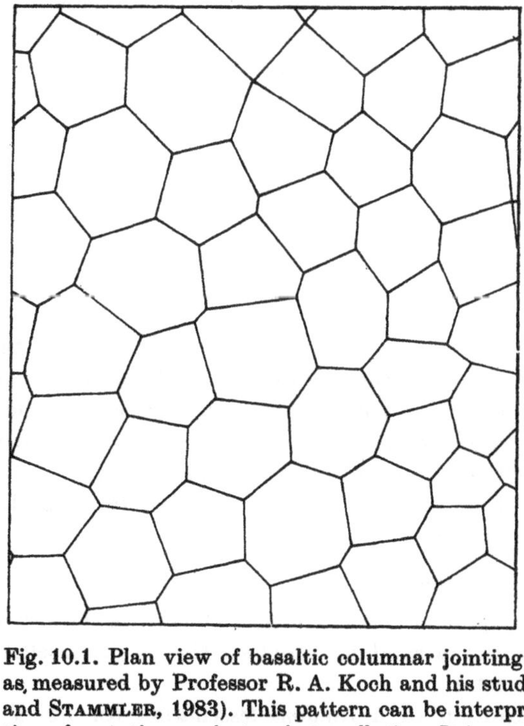

Fig. 10.1. Plan view of basaltic columnar jointing at Burg Stolpen (G.D.R.) as measured by Professor R. A. Koch and his students (see Koch, Pfeiffer, and Stammler, 1983). This pattern can be interpreted as a part of a realization of a stationary isotropic tessellation. It is not a Voronoi tessellation (see Stoyan and Stoyan, 1980).*

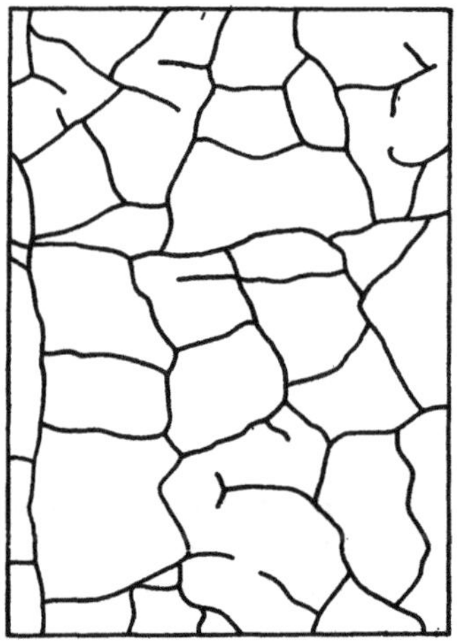

Fig. 10.2. Mudcracks in a ditch. The pattern shows a clear anisotropy caused by the ditch geometry, and can be interpreted as an intermediate stage in the formation of a tessellation with non-convex cells

Planar random tessellations

We follow MECKE (1983b). Let $\boldsymbol{P}$ be the set of all bounded open convex non-empty polygons $\boldsymbol{p}$ in the plane $\mathbb{R}^2$. A subset $\theta \subset \boldsymbol{P}$ of polygons is said to be a *tessellation* if the constituent polygons are pairwise disjoint, the union of their closures fills the plane, and the family is locally finite. These three conditions can be expressed formally by

(a) $\boldsymbol{p}_1 \cap \boldsymbol{p}_2$ is empty if $\boldsymbol{p}_1, \boldsymbol{p}_2 \in \theta$, and $\boldsymbol{p}_1 \neq \boldsymbol{p}_2$,

(b) $\bigcup_{\boldsymbol{p} \in \theta} \boldsymbol{p}^{\text{cl}} = \mathbb{R}^2$,

(c) if B is a bounded planar set then the set

$$\{\boldsymbol{p} \in \theta : \boldsymbol{p} \cap B \text{ is not empty}\}$$

is finite.

The polygons $\boldsymbol{p} \in \theta$ are the *cells* of the tessellation θ. The *nodes* of the tessellation are the vertices of the constituent polygons. If E_θ is the union of the boundaries of polygons in θ then a segment lying in E_θ is an *edge* of the tessellation if its endpoints are nodes of θ and no further nodes of θ lie in the segment. Note that E_θ is a fibre system and θ can be recovered from E_θ.

Let $\mathbb{T}$ denote the class of all tessellations and let $\mathcal{T}$ denote the σ-algebra on $\mathbb{T}$ generated by sets of the form

$$\{\theta \in \mathbb{T} : E_\theta \cap K \text{ is not empty}\}$$

where K runs through all compact subsets of $\mathbb{R}^2$. Thus $\mathcal{T}$ is the restriction to $\mathbb{T}$ of the hit-or-miss σ-algebra of random closed set theory (if tessellations are identified with their edge-sets).

The mathematical definition of a *random tessellation* (or *tessellation* for brevity) is of a random variable Θ with values in $[\mathbb{T}, \mathcal{T}]$. Its *distribution* is the induced probability measure on $[\mathbb{T}, \mathcal{T}]$. The definitions of stationarity and isotropy for random tessellation follow the usual form. A tessellation Θ and its distribution P are said to be *stationary* if for all x in $\mathbb{R}^2$ the translated tessellation

$$\Theta + x = \{p + x : p \in \Theta\}$$

has the same distribution as the original tessellation Θ,

$$P(Y_x) = P(Y) \quad \text{if } Y_x = \{\theta + x : \theta \in Y\} \text{ and } Y \text{ belongs to } \mathcal{T}.$$

If there is invariance with respect to rotations about the origin then Θ and P are called *isotropic*; isotropic stationary tessellations are also called *motion-invariant*.

10.2 Mathematical models for random tessellations

In this section a representative selection of models is discussed and a partial summary of the literature is given. The purpose is to acquaint the reader with various types and methods of construction; however the summary is not complete.

Line and plane tessellations

If Ψ is a line process in $\mathbb{R}^2$ then taking

$$E_\theta = \cup \{l : l \in \Psi\}$$

yields the edge set of a tessellation Θ. Figure 8.2, showing part of a line process, can therefore also be interpreted as part of a *line tessellation*. In an entirely analogous fashion hyperplane processes provide tessellations of $\mathbb{R}^d$.

The distributional properties of Θ depend on those of Ψ. In particular if Ψ is stationary then so is Θ. In general there is little known about general line tessellations (save simple assertions, such as that the mean number of sides of the 'typical' polygon is four if no intersections involve more than two lines). Much work has been done for the special case of *Poisson* line and plane tessellations (built on Poisson line and plane processes). Some important formulae are to be found in Section 10.5.

Voronoi tessellations and related models

DIRICHLET (1850) and VORONOI (1908) considered regular tessellations of planes and higher-dimensional spaces. Their motivations were provided by problems in number theory. The Dirichlet and Voronoi tessellations as applied to random point patterns appear to have arisen independently in various fields of science and technology; see below.

Let φ be a locally finite system of points in $\mathbb{R}^d$. Almost all x in $\mathbb{R}^d$ have a unique *nearest point* $n(x)$ belonging to φ. The *neighbourhood* or *domain* $T(y)$ of a point y of φ is defined by

$$T(y) = \{x \in \mathbb{R}^d : n(x) = y\}.$$

The $T(y)$ are all open convex polytopes but some can be unbounded. If this possible deficiency does not arise then the $T(y)$ constitute a tessellation of $\mathbb{R}^d$, the *Voronoi tessellation* $\mathcal{V}(\varphi)$ relative to φ. If $d = 2$ then some authors refer to it as the Dirichlet or Thiessen tessellation.

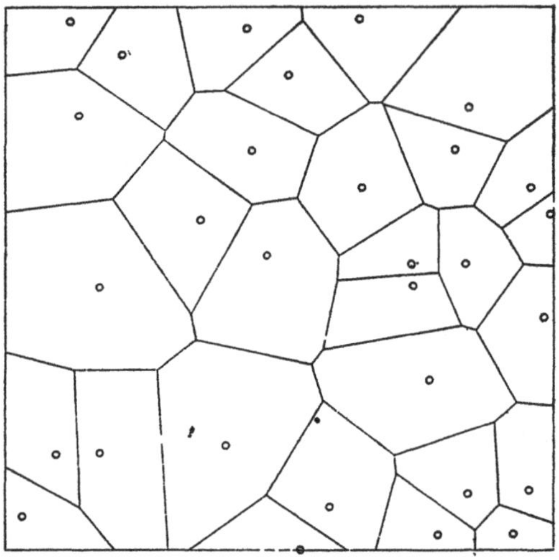

Fig. 10.3. A typical realization of the Dirichlet tessellation $\mathcal{V}$ relative to a stationary Poisson point process. This tessellation was drawn using the TILE program of GREEN and SIBSON (1978)

If Φ is a stationary point process of finite intensity ϱ then almost surely all the $T(y)$ are bounded so $\mathcal{V}(\Phi)$ is indeed a random tessellation. Figure 10.3 shows a typical realization of $\mathcal{V}(\Phi)$ relative to a stationary Poisson process Φ. If Φ is stationary then $\mathcal{V}(\Phi)$ inherits this property. Known formulae for $\mathcal{V}(\Phi)$ are almost entirely confined to the case when Φ is a stationary Poisson process and are described in Section 10.6.

The Voronoi tessellation has been often used as a model for natural phenomena in a variety of fields:

Agriculture and forestry	FISCHER and MILES (1973), KLIER (1969)
Astrophysics	KIANG (1966)
Cell biology	LEWIS (1946), HONDA (1983)
Communication theory	SHANNON (1949)
Crystal growth and Aggregates	MEIJERING (1953), GILBERT (1962)
Geography	MARDIA, EDWARDS, and PURI (1977)

Geology	Smalley (1966), Crain (1976), Gray, Anderson, Devine, and Kwasnik (1976), Stiny (1929), Stoyan and Stoyan (1980b)
Metallography	Boots (1984)
Physics	Ogawa and Tanemura (1974), Rahman (1966), Finney (1979), Boots (1982)
Zoology and ecology	Hamilton (1971), Hasegawa and Tanemura (1976, 1980), Hasegawa, Tanemura, and Takiguchi (1981)

Voronoi tessellations are also used in data analysis of geometrical structures. Finney (1979) used the term 'polyhedral statistics'. Sibson (1980, 1981) described the use of Voronoi tessellations relative to given point patterns as a basis for 'natural neighbour interpolation': interpolating a smooth function for data located at irregularly distributed points. Thiessen (1911) had this in mind when he suggested this tessellation.

The regular square tessellation is a particular example of a Voronoi tessellation. It arises from a regular square grid of points. This tessellation is degenerate; more commonly point patterns give rise to tessellations with nodes at which only three cells meet. If $\mathcal{V}(\Phi)$ has the property that almost surely every node is touched by exactly three cells (in the planar case) or by exactly four cells (in the spatial case) then a further tessellation can be produced, the *Delaunay* tessellation $\mathcal{D}(\Phi)$. In the planar case it is constructed out of the triangles formed by points of Φ whose cells share the same node. Rogers (1964) and Miles (1972a) gave some formulae for $\mathcal{D}(\Phi)$. The construction of Voronoi and Delaunay tessellations has been discussed for example in Bowyer (1981), Finney (1979), Green and Sibson (1978), Lee and Schachter (1980), Ohya, Iri, and Murota (1984), and Suzuki and Iri (1986).

The Voronoi tessellation can be generalized. Miles (1970) defined the *generalized Voronoi tessellation* $\mathcal{V}_n$, which uses for cells the sets of positions sharing the same n nearest neighbours of the generating point pattern (of course $\mathcal{V}_1 = \mathcal{V}$, the original Voronoi tessellation). Details are given in Miles (1970, 1972a) and Miles and Maillardet (1982). The latter paper gave pictures of $\mathcal{V}_n$ for $n = 4$, 16, 64, and 256 relative to a stationary Poisson process.

Another generalization is the *Johnson-Mehl* model which contains a dynamic element; see Meijering (1953), Gilbert (1962), and Miles (1972a). Here 'nuclei' are generated by some birth process, and each nucleus grows so that at time t after its birth it occupies all the previously vacant region within the sphere of radius vt centred on its original point. Consequently, if a nucleus is born at a point in the current cell of another nucleus then it vanishes immediately; those that survive form cells that grow radially in each direction with velocity v, till cells of other nuclei are met. In the standard Johnson-Mehl model, the points in $\mathbb{R}^3 \times [0, \infty)$ giving nucleus birth locations and times form a stationary Poisson process of constant intensity. If all nuclei are born at the same instant then the resulting tessellation is Voronoi; but in the general model the cells of the tessellations are not even convex. So the Johnson-Mehl model in its standard form falls outside the domain of discussion of this chapter. Figure 10.4 shows a part of a realization of a Johnson-Mehl model. Formulae for the model are given in the references above. Its relevance for crystal growth is discussed in MEIJERING (1953), GILBERT (1962), and SALTYKOV (1974).

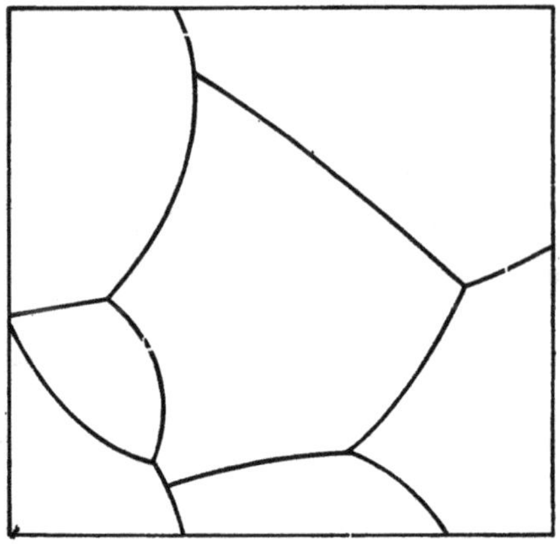

Fig. 10.4. A realization of a Johnson-Mehl model

Crack tessellations

Other models are based on the growth of edges rather than the growth of cells. GRAY, ANDERSON, DEVINE, and KWASNIK (1976) suggested some interesting models of this type. One of them is as follows. A marked point process Φ is constructed on the plane, each point of Φ being marked with the orientation of a line (for example, the mark distribution might give independent and uniform directions). The edges are produced by a growth process; each edge starts at one of the points and grows in the two opposing directions specified by the mark at constant rate. Growth in a particular direction continues until further edges are hit. This process is reminiscent of crack growth — hence the name. Figure 10.5 shows a part of a simulation of such a tessellation. Higher dimen-

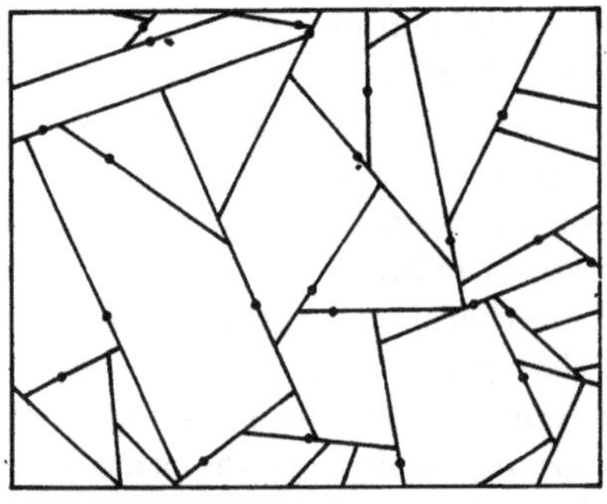

Fig. 10.5. A realization of a crack tessellation relative to a stationary Poisson point process. The directions of growth are uniformly and independently distributed.

sional analogues replace the line-orientation marks by normal directions of hyperplanes.

For this and other models workers have used simulation to determine the distributions of numbers of sides for the 'typical' polygon, and of the number of edges truncated by the 'typical' edge.

Further methods of tessellation construction

More complicated tessellations can be constructed by means of operations on simpler tessellations or by combining the methods given above. We give two examples. Further discussion can be found in SERRA (1982), who also considered the interesting *twin flat model*, constructed from a Poisson plane process by duplicating each plane by a parallel plane separated from it by a random distance.

Superposition

If Θ_1, Θ_2 are two tessellations with edge-sets (regarded as fibre processes) Ψ_1, Ψ_2 then the union of Ψ_1 and Ψ_2 forms a further fibre process which defines the edge-set of the *superposition tessellation*. MECKE (1984a) and SANTALO (1980, 1984) discuss this.

Iterated division of cells

Each cell of a tessellation Θ may itself be divided into further cells by use of one of the procedures described above. Figure 10.6 shows a part of a Voronoi tessellation for which the cells are further divided by a crack process as described above (of course, the crack growth is here terminated on encountering the edge-set of Θ, not just on meeting other edges of the crack process within the cell. See also COWAN (1984)).

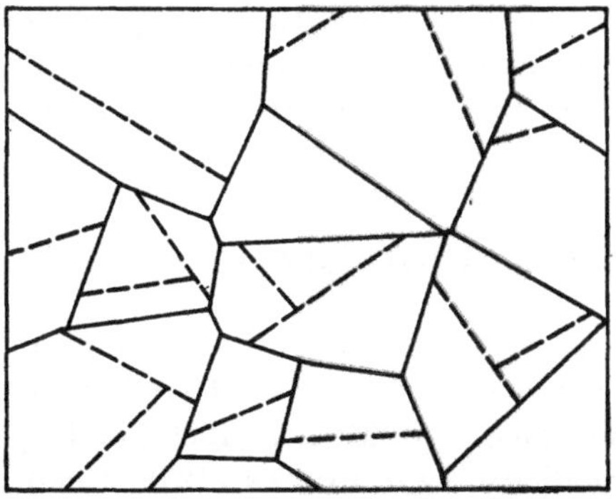

Fig. 10.6. A part of a Voronoi tessellation in which the cells are further subdivided by means of a crack tessellation

10.3 General results for planar tessellations

A planar tessellation Θ automatically produces a fibre process, the edge-set Ψ. Together with Ψ, three planar point processes can be constructed from Θ: the *set of nodes* $\alpha_0(\Theta)$; the *set of edge-midpoints* $\alpha_1(\Theta)$; and the *set of cell-centroids* $\alpha_2(\Theta)$. If Θ is stationary then so are Ψ and the $\alpha_k(\Theta)$. The intensities of the $\alpha_k(\Theta)$ are denoted by λ_k for $k = 0, 1, 2$; and the intensity of Ψ is denoted by L_A; it is assumed that all four intensities are positive and finite.

Further characteristics of the tessellation can be obtained by marking the point processes $\alpha_k(\Theta)$ in various geometrically natural ways. The means of the ensuing mark distributions provide the characteristics.

Marks for a node

If $x \in \alpha_0(\Theta)$ then

$$n_{02} = \#\{\boldsymbol{p} \in \Theta : x \in \boldsymbol{p}^{\text{cl}}\}$$

is the number of polygons touching x (equivalently the number of sides emanating from x); and $l_0(x)$ is the total length of the edges emanating from x.

Marks for an edge-midpoint

If $x \in \alpha_1(\Theta)$ then $l_1(x)$ is the length of the edge through x.

Marks for a cell-centroid

If $x \in \alpha_2(\Theta)$ then $n_{20}(x)$ is the number on nodes of the boundary of the cell containing x

$$l_2(x) = \nu_1(\partial \boldsymbol{p}) \qquad \text{where } x \in \boldsymbol{p} \in \Theta$$

$$a_2(x) = \nu_2(\boldsymbol{p}) \qquad \text{where } x \in \boldsymbol{p} \in \Theta .$$

Each mark leads to a mark distribution. The corresponding means are denoted by $\bar{n}_{02}$, $\bar{l}_0$, $\bar{l}_1$, $\bar{n}_{20}$, $\bar{l}_2$, and $\bar{a}_2$. Each is referred to as the appropriate mean value of the corresponding geometrical quantity. For example, $\bar{l}_1$ is referred to as the mean length of the 'typical' edge, $\bar{n}_{20}$ as the mean number of nodes on the boundary of the 'typical' cell.

Mean-value relationships for stationary tessellations

There are a number of relationships holding between the above mean values in the case of stationary tessellations, see MECKE (1984a). They have been established by means of Palm distributions, as in MECKE (1980), and by means of limiting arguments involving ergodicity, as in COWAN (1978, 1980). They are due to simple geometric reasoning (which has to be made precise in one of the two ways above).

For example, the Euler formula for *finite* polyhedra in $\mathbb{R}^3$ states that

$$v - e + f = 2$$

where v is the number of vertices, e the number of edges, and f the number of faces. A tessellation can be interpreted as the surface of an *infinite* polyhedron, and limiting arguments based on this idea show that

$$\lambda_1 = \lambda_0 + \lambda_2. \tag{10.3.1}$$

This basic formula exhibits a duality between nodes and cells, that is between the indices $_0$ and $_2$. Any tessellation is associated with a *dual* tessellation obtained by taking the original cell-centroids as new nodes and connecting pairs of new nodes if they are centroids of neighbouring old cells. This association helps to explain the duality between indices which underlies several of the formulae given below

$$\begin{aligned} \bar{n}_{02} &= 2 + 2\lambda_2/\lambda_0, \qquad \bar{n}_{20} = 2 + 2\lambda_0/\lambda_2 \\ \bar{l}_0 &= 2L_A/\lambda_0, \qquad \bar{l}_2 = 2L_A/\lambda_2 \\ \bar{a}_2 &= 1/\lambda_2 \\ \bar{l}_1 &= L_A/(\lambda_0 + \lambda_2) = L_A/\lambda_1. \end{aligned} \tag{10.3.2}$$

From the above formulae others can be derived

$$\frac{1}{\bar{n}_{02}} + \frac{1}{\bar{n}_{20}} = \frac{1}{2} \tag{10.3.3}$$

$$\begin{aligned} &\bar{l}_0 = \bar{n}_{02}\bar{l}_1, \qquad \bar{l}_2 = \bar{n}_{20}\bar{l}_1 \\ &3 \leqq \bar{n}_{02}, \qquad \bar{n}_{20} \leqq 6. \end{aligned} \tag{10.3.4}$$

As Mecke (1984a) showed, all these mean values can be expressed by the three parameters λ_0, λ_2 (or $\bar{n}_{02}$) and L_A.

The ordinary equilibrium state is an important special case. It holds when the lower bound above is attained and

$$\bar{n}_{02} = 3.$$

Since $n_{02} \geqq 3$ by construction this implies that the number of edges emanating from each node is always three, as in Figures 10.1 and 10.7. In this case, simpler relationships hold

$$\bar{n}_{02} = 3, \qquad \bar{n}_{20} = 6, \qquad 2\lambda_2 = \lambda_0, \qquad 2\lambda_1 = 3\lambda_0.$$

Thus in the ordinary equilibrium state the mean number of edges (or nodes) of the 'typical' polygon is 6.

Formulae for the particular case of tessellations formed by line processes can be found in Section 10.5. In this case, four edges emanate from each node so the ordinary equilibrium state cannot hold.

The following gives proofs for two examples of the general mean-value formulae. Proofs are taken from Mecke (1980) and are good examples of the use of the theorems concerning Palm distribution calculations as given in Chapter 7. To begin with, some notation is introduced.

If Θ is a stationary planar tessellation with distribution P, then the $\alpha_k(\Theta)$ ($k = 0, 1, 2$) and also $\beta(\Theta) = \alpha_0(\Theta) \cup \alpha_1(\Theta) \cup \alpha_2(\Theta)$ are stationary point processes. The intensity of $\beta(\Theta)$ is $\lambda' = \lambda_0 + \lambda_1 + \lambda_2$, and (under our general assumptions on tessellations) λ' is positive and finite.

The Palm distribution P' of $\beta(\Theta)$ is

$$\lambda' P'(Y) = \int \sum_{x \in \beta(\theta) \cap [0,1]^2} \mathbf{1}_Y(\theta - x)\, P(\mathrm{d}\theta) \quad \text{for } Y \text{ in } \mathcal{T}.$$

This may be regarded intuitively as the probability distribution of a tessellation configuration given that a point of $\beta(\Theta)$ is at o. It is useful to further distinguish which of the $\alpha_k(\Theta)$ contains this 'typical' point. If $\boldsymbol{T}_k = \{\theta \in \boldsymbol{T} : o \in \alpha_k(\theta)\}$, then let

$$E_k g(\Theta) = \frac{\lambda'}{\lambda_k} \int_{\boldsymbol{T}_k} g(\theta)\, P'(\mathrm{d}\theta) \quad \text{for } k = 0, 1, 2$$

where $g : \boldsymbol{T}_k \to [0, \infty)$ is measurable. We call $E_k g(\Theta)$ the mean of $g(\Theta)$ under the condition that o is a point of $\alpha_k(\Theta)$.

Proposition. *Let $C_0(\Theta)$ be the cell of the tessellation Θ containing o. Then*

$$\mathsf{E} f(\Theta) = \lambda_2 E_2 \int_{C_0(\Theta)} f(\Theta - x)\, \mathrm{d}x \tag{10.3.5}$$

for every measurable non-negative function f on $\mathbb{T}$.

Proof. The right-hand side, under interpretation of the conditional expectation E_2 and change of co-ordinates $x \to -x$, yields

$$\lambda_2 E_2 \int_{C_0(\Theta)} f(\Theta - x)\, \mathrm{d}x = \lambda' \iint \mathbf{1}_{\boldsymbol{T}_2}(\theta)\, \mathbf{1}_{C_0(\theta)}(-x)\, f(\theta + x)\, \mathrm{d}x P'(\mathrm{d}\theta).$$

A variation of Theorem 7.1(a) yields

$$\lambda' \iint v(x, \theta)\, \mathrm{d}x P'(\mathrm{d}\theta) = \int \sum_{x \in \beta(\theta)} v(x, \theta - x)\, P(\mathrm{d}\theta)$$

and if $v(x, \theta) = \mathbf{1}_{\boldsymbol{T}_2}(\theta) \cdot \mathbf{1}_{C_0(\theta)}(-x) \cdot f(\theta + x)$ then the result can be applied to show that the right-hand side equals

$$\int \sum_{x \in \beta(\theta)} \mathbf{1}_{\boldsymbol{T}_2}(\theta - x)\, \mathbf{1}_{C_0(\theta - x)}(-x)\, f(\theta)\, P(\mathrm{d}\theta)$$

$$= \int f(\theta)\, \#\{x \in \alpha_2(\theta) : -x \text{ belongs to the cell of } \theta - x \text{ containing } o\} P(\mathrm{d}\theta).$$

If o is in the edge-set of θ then there can be no cells containing it. Otherwise there can be only one such cell, and so only one corresponding cell-centroid x in $\alpha_2(\theta)$. But also by stationarity

$$\mathsf{P}(o \in E_\Theta) = 0.$$

Hence, the right-hand side of the equation in the proposition must equal

$$\int f(\theta)\, P(\mathrm{d}\theta)$$

which proves the proposition.

Taking $f \equiv 1$, formula (10.3.5) yields

$$1 = \lambda_2 E_2 \int_{C_0(\Theta)} \mathrm{d}x = \lambda_2 \bar{a}_2$$

which proves (10.3.2). With $f(\theta) = \left(\nu_2(C_0(\theta))\right)^{-1}$, the formula yields

$$\mathsf{E}\left(\nu_2(C_0(\Theta))\right)^{-1} = \lambda_2 = 1/\bar{a}_2 .$$

An application of Jensen's inequality shows that

$$\bar{a}_2 \leqq \mathsf{E}\nu_2\left(C_0(\Theta)\right)$$

which means that the mean area of the 'typical' polygon is smaller than the mean area of the polygon containing o. This accords with intuition since the latter polygon is selected by an area-biased method, so larger polygons are more likely to contain o.

Formula (10.3.5) can be explained intuitively as follows. The left-hand side is a kind of average of $f(\theta - x)$ over all x (using an ergodic-theory understanding of expectation and the stationary nature of Θ). The right-hand side is a kind of sum over all cell-centroids (hence the λ_2) of integrals of $f(\theta - x)$ for x in the corresponding cell. It is natural to suppose that the two sides should be equal. The proof above uses Palm distributions to make this argument rigorous.

Proposition. *For a tessellation θ let $S(\theta)$ be the set of all cell-centroids of cells containing o in their closure. Then*

$$\lambda_2 E_2 \sum_{x \in \alpha_0(\Theta) \cap C_0(\Theta)^{\mathrm{cl}}} h(x, \Theta) = \lambda_0 E_0 \sum_{x \in S(\Theta)} h(-x, \Theta - x) \tag{10.3.6}$$

for any non-negative measurable function h on $\mathbb{R}^2 \times \mathbb{T}$.

Proof. A variation of Theorem 7.1(c) yields

$$\lambda' \int \sum_{x \in \beta(\theta)} v(x, \theta)\, P'(\mathrm{d}\theta) = \lambda' \int \sum_{x \in \beta(\theta)} v(-x, \theta - x)\, P'(\mathrm{d}\theta)$$

for non-negative measurable functions v on $\mathbb{R}^2 \times \mathbb{T}$. Taking

$$v(x, \theta) = \mathbf{1}_{T_2}(\theta) \cdot \mathbf{1}_{\alpha_0(\theta) \cap C_0(\theta)^{\mathrm{cl}}}(x) \cdot h(x, \theta)$$

and applying the theorem to the left-hand side of (10.3.6)

$$\lambda_2 E_2 \sum_{x \in \alpha_0(\Theta) \cap C_0(\Theta)^{\mathrm{cl}}} h(x, \Theta) = \lambda' \int_{T_2} \sum_{x \in \alpha_0(\theta) \cap C_0(\theta)^{\mathrm{cl}}} h(x, \theta)\, P'(\mathrm{d}\theta)$$

$$= \lambda' \int \sum_{x \in \beta(\theta)} v(x, \theta)\, P'(\mathrm{d}\theta) = \lambda' \int \sum_{x \in \beta(\theta)} v(-x, \theta - x)\, P'(\mathrm{d}\theta)$$

$$= \lambda' \int \sum_{x \in \alpha_2(\theta)} h(-x, \theta - x)\, \mathbf{1}_{\alpha_0(\theta - x)}(-x)\, \mathbf{1}_{C_0(\theta - x)^{\mathrm{cl}}}(-x)\, P'(\mathrm{d}\theta).$$

If x is in $\alpha_2(\theta)$ then the summand above is zero unless o is a node (in which case $\theta \in T_0$) and x is the cell-centroid of one of the cells $\boldsymbol{p}$ in θ with o in $\boldsymbol{p}^{\mathrm{cl}}$. Thus the integrand above is equal to

$$\mathbf{1}_{T_0}(\theta) \sum_{x \in S(\theta)} h(-x, \theta - x)$$

from which (10.3.6) follows.

As with (10.3.5), formula (10.3.6) can be understood intuitively by an 'averaging' argument. The left-hand side averages over all cell-centroids and counts the quantity $h(x, \theta)$ for x running through the nodes of a 'typical' cell. The right-hand side averages over all nodes and counts the appropriate quantity for x running through the cell-centroids of cells meeting at the 'typical' node.

With $h = 1$ formula (10.3.6) yields

$$\lambda_2 \bar{n}_{20} = \lambda_0 \bar{n}_{02} .$$

If $h(x, \theta)$ is defined as the angle at the node x in $C_0(\theta)$ then the left-hand side of (10.3.6) involves a sum of all angles of the convex polygon $C_0(\theta)$. This sum equals

$$\pi\big(\#\{\text{nodes of } C_0(\theta)\} - 2\big)$$

and so the left-hand side of (10.3.6) is $\lambda_2(\bar{n}_{20} - 2)\,\pi$. The right-hand side is $2\pi\lambda_0$ and so

$$\lambda_2 \bar{n}_{20} - 2\lambda_2 = 2\lambda_0 .$$

Together, these two derived equations yield (10.3.3).

MECKE (1984a) also gave mean-value relationships for superpositions of independent tessellations and studied 'extremal' tessellations: those that minimize L_A when all cells either are of the same fixed cell area or of the same fixed perimeter. Such tessellations turn out to be regular, based on equilateral triangles, squares, or regular hexagons (MECKE, 1984b).

AMBARTZUMIAN (1974) studied intersections of motion-invariant tessellations with lines, establishing relationships between the intersection structure (the 'typical' chord and adjacent angles of intersection) and tessellation characteristics. For example, if the 'typical' chord has length independent of the two adjacent angles then the distribution of the length of the 'typical' edge l_1 and the 'typical' chord χ are connected by

$$\mathsf{P}(l_1 \leqq x) = 1 - \bar{l}_1 \frac{\mathrm{d}}{\mathrm{d}x} \mathsf{P}(\chi \leqq x)$$

where $\bar{l}_1$ is the mean length of l_1. This gives the familiar length-biased distribution for χ.

10.4 Mean-value formulae for stationary spatial tessellations

Mean-value relationships hold for stationary spatial tessellations just as for stationary planar tessellations. The relationships are due to MECKE (1984a) and RADECKE (1980) and are displayed below. Some additional notation is helpful. To aid description vertices, edges, faces, and cells are called 0-facets, 1-facets, 2-facets, and 3-facets respectively. Let λ_k be the intensity of the point process of centroids of k-facets ($k = 0, 1, 2, 3$), L_V the intensity of the fibre process that is the union of all 1-facets, and S_V the intensity of the surface process that is the union of all 2-facets. If the vertices and edges are weighted according to the numbers of adjacent cells then the vertex point process and edge fibre process can be used to produce stationary random measures (multiplying by the weights) whose corresponding intensities are denoted by T_V and Z_V. Finally,

$\bar{n}_{kl}$	denotes the mean number of l-facets adjacent to the 'typical' k-facet, for k and l in $\{0, 1, 2, 3\}$,
$\bar{l}_1$	denotes the mean length of the typical 1-facet or edge,
$\bar{l}_2$	the mean perimeter,
$\bar{A}_2$	the mean area of the 'typical' 2-facet or face,
$\bar{l}_3$	the mean total edge length,
$\bar{S}_3$	the mean surface area,
$\bar{V}_3$	the mean volume,
$\bar{b}_3$	the mean of the average breadth of the 'typical' 3-facet or cell.

MECKE (1984a) showed that all these means can be expressed by the parameters

$$\lambda_0, \lambda_3, \lambda = \lambda_1 + \lambda_2, L_V, S_V, T_V, \text{ and } Z_V.$$

The relevant formulae are as follows:

$$2\lambda_1 = \lambda + \lambda_0 - \lambda_3; \qquad 2\lambda_2 = \lambda + \lambda_3 - \lambda_0$$
$$\lambda_0\bar{n}_{03} = T_V; \qquad \lambda_3\bar{n}_{30} = T_V$$
$$\lambda_0\bar{n}_{01} = \lambda + \lambda_0 - \lambda_3; \qquad \lambda_3\bar{n}_{32} = \lambda + \lambda_3 - \lambda_0$$
$$\lambda_0\bar{n}_{02} = T_V + \lambda - \lambda_0 - \lambda_3; \qquad \lambda_3\bar{n}_{31} = T_V + \lambda - \lambda_0 - \lambda_3$$

$$\bar{n}_{12} = \bar{n}_{13} = 2(T_V + \lambda - \lambda_0 - \lambda_3)/(\lambda + \lambda_0 - \lambda_3)$$
$$\bar{n}_{21} = \bar{n}_{20} = 2(T_V + \lambda - \lambda_0 - \lambda_3)/(\lambda + \lambda_3 - \lambda_0)$$

$$\bar{l}_1(\lambda + \lambda_0 - \lambda_3) = 2L_V$$
$$\bar{l}_2(\lambda + \lambda_3 - \lambda_0) = 2Z_V$$
$$\bar{A}_2(\lambda + \lambda_3 - \lambda_0) = 2S_V$$

$$\lambda_3\bar{l}_3 = Z_V$$
$$\lambda_3\bar{S}_3 = 2S_V; \qquad \lambda_3\bar{V}_3 = 1$$
$$4\lambda_3\bar{b}_3 = Z_V - 2L_V.$$

These formulae imply the relationships

$$\lambda_0 - \lambda_1 + \lambda_2 - \lambda_3 = 0$$
$$\bar{n}_{01} - \bar{n}_{02} + \bar{n}_{03} = 2; \qquad \bar{n}_{30} - \bar{n}_{31} + \bar{n}_{32} = 2$$
$$\bar{n}_{12}\bar{n}_{01} = 2\bar{n}_{02}; \qquad \bar{n}_{21}\bar{n}_{32} = 2\bar{n}_{31}$$
$$\bar{n}_{02}\bar{n}_{30} = \bar{n}_{31}\bar{n}_{03}$$

$$\bar{S}_3 = \bar{n}_{32}\bar{A}_2.$$

Some of these formulae display symmetries between the indices $_0$ and $_3$ and between the indices $_1$ and $_2$.

Again the *ordinary equilibrium state* is an important special case; it is similar to that of the planar tessellation and holds when

$$\bar{n}_{03} = 4, \qquad \bar{n}_{13} = 3, \qquad \bar{n}_{23} = 2.$$

The intensities then satisfy

$$\lambda_1 = 2\lambda_0$$
$$\lambda_2 = \lambda_0 + \lambda_3.$$

The particular case of tessellations formed by random planes is discussed in the next section.

Note that Chapter 11 contains a section on stereology for spatial tessellations.

10.5 Poisson line and plane tessellations

Poisson line and plane processes produce tessellations in the plane and in space respectively. Formulae for these tessellations are presented here; some results for higher dimensional Poisson hyperplane tessellations can be found in MATHERON (1975) and MILES (1971, 1974a).

10.5.1 Poisson line tessellations

These are also called *Crofton tessellations*. Let Ψ be a motion-invariant Poisson line process of intensity L_A; we consider the motion-invariant tessellation with Ψ as edge-set.

The quantity $\varrho = P_L = 2 \cdot L_A/\pi$ (the mean number of lines intersected by a test line segment of unit length) is a convenient parameter. The intensity of the point process of intersection points on a fixed line g in Ψ is given by ϱ (using Slivnyak's theorem). Hence

$$\bar{l}_1 = 1/\varrho.$$

With probability one the Poisson line process is bundlefree (has no triplets of lines meeting in a single point) so every node is formed by the intersection of precisely two lines. Thus

$$\bar{n}_{02} = 4$$

and so the formulae in Section 10.3 yield

$$\lambda_0 = \pi\varrho^2/4$$
$$\bar{l}_0 = 4/\varrho$$
$$\lambda_1 = \pi\varrho^2/2$$
$$\lambda_2 = \pi\varrho^2/4.$$

Suppose the cell-centroid process is marked by the corresponding polygons. The mark distribution is then the distribution of the 'typical' polygon of the tessellation process, also known as the *Poisson polygon* or *Crofton polygon*. Several formulae are known for the distribution of this polygon. They were mostly found by MILES (1964, 1973).

Table 10.1. Poisson polygon formulae

First-order moments of area, perimeter, edge number

A	U	N
$4/(\pi\varrho)^2 = \bar{a}_2$	$4/\varrho = \bar{l}_2$	$4 = \bar{n}_{20}$

Second-order moments

	A	U	N
A	$8/\varrho^4$		
U	$4\pi/\varrho^3$	$2(\pi^2 + 4)/\varrho^2$	
N	$2\pi/\varrho^2$	$(\pi^2 + 8)/\varrho$	$(\pi^2 + 24)/2$

The first-order moment for N was found by GOUDSMIT (1945). Further formulae, including higher order moments, are given in MILES (1964, 1973). Information about density functions of area, perimeter, and breadth can be found in CRAIN and MILES (1976), MATHERON (1975), and MILES (1964, 1973). The distribution of the number of sides is partially known and tabulated in Table 10.2; see TANNER (1983b) who also gave the values of the corresponding third and fourth moments.

Table 10.2. p_n the probability that the Poisson polygon has n sides

n	p_n	
3	$2 - \pi^2/6 \simeq 0.357$	MILES (1964)
4	$\pi^2 \ln 2 - 1/3 - 7\pi^2/36 - (7/2)(1 + 2^{-3} + 3^{-3} + \cdots) \simeq 0.381$	TANNER (1983a)
5	0.192	
6	0.059	
7	0.013	
8	0.002	
...		

CRAIN and MILES (1976) note the empirical approximation of the distribution $\{p_n\}$ by a shifted Poisson distribution of parameter 1:

$$p_n \simeq e^{-1}/(n-3)! \quad \text{for } n = 3, 4, \ldots$$

They found the values $p_5, p_6, \ldots$ and much other distributional information by using Monte Carlo simulation of the Poisson polygon. The method was to generate the tessellation in a circle and to determine the desired quantities by statistical estimation. Since their estimation procedure allow for edge-effects they are of some methodological interest.

MATHERON (1975) discusses some formulae for the tessellation polygon containing the origin. This is in some sense greater than the Poisson polygon; for example the mean area and mean number of sides are both greater. Matheron also showed a 'lack of memory' property for the Poisson polygon.

In conclusion we note that mean-value formulae can also be derived in the anisotropic case. Let $\boldsymbol{R}$ be the rose of direction of the anisotropic Poisson line process generating the tessellation. Then

$$\bar{n}_{02} = 4$$

and

$$\lambda_0 = L_A{}^2\zeta/2$$

where

$$\zeta = \int_{(0,\pi]} \int_{(0,\pi]} |\sin(\alpha - \beta)| \, \boldsymbol{R}(\mathrm{d}\alpha) \, \boldsymbol{R}(\mathrm{d}\beta)$$

(see MATHERON, 1975, and THOMAS, 1984). From this one may conclude (using the mean value formulae in 10.3) that

$$\begin{aligned} \bar{n}_{20} &= 4 \\ \lambda_1 &= L_A{}^2\zeta \quad \text{and} \quad \lambda_0 = \lambda_2 = L_A{}^2\zeta/2 \\ \bar{l}_1 &= 1/(L_A\zeta) \\ \bar{l}_0 &= \bar{l}_2 = 4/(L_A\zeta) \\ \bar{a}_2 &= 2/(L_A{}^2\zeta) \end{aligned} \qquad (10.5.1)$$

The epuation $\lambda_0 = \lambda_2$ is true for general stationary bundlefree line tessellations; the Poisson assumption is not necessary (see MECKE, 1983b).

10.5.2 Poisson plane tessellations

Let Ψ now be a motion-invariant Poisson plane process with intensity S_V. As before, the quantity $\varrho = P_L = S_V/2$ (the mean number of planes intersected by a test line segment of unit length) is used as a parameter. With probability one there is no point of $\mathbb{R}^3$ in which more than three planes intersect and so

$$\begin{aligned} \bar{n}_{01} &= 6 \\ \bar{n}_{02} &= 12 \\ \bar{n}_{03} &= 8 \\ \bar{n}_{12} &= 4. \end{aligned}$$

It is known that

$$\begin{aligned} \bar{n}_{30}' &= 8 \\ \bar{n}_{31} &= 12 \\ \bar{n}_{32} &= 6 \\ \bar{n}_{21} &= 4, \end{aligned}$$

Table 10.3. Poisson polyhedron formulae ; see Santalo (1976)

First order moments

V	S	$\bar{b}$	N (faces)
$\bar{V}_3 = 6/(\pi\varrho^3)$	$\bar{S}_3 = 24/(\pi\varrho^2)$	$\bar{b}_3 = 3/(2\varrho)$	$\bar{n}_{32} = 6$

Second order moments

V	$48/\varrho^6$			
S	$96/\varrho^5$	$240/\varrho^4$		
$\bar{b}$	$2(\pi^2 + 3)/(\pi\varrho^4)$	$(7\pi^2 + 12)/(\pi\varrho^3)$	$(13\pi^2 + 48)/(48\varrho^2)$	
N	$4(\pi^2 + 3)/(\pi\varrho^3)$	$2(7\pi^2 + 24)/(\pi\varrho^2)$	$(13\pi^2 + 120)/(12\varrho)$	$(13\pi^2 + 336)/12$

see Miles (1974b) and Mecke (1984c). These are the same values as for a regular cubic tessellation. In the sense of 10.4 basic parameters are

$$\lambda_0 = \lambda_3 = \pi\varrho^3/6$$

$$\lambda = \pi\varrho^3$$

$$L_V = \pi\varrho^2/2$$

$$S_V = 2\varrho$$

$$T_V = 4\pi\varrho^3/3$$

$$Z_V = 2\pi\varrho^2.$$

Some formulae are known for the 'typical' polyhedron (the *Poisson polyhedron* in the sense of the 'number law' of Matheron, 1975); these are mainly due to Miles (1972a). Miles (1972a) and Matheron (1975) give further formulae. The latter proves a 'lack of memory' property for the Poisson polyhedron Π.

The Poisson polyhedron finds useful employment as a model component, for example, as the primary grain for Boolean models, as in Serra (1982, p. 499) and Example 3.3 in this book.

In conclusion it should be noted that the planar section of a Poisson plane tessellation is itself a Poisson line tessellation defined on the plane of intersection with the same parameter ϱ as the plane tessellation (this allows application of the formulae of Section 10.5.1). Mecke (1984d) showed that for general stationary bundlefree[1]) hyperplane tessellations in $\mathbb{R}^d$ the intensities satisfy

$$\lambda_k = \binom{d}{k} \cdot \lambda_0$$

[1]) A hyperplane process in $\mathbb{R}^d$ is said to be *bundlefree*, if every point or $\mathbb{R}^d$ is contained in at most d hyperplanes of the process.

10.6 Planar and spatial Poisson-Voronoi tessellations

10.6.1 Planar Poisson-Voronoi tessellations (Poisson-Dirichlet tessellations)

A planar Poisson-Voronoi tessellation is the Voronoi tessellation of a stationary Poisson process. Let τ be the intensity of the point process. Then, as MEIJERING (1953) and MILES (1970) establish, the mean-value parameters defined in Section 10.3 take on the values

$$\lambda_0 = 2\tau \tag{10.6.1}$$

$$\lambda_2 = \tau$$

$$L_A = 2\sqrt{\tau}. \tag{10.6.2}$$

All other mean-values follow from these by (10.3.2). For example $\bar{n}_{20} = 6$ so the 'typical' cell of the Poisson-Voronoi tessellation has an average of six vertices.

It would be extremely convenient if one could identify tessellations that were in the ordinary equilibrium state as always derivable from a point pattern as a Voronoi tessellation. This would enable one to adopt a model-fitting approach to the analysis of tessellations. Unfortunately, a simple degrees-of-freedom argument shows that this cannot be the case. So we must content ourselves with using the Poisson-Voronoi tessellation as a kind of 'bench-mark'; the mean-value quantities extracted here provide indices with which to compare those obtained by empirical measurement from observed tessellations.

Table 10.4. The probability p_n that the 'typical' Poisson-Voronoi polygon has n sides

n	p_n
3	0.011
4	0.107
5	0.259
6	0.294
7	0.199
8	0.090

Moments of characteristics of the 'typical' cell of the Poisson-Voronoi tessellation

	A	U	N
Variance:	$0.280/\tau^2$	$0.947/\tau$	1.782
Correlation coefficient:			
A			
U	0.953		
N	0.568	0.502	

Gilbert (1962) established the second moment of the area of the 'typical' polygon of the Poisson-Voronoi tessellation. He obtained the value as $1.280\tau^{-2}$. In principle all characteristics of this andom polygon can be obtained by numerical integration. For example Miles and Maillardet (1982) derived integral formulae for the probabilities p_n of the polygon having exactly n vertices. These formulae are very complicated and hence it is helpful to determine such characteristics by Monte Carlo methods. Hinde and Miles (1980) obtain a series of estimate by simulating the 'typical' cell directly rather than by dealing with the whole tessellation. The results are tabulated on p. 274 together with variances and correlations for geometrical characteristics of the 'typical' cell.

Hinde und Miles gave histograms for the distributions of perimeter, area, and iuner arries (see also Quine and Watson, 1984) and suggested an approximation for perimeter and area distribution using the so called generalized gamma densities

$$f(x) = \gamma \cdot x^{\nu-1} \cdot \exp(-\chi \cdot x^{\kappa} \quad \text{for } x \geqq 0 .$$

10.6.2 Spatial Poisson-Voronoi tessellations

Here again τ is the intensity of the underlying Poisson point process. Following Meijering, 1953, and Miles, 1972a the mean-value parameters of section 10.3 take the values

$$\lambda_0 = 24\pi^2\tau/35 \simeq 6.768\tau$$

$$\lambda_3 = \tau$$

$$\lambda = 3\lambda_0 + \tau$$

$$L_V = (16/15)\,(3/4)^{1/3}\,\pi^{5/3}\Gamma(4/3)\,\tau^{2/3} \simeq 5.832\tau^{2/3}$$

$$S_V = 4(\pi/6)^{1/3}\,\Gamma(5/3)\,\tau^{1/3} \simeq 2.910\tau^{1/3}$$

$$T_V = 4\lambda_0$$

$$Z_V = 3L_V .$$

All other mean values follow from these results. For example, the mean volume of the 'typical' cell is given by

$$\overline{V}_3 = \tau^{-1}$$

the mean surface area by

$$\overline{S}_3 = (256\pi/3)^{1/3}\,\Gamma(5/3)\,\tau^{-2/3} \simeq 5.821\tau^{-2/3}$$

the mean average breadth by

$$\overline{b}_3 = 3^{-5/3} \cdot 4^{2/3} \cdot 5^{-1} \cdot \pi^{5/3}\Gamma(1/3) \cdot \tau^{-1/3} \simeq 1.458\tau^{-1/3} .$$

With probability one the Poisson-Voronoi tessellation is in the natural equilibrium state.

GILBERT (1962) used numerical integration to estimate the second moment of volume as

$$1.180\tau^{-2}.$$

QUINE and WATSON (1984) used simulation to obtain histograms of volume, surface, and total edge length.

Random intersections of Poisson-Voronoi tessellations are considered in Section 11.5.

10.7 Statistics for planar stationary tessellations

We confince ourselves here to planar tessellations. The spatial case is considered briefly in the spirit of stereology in Section 11.5.4. In general, if a tessellation is interpreted as a germ-grain model then the random set methods of Chapter 6 can be applied (the grains being the cells *exclusive* of their boundaries).

In the first place the problem is to estimate the fundamental parameters. By Section 10.3 these are L_A, λ_0, and λ_2. STOYAN and HERMANN (1986) considered the estimation of second-order characteristics.

The characteristic L_A can be estimated by applying the methods for planar fibre processes as in Chapter 9. Thus direct measurement can be made via an image analyzer, or else intersection with test line segments can be observed. In the case of isotropy one can use

$$L_A = (\pi/2)\, P_L$$

where P_L is the intensity of the intersection point pattern on the test lines.

Determination of the mean number of nodes per unit area λ_0 is a matter of applying an estimation method for the intensity of a planar stationary point process. Given a window of observation W an unbiased estimator is

$$\hat{\lambda}_0 = \frac{\#\{\text{nodes observed in } W\}}{\nu_2(W)}.$$

Unbiased estimation of the number of cells per unit area λ_2 is complicated by the presence of edge effects. In the case of a tessellation with $\bar{n}_{02} = 3$, which is to say when each node of the tessellation is formed by precisely three edges, the formula on page 265 can be used

$$\hat{\lambda}_2^{(1)} = \hat{\lambda}_0/2. \tag{10.7.1}$$

So the problem is reduced to the estimation of λ_0. In the case of isotropy the formulae of 6.4 apply

$$\hat{\lambda}_2^{(2)} = \left(N(W) - 1 - \frac{1}{2}\, NE(W)\right) \Big/ \nu_2(W) \tag{10.7.2}$$

for a convex window of observation W. Here $N(W)$ is the number of cells intersecting the window W while $NE(W)$ is the number of edges intersecting the boundary of W.

If W is a rectangle and if the cells are so small that each cell produces at most two intersections of edges with the boundary of W then a formula of SALTYKOV (1974) gives

$$\hat{\lambda}_2^{(3)} = \big(z(W) + 0.5w(W) + 0.25u(W)\big)/\nu_2(W) \tag{10.7.3}$$

where

$z(W)$ is the number of cells completely in W,
$w(W)$ is the number of cells that intersect the sides of W but *not* its corners,
$u(W)$ is the number of cells containing the corners of W.

Formula (10.7.2) is a consequence of (6.4.6) and $L_A = \pi P_L/2$, which yield

$$\begin{aligned}[\text{Intensity of cells}] \cdot \nu_2(W) &= \overline{N(W)} - 1 - \frac{L_A}{2\pi} U(W) \\ &= \overline{N(W)} - 1 - \overline{NE(W)}/2.\end{aligned}$$

Here the upper bars denote means. Note also that L_A here is the quantity given as $L_A/2$ in Section 6.4.

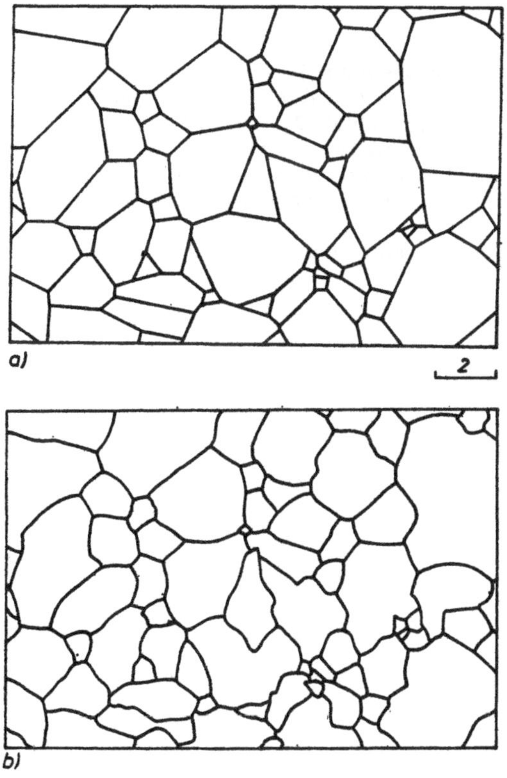

Fig. 10.7. Planar section through a specimen of steel. (a) Tellesation obtained by linearizing the edges of Austenite grains in (b) Original boundaries of Austenite grains.

If $\bar{n}_{02} = 3$ then the question may arise as to whether the tessellation is Poisson-Voronoi. Of course the comments at the head of Section 10.6.1 mean that one should first ask whether the tessellation is one that could be Voronoi in any sense; if so, then one might attempt to reconstruct the points of the generating pattern and test this pattern to see if it could be considered to be a reasonable Poisson process pattern. A quicker approach, and one that at any rate assesses the tessellation against the Poisson-Voronoi tessellation *qua* bench-mark, estimates λ_0, λ_2, and L_A, and compares these values with the Poisson-Voronoi tessellation relationships

$$\lambda_0 = 2\lambda_2; \; L_A = 2\sqrt{\lambda_2}.$$

Example 10.1. *Austenite grain boundaries in steel.*

Figure 10.7 shows a planar tessellation which results from a 'linearization' of an image obtained by planar section through a sample of steel. The edges correspond to the Austenite grain boundaries. From the tessellation the following values are derived

Total edge length	172.2
Number of nodes	124
$N(W)$	78
$NE(W)$	30

This yields

$$\hat{L}_A = 0.96, \qquad \hat{\lambda}_0 = 0.69.$$

(The unit of length is as indicated on the figure.)
Since $\bar{n}_{02} = 3$, the parameter λ_2 can be estimated in different ways. By formula (10.7.1) we obtain $\hat{\lambda}_2 = 0.34$, while by (10.7.2) we obtain $\hat{\lambda}_2 = 0.35$. Formula (10.7.3) is inappropriate here, as there are two grains hitting two sides each.

These estimates differ considerably from those expected for a Poisson-Voronoi tessellation: the estimate $\hat{\lambda}_2 = 0.23$ is derived from L_A, which differs from the direct estimates.

Further discussion of this example can be found in Example 11.7.

11. Stereology

11.1 Introduction

The subject of stereology represents a major part of the applications of stochastic geometry. Here we introduce some of the results and ideas of stereology, but in no way is this chapter a replacement for textbooks on the subject such as CHERNYAVSKIJ (1977), COLEMAN (1979), DE HOFF and RHINES (1968), SALTYKOV (1974), UNDERWOOD (1970), and WEIBEL (1980) (readers wanting further information should also consult surveys on stereology such as BADDELEY (1977), COLEMAN (1981b), JENSEN, BADDELEY, GUNDERSEN, and SUNDBERG (1985), MORAN (1972), SAXL (1981, 1984, 1985), WEIL (1983)). Furthermore, this chapter hardly mentions problems arising in the practical employment of stereology (such as measurement errors). Nevertheless, the reader will find a selection of important formulae and methods, and some material that has not previously appeared in book form.

The objective of stereology is to draw inferences about the geometrical properties of three-dimensional structure when information is only available in some lower-dimensional form via linear probes, planar sections, or projections of thick slices. Its application arises in the study of geometrical structure of inclusions or pores in opaque bodies such as metals, minerals, synthetic materials, or biological tissues; in these cases the available information must come from linear probes or planar sections. Projection methods are applied in X-ray analysis and in computerized tomography; see HERMAN and NATTERER (1981). In general, stereological problems are very special cases of the general problem of the reconstruction of geometrical structure from transformed images (GRENANDER, 1976, 1978, 1981).

The methods and formulae of stereology relate characteristics of three-dimensional structures to quantities arising from measurements of planar sections (such as slices through mineral material), projections of thick sections (such as foils or sheets studied by light or electron microscopy), or linear probes (such as might arise in biopsy). The step from spatial structures to their sections involves a great loss of information and so stereological methods commonly yield only 'global' information of a statistical character. A further consequence of this loss of information is that many of the applications require the solution of *ill-posed* mathematical problems: numerical problems in which small deviations due to measurement error can lead to large discrepancies in the final solution. For example, this is the case for the unfolding problem for sphere diameters sampled by a plane in Section 11.4.

Stereological formulae depend on some assumption of 'randomness'. This can come in one of two forms: either the structure itself is assumed random or the section process is taken to involve a randomising element. One of the authors collaborates a great deal with scientists using stereology in metallography and mineralogy, where it is natural to assume some kind of random structure. So we assume that the geometrical structures are random in some way and that the section plane or line is fixed (for example the section plane is taken to be the (x, y)-plane). The alternative approach typically assumes the sectioning planes or lines to form motion-invariant Poisson plane or line processes, and is more natural in biological applications. It is not considered in this book. The resulting formulae are generally similar for both approaches, differing only in the statistical theory and underlying interpretation. The random section approach is described in detail in the papers of MILES and DAVY (1977), MILES (1978), JENSEN (1984), and in the books by WEIBEL (1980) and SERRA (1982).

A particular kind of stereology has already been introduced in this book in the study of the determination of characteristics of planar structures by analysis of linear sections (see Section 9.3). In the present chapter we do not consider stereology in dimensions greater than three or in non-Euclidean space.

Historically DELESSE (1847) and ROSIWAL (1898) were probably the first to work in the spirit of stereology. They recognized that under certain assumptions the volume fraction of a component in a body equals the area or linear fraction in a planar or linear section (see Sections 11.2 and 6.3). WICKSELL (1925) solved the problem of inferring the size distribution for systems of spheres given planar sections (see Section 11.4). The great importance of the subject has been reflected in a vast and flourishing literature. We do not attempt a complete coverage, nor complete accuracy in the attribution of ideas. The reader is referred to textbooks such as those mentioned above, to the proceedings of frequent conferences on stereology, and to journals (such as the *Journal of Microscopy*) that publish many papers on stereology. In the main body of the chapter, many workers of considerable importance in the development of stereology are not adequately credited or even not named at all. Some of them (in alphabetical order) are: Bach, Giger, Hilliard, Moran, and Saltykov. This list omits some important scientists who worked on significant applications of stereology in material sciences and biology. A detailed history of stereology can be found in WEIBEL (1980).[1)]

The mathematical basis of modern stereology is due to Ambartzumian, Coleman, Cruz-Orive, Davy, Gundersen, and Miles (in alphabetical order) and others, with important impulses for its development coming repeatedly from the needs of practical applications. The modern theory is a synthesis of integral geometry, point processes and random measure theory. The proof on page 291 demonstrates the use of point processes; random measure theory occurs in Section 9.3 in the proofs of sectioning formulae for fibre processes.

1) The term "stereology" (in German: Stereologie) had another meaning in the past: In the "Handbuch der Fremdwörter", 20th edition, by F. E. PETRI (C. B. Griesbachs Verlag, Gera 1897), Stereologie is "Kreuzlehre; Erklärung aller Arten von Kreuzen auf Münzen, in Wappen und Urkunden" (science of crosses; explanation of all kinds of crosses on coins, in coat-of-arms and records).

11.2 The fundamental mean-value formulae of stereology

11.2.1 Notation

Stereology has for its basis a small number of mean-value formulae concerning intensities of random measures. These quantities, often called *densities* or *specific values* in the subject, arise here as intensities of stationary (often isotropic) random measures. They are denoted systematically in Table 11.1, using a system described for example in UNDERWOOD (1976a, 1981) and WEIBEL (1980). Each mean-value quantity is denoted by one symbol sufficed by another. The main symbol is the quantity measured while the sufficed symbol indicates the dimensionality of the section employed in the measurement.

Thus the notation A_A denotes the mean area (of some phase or structure being considered) per unit area. It is therefore the area density of the phase. The notation S_V stands for the mean surface content per unit volume; the surface density or specific surface area. It is measured in units of m^2/m^3.

This section is confined to general formulae. Specific results for germ-grain models such as the determination of N_V and related quantities, are considered in Section 11.3.

Table 11.1. Symbols in common use in stereology

Interpretation of main symbol	
V	Volume (in m^3)
S	Surface area (in m^2)
J	Length (of linear elements in the space) (in m)
M	Integral of mean curvature (in m)
M^+	Integral of mean curvature taken over the region in which all curvatures are non-negative (see Section 7.3)
A	Planar area (of profiles or section structures) (in m^2)
L	Length (of linear elements in the plane) (in m)
N	Number of grains or particles (for non-overlapping germ-grain models) or more generally the connectivity number
P	Number of points (in a section)
Interpretation of symbol in suffix	
V	per unit volume
A	per unit area
L	per unit length
P	per point

11.2.2 Planar and linear sections

Here we cover the general formulae that arise when the structure is sectioned by a plane or line. Each formula will be stated and either a brief indication of its proof will be given or a reference will be made.

Volume fraction

The volume fraction $p = V_V$ of a three-dimensional stationary random closed set Ξ_V satisfies

$$V_V = A_A = L_L = P_P. \tag{11.2.1}$$

This is interpreted as follows. The volume fraction V_V is equal to the quantity obtained by intersecting Ξ_V by an arbitrary test plane, line, or point system; and measuring respectively the area fraction, linear fraction, or fraction of test points in Ξ_V. Section 6.3 discusses this further, and considers the question of estimating p.

For example, let Ξ_A be the planar random closed set resulting from intersection of Ξ_V with a test plane. Then Ξ_A is also stationary and so,

$$A_A = \mathsf{E}\nu_2(\Xi_A \cap [0, 1]^2) = \int_{[0,1]^2} \mathsf{P}(x \in \Xi_A)\, dx = \mathsf{P}(o \in \Xi_V) = p = V_V$$

as discussed in Section 6.2.

Specific surface area

For a stationary isotropic surface process, in particular one arising as the boundary of a motion-invariant grain model or random closed set, the following formulae hold:

$$S_V = 4L_A/\pi \tag{11.2.2}$$

$$S_V = 2P_L. \tag{11.2.3}$$

This means that S_V can be determined from planar sections by measuring line lengths and determining the intensity L_A, or from linear sections by counting the points formed by intersection of the linear probe with the surface of interest.

Remarks on the proofs of these two formulae and their generalizations can be found in Section 9.5. The anisotropic case requires either additional angle measurements or assumptions on the orientation distribution (but see also Baddeley, 1983a, b).

Specific curve length

For a stationary and isotropic fibre process the specific curve length satisfies

$$J_V = 2P_A. \tag{11.2.4}$$

This means that J_V can be determined from a planar section by counting the number of points formed by intersection of the planar section with the fibre process. Remarks on the proof are to be found in Section 9.4. Section 11.6.4 discusses the case of anisotropic fibre processes.

Specific mean curvature

For a stationary and isotropic surface process the specific mean curvature or mean curvature density M_V can be defined as the intensity of the mean curvature measure associated with the process. It satisfies

$$M_V = C_A. \tag{11.2.5}$$

Here C_A denotes the density of curvature of fibres in the planar section. In the terminology of Section 7.3 it is the intensity of the planar curvature measure. Equations (7.3.2), and (11.2.5) together yield

$$M_V = 2\pi N_A. \tag{11.2.6}$$

In the case of non-overlapping grains this formula is due to BODZIONY (1965).

Analogously the intensity M_V^+ of the positive curvature measure (as described in Section 7.3) satisfies

$$M_V^+ = 2\pi N_A^+. \tag{11.2.7}$$

WEIBEL (1980) gives proofs and details. The particular case of germ-grain models is discussed in Section 11.3.

The intensity of the absolute curvature measure K_V satisfies

$$K_V = 2\pi(N_A^+ + N_A^-) \tag{11.2.8}$$

see DEHOFF (1978) and BADDELEY and AVERBACK (1983). The latter give an instructive example of the use of K_V.

In the literature the specific total (or Gauss) curvature has also been used (WEIBEL, 1980).

Formulae above which contain mean values of numbers or of intensities of point processes can sometimes be transformed into confidence intervals for S_V, L_V, and M_V. This is possible if the corresponding point process is approximately a Poisson point process. For then one can use the asymptotic confidence interval for the parameter of a Poisson distribution, as in Section 2.7, and so obtain confidence intervals for S_V, L_V, and M_V. Thus, for example, an approximate confidence interval for S_V with confidence level $1 - \alpha$ is given by

$$2(\sqrt{n} - z_\alpha/2)^2/l \leq S_V \leq 2(\sqrt{(n+1)} + z_\alpha/2)^2/l \tag{11.2.9}$$

where l is the total length of the test lines employed, while n is the number of points of intersection of the surface with the test system. However this depends on the somewhat dubious assumption that the points of intersection can be approximated by a Poisson point process.

There are a large number of papers on the statistics of stereology, including the topic of experimental design. A representative list is: CLARKE (1981), CRUZ-ORIVE (1980a), CRUZ-ORIVE and WEIBEL (1981), MASE (1982a), MILES (1978), NICHOLSON (1970, 1976, 1978), HERMANS and RASSON (1981), and WATSON (1978). References to papers on some specific statistical questions are given in the following sections.

11.2.3 Thick sections

To introduce this section we describe the process of taking a thick section (illustrated by Figure 11.2, page 299). The random closed set Ξ_V under study is intersected by two parallel planes separated by a distance t. For the sake of definiteness one of these planes is taken to be the (x, y)-plane. The 'slice' that is to be examined is the set

$$T^t = \{\boldsymbol{x} = (x, y, z) : 0 \leqq z \leqq t\}$$

lying between these two planes. The experimental data is the image Ξ^t of the projection of $\Xi_V \cap T^t$ onto the (x, y)-plane. The projected thick section Ξ^t inherits the properties of stationarity and isotropy from Ξ_V.

There seem to be no exact general formulae relating V_V and S_V to intensities that are measurable on Ξ^t. Formulae are known for particular cases. Section 11.3.3 gives examples. In the motion-invariant case an accurate approximate formula is given by

$$V_V \simeq A_A(t) - tL_A(t)/\pi \tag{11.2.10}$$

where $A_A(t)$ is the area fraction and $L_A(t)$ is the mean boundary length per unit area of Ξ^t. This formula was discussed by STOYAN (1985b).

An approximation for S_V is

$$S_V \simeq (4/\pi)\, L_A(t)\,. \tag{11.2.11}$$

This is not as accurate as (11.2.10).

Let Ξ_V be a motion-invariant fibre process. Then Ξ^t is a planar motion-invariant fibre process. The specific length J_V of Ξ_V satisfies

$$J_V = 4L_A(t)/(\pi t) \tag{11.2.12}$$

where $L_A(t)$ is the specific curve length of Ξ^t. NAGEL (1982) gives a proof using the theory of fibre processes.

11.3 Stereological mean-value formulae for germ-grain models

11.3.1 Planar sections

Let Ξ_V be a spatial stationary germ-grain model and let $\Psi = \{[x_n; \Xi_n]\}$ be the corresponding marked point process, following the notation of Section 6.4. The distribution of the 'typical' primary grain Ξ_0 is assumed to be rotation-invariant (but Ξ_V itself need not be isotropic!). We consider the intersection Ξ_A of Ξ_V with an arbitrary test plane T

$$\Xi_A = \Xi \cap T\,.$$

The position and orientation of T can be chosen for convenience (because of the assumptions on Ξ) so we take T to be the (x, y)-plane. The intersection Ξ_A is itself a germ-grain model. We seek to relate the mean-value characteristics and germ-process intensity λ_V of Ξ_V to the mean-value characteristics and germ-process intensity λ_A of Ξ_A. The following Sections (11.4 and 11.5) consider further the problem of estimating shape and size distributions for the grains of Ξ_V (the so-called *unfolding problem*).

Throughout this section we assume that the grains of Ξ_V are convex. This assumption is frequently made; it seems to be realistic in many cases and measurement difficulties arise when it does not hold. However the basic formula (6.4.5) has a generalized version holding for non-convex particles so generalization is possible. Miles (1983, 1985a) discussed the non-convex case. It is necessary to be able to decide whether different components of a section actually arise from the same grain.

Under the assumption of convex grains it is clear that the intersection Ξ_A is also a stationary germ-grain model with convex primary grains. Because the grains of Ξ_V have rotation-invariant distribution so has the 'typical' grain $\Xi_{A,0}$. If Ξ_V is a Boolean model then Ξ_A is as well; the intersection with the test plane can then be interpreted as a position-dependent, independent thinning procedure.

Theorem 11.1. *The mean-value characteristics of Ξ_V and Ξ_A are connected by*

$$\lambda_V \bar{V} = \lambda_A \bar{A} \tag{11.3.1}$$

$$\lambda_V \bar{S} = 4\lambda_A \bar{U}/\pi \tag{11.3.2}$$

$$\lambda_V \bar{b} = \lambda_V \bar{M}/2\pi = \lambda_A \tag{11.3.3}$$

where $\bar{b}$ is the mean average breadth of the 'typical' grain.

Proof (Stoyan, 1979c). Consider the disc

$$c_r = \{\boldsymbol{x} = (x, y, z) : x^2 + y^2 \leq r^2, z = 0\}.$$

By construction the numbers of primary grains hitting c_r must be the same for Ξ_V and Ξ_A, for all r. The corresponding mean values are given by formula (6.4.7) as

$$\lambda_V \mathsf{E}\nu_3(c_r \oplus \check{\Xi}_0) = \lambda_V[\bar{V} + \pi\bar{S}r/4 + \pi\bar{b}r^2]$$

$$\lambda_A \mathsf{E}\nu_2(c_r \oplus \check{\Xi}_{A,0}) = \lambda_A[\bar{A} + \bar{U}r + \pi r^2]$$

for all $r \geq 0$. The area in the second formula has been calculated by means of the Steiner formula (p. 173); the volume in the first by means of the generalization (6.1.1). Comparison of the coefficients of powers of r leads to formulae (11.3.1), (11.3.2), (11.3.3).

The formulae do not depend on assumptions such as independence of the grains or particular distributions for the germ process. If the primary grains are non-overlapping then $\lambda_V = N_V$ and so

$$V_V = \lambda_V \bar{V}$$

$$S_V = \lambda_V \bar{S}$$

$$M_V = \lambda_V \bar{M} = 2\pi\lambda_V \bar{b}.$$

The fundamental stereological formulae (11.2.1), (11.2.2) and (11.2.6) are obtained by transformation of (11.3.1), (11.3.2), and (11.3.3).

The primary grain of the plane section set Ξ_A has mean characteristics (mean Minkowski functionals) given by

$$\bar{A} = \bar{V}/\bar{b} \tag{11.3.4}$$

$$\bar{U} = \pi\bar{S}/(4\bar{b}). \tag{11.3.5}$$

The spatial characteristics λ_V, $\bar{V}$, $\bar{S}$, and $\bar{b}$ cannot be determined directly from the formulae in the theorem; there are only three equations available for four unknowns. The usual procedure adopted to overcome this difficulty is to make use of *form factors* such as Blaschke's coefficients

$$f = 4\pi\bar{S}/\bar{M}^2$$

$$g = 48\pi^2\bar{V}/\bar{M}^3.$$

Voss (1980) gives an effective and well-thought-out form factor method. Form-factors are discussed in Weibel (1980) and Underwood (1976b).

Miles (1983) suggested a parametric method in which a one-parameter family of distribution functions is assumed for the grain-volume. The actual parameter is estimated using the volume moment ratio $\zeta = \overline{V^2}/\bar{V}$. This ratio can be estimated from linear sections as can be seen by

$$\overline{l^4}/\bar{l} = 3\overline{V^2}/(\pi\bar{V}). \tag{11.3.6}$$

Here l is the chord length of a typical grain intersecting the section line; $\bar{l}$ and $\overline{l^4}$ are the first and fourth moments respectively. Miles (1983, 1985a) proves this formula and also a generalization for the non-convex case. Formula (11.3.6) also yields the mean grain volume $\bar{V}_{(v)}$ if the mean is taken with respect to the volume-weighted grain distribution. (Note that $\bar{V}_{(v)} = \bar{V}$ if all grains have the same volume.) Equation (6.4.2) gives

$$\bar{V}_{(v)} = \overline{V^2}/\bar{V} = \zeta. \tag{11.3.7}$$

An alternative formula for $\bar{V}_{(v)}$ is

$$\bar{V}_{(v)} = \frac{\pi}{3}\,\overline{l_0^3}$$

where $\overline{l_0^3}$ is the third moment of the linear intercept in a random direction through a random test point hitting a grain (see also 11.4.4). Care must be taken to correctly sample the intercepts; see Gundersen and Jensen (1985).

Example 11.1. *Planar section of a Boolean model.*

The mean characteristics of a spatial stationary Boolean model Ξ_V with convex isotropic primary grains are given by

$$\lambda_V,\ \bar{V},\ \bar{S} \quad \text{and} \quad \bar{M} = 2\pi\bar{b}.$$

The planar section Ξ_A is again a Boolean model with characteristics

$$\lambda_A,\ \bar{A} \quad \text{and} \quad \bar{U}$$

given by (11.3.3), (11.3.4), and (11.3.5).

The area fraction A_A, specific boundary length L_A, and specific convexity number N_A^+ of Ξ_A are therefore obtained by using the formulae in Section 3.2

$$A_A = 1 - \exp(-\lambda_A \bar{A}) = 1 - \exp(-\lambda_V \bar{V})$$

$$L_A = \lambda_A \bar{U} \exp(-\lambda_A \bar{A}) = \lambda_V \pi (\bar{S}/4) \exp(-\lambda_V \bar{V})$$

and

$$N_A^+ = \lambda_A \exp(-\lambda_A \bar{A}) = \lambda_V \bar{\bar{b}} \exp(-\lambda_V \bar{V}).$$

11.3.2 Sections of d-dimensional germ-grain models by k-dimensional subspaces

The proofs of formulae for the general case follow that of 11.3.1 for the three-dimensional case. As in 6.4. λ denotes the intensity of the point process of germs while $\overline{W}_l$ denotes the mean lth Minkowski functional of the 'typical' grain Ξ_0. The corresponding quantities in the k-dimensional subspace are λ_k and $\overline{W}_l^k$. The formulae are

$$\lambda_k = \lambda \frac{\omega_{d-k}}{\omega_d} \overline{W}_k \tag{11.3.8}$$

and

$$\overline{W}_l^k = \frac{\omega_k \omega_{d-l}}{\omega_{k-l} \omega_{d-k}} \overline{W}_l / \overline{W}_k . \tag{11.3.9}$$

These formulae are taken from MATHERON (1975) who gave them for the particular case of a Boolean model.

11.3.3 Thick sections of spatial germ-grain models

The mean-value characteristics of Ξ_V can be related to those of Ξ_t when Ξ_V is a germ-grain model. The formulae are

$$\lambda_V(\bar{V} + \bar{S}t/4) = \lambda_A \bar{A} \tag{11.3.10}$$

$$\lambda_V(\pi \bar{S}/4 + \pi \bar{b} t) = \lambda_A \bar{U} \tag{11.3.11}$$

$$\lambda_V(\bar{b} + t) = \lambda_A . \tag{11.3.12}$$

VOSS and STOYAN (1985) gave the proof which follows that of theorem 11.1; the disc c_r is replaced by the cylinder

$$C_r = \{\boldsymbol{x} = (x, y, z) : x^2 + y^2 \leq r^2, 0 \leq z \leq t\}.$$

Formula (11.3.10) is due to CAHN and NUTTING (1959) and formula (11.3.11) to DE HOFF (1968).

The mean characteristics of the 'typical' projected grain of Ξ_t are

$$\bar{A} = \frac{\bar{V} + \bar{S}t/4}{\bar{b} + t} = \frac{4\pi \bar{V} + \pi \bar{S} t}{2\overline{M} + 4\pi t} \tag{11.3.13}$$

$$\bar{U} = \frac{\pi \bar{S}/4 + \pi \bar{b} t}{\bar{b} + t} = \frac{\pi^2 \bar{S} + 2\pi \overline{M} t}{2\overline{M} + 4\pi t} . \tag{11.3.14}$$

See, for example, Voss (1980). The formulae for the Boolean model in Section 3.2 give formulae for characteristics of Ξ_t when Ξ_V is a Boolean model. The formulae concern the area fraction $A_A(t)$, specific boundary length $L_A(t)$, and specific connectivity number $N_A(t)$

$$A_A(t) = 1 - \exp\left(-\lambda_V(\bar{V} + \bar{S}t/4)\right) \tag{11.3.15}$$

$$L_A(t) = \pi\lambda_V(\bar{S}/4 + \bar{b}t) \exp\left(-\lambda_V(\bar{V} + \bar{S}t/4)\right) \tag{11.3.16}$$

$$N_A(t) = \left[\lambda_V(\bar{b} + t) - \frac{\pi}{4}\lambda_V^2(\bar{b}t + \bar{S}/4)^2\right] \exp\left(-\lambda_V(\bar{V} + \bar{S}t/4)\right) \tag{11.3.17}$$

as in MILES (1976).

The equation systems (11.3.10) to (11.3.12), as well as (11.3.15) to (11.3.17), have properties similar to those of the system (11.3.1) to (11.3.3). Again there are three equations and four unknown variables λ_V, $\bar{V}$, $\bar{S}$, and $\bar{b}$. One way of resolving this problem is to use two probes of different thicknesses t_1 and t_2 as proposed and discussed by EBBESON and TANG (1965), GIGER (1967) and MILES (1976) (see also WEIBEL (1980)). This approach has a disadvantage in that the resulting equations are ill-conditioned and sensitive with respect to measurement errors.

Another possibility is to use a whole sequence of different probes with thicknesses $t_1, t_2, \ldots, t_n$, to measure the corresponding planar quantities and to determine the parameters of the spatial model by regression. This method can be used to analyze germ-grain models that are not necessarily Boolean as in the following example.

Example 11.2. *Object counting method of Voss applied to liver cell nuclei of rats.* See Voss (1982b), WASSILEW and Voss (1982), and Voss and STOYAN (1985).

The objective of the work referred to here was to determine N_V for the spatial process of liver cell nuclei in rat liver tissue. Slices of tissue were taken and their approximate thicknesses were determined by micrometer. The thicknesses were grouped as $t \simeq 2, \ldots, 10$ μm. Counts were taken of Feulgen-stained liver cell nuclei from rectangular windows of observation of dimensions 97 × 70 μm. The results are graphed in Figure 11.1. No nuclei overlapped. The regression line for the data is

$$N(t) = 11.1 + 1.34t \quad \text{for } t \text{ in μm}.$$

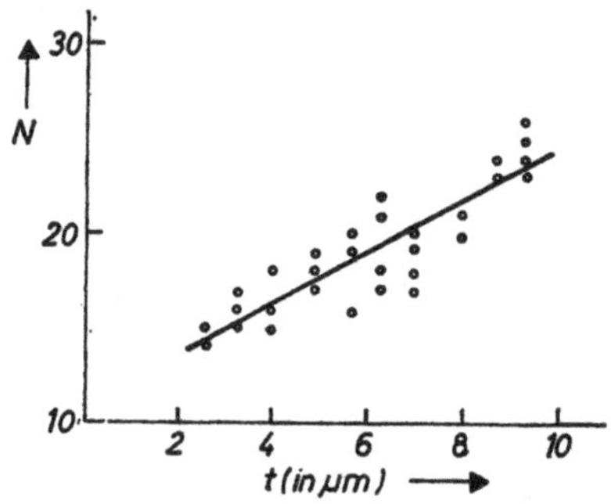

Fig. 11.1. The number of rat liver cell nuclei counted in an observation window, plotted against the section thickness t. See Example 11.2

and so a formula is obtained for the thickness-dependent intensity $N_A(t)$

$$N_A(t) = 0.00163 + 0.000197\, t.$$

Equation (11.3.12) yields

$$\lambda_V = N_V = 0.000197\ \mu\text{m}^{-3}$$

$$\bar{\bar{b}} = 8.2\ \mu\text{m}$$

$$\bar{M} = 51.8\ \mu\text{m}.$$

Additional measurements of $\bar{A}$ and $\bar{U}$ for one value of t would allow the estimation of $\bar{V}$ and $\bar{S}$.

The determination of N_V is straightforward if information can be obtained about the structure between the parallel planes (for example, whether a particle intersects only one of the planes; STERIO, 1984).

Further approaches to the determination of N_V are the '$2^1/_2$-dimensional' method using wedge sections, after MILES and DAVY (1977); and the use of parallel serial sections as described by CRUZ-ORIVE (1980b), DE HOFF (1983), and STERIO (1984). For stationary and isotropic fibre processes ZÄHLE (1984b) showed how to determine the mean length of fibre per unit volume and the mean length of the 'typical' fibre.

11.3.4 Tubular structures and thin surfaces

It should be noted that there are stereological methods for models that relate to spatial fibre and surface processes in the same manner as germ-grain models relate to point processes. These fibre and surface based models are systems of tubes and thin surfaces (for example tissue membranes). Stereological techniques yield V_V, S_V, L_V and so forth; see GUNDERSEN (1979), WEIBEL (1980), and BADDELEY and AVERBACK (1983). The latter authors demonstrate the use of K_V for the characterisation of the 'tortuosity' of tubular structures. Methods for the determination of the distribution of the thickness of membranes are discussed by JENSEN, GUNDERSEN, and ØSTERBY (1979), and CRUZ-ORIVE (1979); see also WEIBEL (1980).

11.4 Stereological methods for systems of spheres

This section presents methods for a particular case of a stationary germ-grain model: that of a random system of spheres in $\mathbb{R}^3$. The case of planar sections is studied in detail while thick and linear sections are touched on briefly. These problems have provoked a vast literature and no attempt is made here to cover all references. Instead we refer the reader to the surveys of GOLDMAN and VISSCHER (1978), WEIBEL (1980), and CRUZ-ORIVE (1983a).

11.4.1 Planar sections and Wicksell's problem

The following quantities can be determined from a planar section by means of the methods discussed in Section 6.4:

- N_A the mean number of section circles per unit area,
- D_A the distribution function of diameters of section circles.

The distribution function D_A always has a density function denoted here by δ_A. The aim of stereology is to determine the following quantities:

N_V the mean number of spheres per unit volume,

D_V the distribution function of the diameters of spheres,

d_V the mean sphere diameter.

Procedures for solving the important problem of determining D_V from knowledge of D_A are called *unfolding* procedures. The unfolding problem was first posed and solved by WICKSELL (1925, 1926). However, as will become clear there are statistical and numerical difficulties in implementing the solution.

The following formulae relate these quantities to each other.

$$d_V = \frac{\pi}{2}\left[\int_0^\infty \frac{1}{r}\,\delta_A(r)\,\mathrm{d}r\right]^{-1} \tag{11.4.1}$$

$$N_V = N_A/d_V \tag{11.4.2}$$

$$D_V(r) = 1 - \frac{2d_V}{\pi}\int_r^\infty \frac{\delta_A(x)}{\sqrt{(x^2 - r^2)}}\,\mathrm{d}x,\ r \geqq 0 \tag{11.4.3}$$

$$D_A(r) = 1 - \frac{1}{d_V}\int_0^\infty \left[1 - D_V\left(\sqrt{(r^2 + x^2)}\right)\right]\mathrm{d}x \quad \text{as } r \geqq 0. \tag{11.4.4}$$

Formula (11.4.2) is a particular case of (11.3.3). If D_V has a density δ_V then

$$\delta_A(r) = \frac{r}{d_V}\int_r^\infty \frac{\delta_V(x)}{\sqrt{(x^2 - r^2)}}\,\mathrm{d}x.$$

The solution of this Abel integral equation is

$$\delta_V(r) = -\frac{2d_V r}{\pi}\int_r^\infty \frac{1}{\sqrt{(x^2 - r^2)}}\frac{\mathrm{d}}{\mathrm{d}x}\left[\frac{\delta_A(x)}{x}\right]\mathrm{d}x \quad \text{for } r \geqq 0. \tag{11.4.5}$$

There are nice relations between the moments of D_A and D_V which can be found on page 300 in the context of thick sections.

Example 11.3. *Particular forms of D_V.*

(a) Let the diameters of the spheres be constant. Then (11.4.4) shows

$$D_A(r) = 1 - \frac{1}{d_V}\sqrt{(d_V^2 - r^2)},$$
$$\delta_A(r) = r/\left(d_V\sqrt{(d_V^2 - r^2)}\right) \quad \text{for } 0 \leqq r \leqq d_V. \tag{11.4.6a}$$

The mean and variance of the section circles are

$$d_A = \pi d_V/4, \qquad \sigma_A^2 = (32 - 3\pi^2)\,d_V^2/48. \tag{11.4.6b}$$

These equations can be used to estimate d_V and to test whether the diameters are constant.

(b) If the diameters of the spheres have a Rayleigh distribution, that is if

$$\delta_V(r) = \frac{r}{a^2} \exp\left(-r^2/(2a^2)\right) \quad \text{for } r \geqq 0, d_V = a \sqrt{\frac{\pi}{2}}$$

then the distribution reproduces itself

$$\delta_V(r) = \delta_A(r).$$

This property makes the use of the Rayleigh distribution attractive for stereology. The paper of Rysz and Wiencek (1980) gives an interesting example.

Before turning to the statistical problems in the determination of N_V, d_V, and D_V by the above formulae we present a proof (following Mecke and Stoyan, 1980b) of formulae (11.4.1) to (11.4.4).

Proof of formulae (11.4.1) *to* (11.4.4). The system of spheres can be described by a marked point process Ψ_V;

$$\Psi_V = \{[x_n, y_n, z_n; d_n]\}.$$

Here $\boldsymbol{x}_n = (x_n, y_n, z_n)$ is the centre of the nth sphere and the mark d_n gives its diameter. The corresponding intensity measure Λ_V has the form

$$\Lambda_V = N_V \nu_3 \boldsymbol{D}_V.$$

where $\boldsymbol{D}_V$ denotes the Palm mark distribution. Thus $\boldsymbol{D}_V$ is the distribution of the diameter of a 'typical' sphere

$$\boldsymbol{D}_V([0, r]) = D_V(r) \quad \text{for } r \geqq 0.$$

All that is observed of a sphere is its intersection with the (x, y)-plane. The condition that a section is not empty is precisely

$$4z^2 \leqq d^2,$$

and the resulting section diameter s is given by

$$s = \sqrt{(d^2 - 4z^2)_+}.$$

Consider the system of all section circles. This can be described by another marked point process which this time lies in the plane:

$$\Psi_A = \{[x_m, y_m; s_m]\}.$$

Its intensity measure is Λ_A given by

$$\Lambda_A = N_A \nu_2 \boldsymbol{D}_A.$$

and $\boldsymbol{D}_A$ here is the diameter distribution of the 'typical' section circle (ignoring those of null section) so

$$\boldsymbol{D}_A([0, r]) = D_A(r).$$

The Λ_A measure of the set $C = [0, 1) \times [0, 1) \times (r, \infty)$ can be expressed in two ways. Firstly the formula for Λ_A yields

$$\Lambda_A(C) = N_A\left(1 - D(r)\right).$$

Secondly since the points of Ψ_A result from points of Ψ_V whose spheres hit the (x, y)-plane such points correspond to points $[x, y, z; d]$ of Ψ_V with

$$\left(x, y, \sqrt{(d^2 - 4z^2)_+}\right) \in C.$$

Therefore the expected number $\Lambda_A(C)$ of Ψ_A points in C must equal the expected number of Ψ_V points for which the above is true:

$$\Lambda_A(C) = \mathsf{E} \sum_{[x,y,z,d] \in \Psi_V} \mathbf{1}_C(x, y, w)$$

where $w = \sqrt{(d^2 - 4z^2)_+}$. Therefore

$$\Lambda_A(C) = \int_{R^3 \times [0,\infty)} \mathbf{1}_C(x, y, w)\, \Lambda_V\big(d(x, y, z, d)\big)$$

$$= N_V \int_0^\infty \boldsymbol{D}_V\left(\sqrt{(r^2 + t^2)}, \infty\right) \mathrm{d}t\,.$$

This follows from the formula for Λ_V and the Campbell theorem. Formula (11.4.2) follows from equating both expressions for $\Lambda_A(C)$, expressing

$$\boldsymbol{D}_V\big([u, \infty)\big) = 1 - D_V(u)$$

and setting r equal to zero.

The formula (11.4.4) follows directly from (11.4.2). Integration shows that (11.4.4) has as solution the result summarized in (11.4.3). Setting $r = 0$ in (11.4.3) gives (11.4.1).

Statistical methods

Formula (11.4.1) gives a basis for the direct estimation of d_V or N_V; an estimator $\hat{d}_V$ for d_V is given by

$$\hat{d}_V = \frac{n\pi}{2} \Big/ \sum_{i=1}^{n} d_i^{-1}. \tag{11.4.7}$$

This estimator has been suggested by Saltykov (1950) and Fullman (1953). It has been studied by Watson (1971) and Franklin (1977, 1981); the latter constructed a confidence interval for d_V. It has been criticized on the grounds that its variance is infinite; see for example, Ripley (1981, p. 210). Estimators for other moments of D_V have been studied in Watson (1971), Nicholson (1970, 1976, 1978), and Jakeman and Schaeffer (1978).

The determination of the distribution function D_V is an ill-posed problem of a well-known kind. Many statistical procedures have been developed to overcome the ill-posed nature of the problem; they are described in books on stereology and in Goldman and Visscher (1978), Cruz-Orive (1983a) and Kanatani and Ishikawa (1985). Following Blödner, Mühlig, and Nagel (1984), they may be classified in one of six categories.

(a) Discretization of (11.4.4) by discretizing D_V, which corresponds to grouping the sample. The integral is then approximated by a finite sum, and a finite system of linear equations is produced. Methods of this kind have been proposed by Wicksell, and by Scheil-Schwartz-Saltykov; see LEWIS, WALTHERS, and JOHNSON (1973), SALTYKOV (1974), and WEIBEL (1980).

(b) Application of (11.4.3), replacing D_A by a empirical distribution function; see BLÖDNER, MÜHLIG, and NAGEL (1984) and KANATANI and ISHIKAWA (1985).

(c) Applying (11.4.5) and using spectral differentiation; ANDERSSEN and JAKEMAN (1975a, b) or kernel density estimators as in TAYLOR (1983).

(d) Parametric methods. These involve making a parametric assumption about the form of D_V and then estimating the parameters; see DE HOFF (1965), GIGER and RIEDWYL (1970), LIKEŠ (1963), KEIDING, JENSEN, and RANEK (1972), SCALES *et al.* (1982), and SUWA *et al.* (1976).

(e) Trial and error methods, which consist of guessing a diameter distribution $D_V^{(1)}$, computing the corresponding $D_A^{(1)}$, comparing this with the empirical D_A, and then making a second guess at the diameter distribution, and so forth; see Voss and SCHUBERT (1975).

(f) Regularization, which consists of replacing the original integral equation by a simpler one to produce approximate solutions; see NYCHKA (1983a, b). This author also used spline methods.

No method has clear advantages over all others. Statistical and numerical difficulties interact in a delicate fashion; see ANDERSSEN and JAKEMAN (1975a). Possibly the most appealing approach is to use a non-parametric method for pilot studies, and then to conduct a final analysis parametrically using whatever family of distributions is suggested by the preliminary stage.

These methods have been compared by simulation studies. See for example that of BLÖDNER, MÜHLIG and NAGEL (1984), which provides evidence to favour the methods of Scheil-Schwartz-Saltykov and Blödner-Mühlig-Nagel over other methods of types (a) and (b). These two methods are discussed briefly below.

The conclusions of simulation studies naturally depend on the diameter distribution D_V that is used. In contrast GERLACH and OHSER (1985) and KANATANI and ISHIKAWA (1985) determined *condition numbers* (see e.g. HOUSEHOLDER (1964)) for matrices derived from the linear systems corresponding to the various methods. They found the two methods mentioned above to have smaller condition numbers and therefore to be less vulnerable than the other methods to *statistical* fluctuations.

To establish notation let $d_1, \ldots, d_n$ be a sample of diameters of section circles in a planar section. An interval $(0, d_u)$ is chosen, with suitable upper limit d_u exceeding all the d_i. This interval is divided into k cells of width $\Delta = d_u/k$. The number n_i of circles with diameters in $C_i = ((i-1)\,\Delta, i\Delta]$ is recorded, for each i. The aim is to provide good estimates $\hat{N}_j$ for the expected number of spheres per unit volume having diameters in C_j for $j = 1, \ldots, k$. The density of the distribution D_V can then be estimated as the histogram density $\hat{N}_j/\sum_l \hat{N}_l$ in the cell C_j.

Scheil-Schwartz-Saltykow method

If N_j is the expected number of spheres per unit volume with diameters in C_j then in a plane section these spheres give contributions to the expected numbers n_i of section circle diameters per unit area as follows

$$n_i = \Delta \sum_{j=1}^{n} b_{ij} N_j, \qquad i = 1, \ldots, k$$

where the coefficients are

$$b_{ij} = \sqrt{(j^2 - (i-1)^2)} - \sqrt{(j^2 - i^2)} \quad \text{if } i \leqq j,\ j = 1, \ldots, k$$
$$= 0 \text{ otherwise.}$$

Here, the assumption is made that all spheres contributing to C_j actually have diameter $j\Delta$. We note in passing that the studies of Blödner-Mühlig–Nagel and Gerlach–Ohser indicate that other assumptions, such as perhaps that the diameters are $j\Delta - \Delta/2$, or that they are uniformly distributed in $\big((j-1)\,\Delta, j\Delta\big)$, do not give better results.

The estimates $\hat{N}_j$ can now be obtained by solving for N_j in these equations. The solutions have the form

$$\hat{N}_j = \frac{1}{\Delta} \sum_{i=1}^{k} c_{ij} n_i \tag{11.4.8}$$

where the c_{ij} may be obtained by back-substitution in the linear equations for n_i above. A table of the c_{ij} coefficients can be found for example in WEIBEL (1980).

As a by-product we obtain estimators for N_V and d_V:

$$\hat{N}_V = \sum_{j=1}^{k} \hat{N}_j \quad \text{and} \quad \hat{d}_V = 2n/\hat{N}_V. \tag{11.4.9}$$

Blödner-Mühlig-Nagel method

This method starts from formula (11.4.3) and replaces D_A by an empirical distribution namely

$$\hat{D}_A(r) = \sum_{i:\left(i-\frac{1}{2}\right)\Delta<r} n_i/n.$$

The corresponding N_j again satisfy the equation system (11.4.8) but this time with coefficients c_{ij} given by

$$c_{ij} = 0 \quad \text{if } i < j$$

$$c_{ij} = \frac{2}{\pi} \frac{1}{\sqrt{\left(\left(i - \frac{1}{2}\right)^2 - (i-1)^2\right)}} \quad \text{if } i = j$$

$$c_{ij} = \frac{2}{\pi} \left[\frac{1}{\sqrt{\left(\left(i - \frac{1}{2}\right)^2 - (j-1)^2\right)}} - \frac{1}{\sqrt{\left(\left(i - \frac{1}{2}\right)^2 - j^2\right)}}\right] \quad \text{if } i > j.$$

Blödner, Mühlig, and Nagel report on their experience with these methods. They suggest the use of large samples (of the order of 1000 section circles) and k in the range 7, ..., 10. The methods can often produce negative $\hat{N}_j$ and these values should not be replaced uncritically by zero as they contain information about the unknown distribution. The authors suggest amalgamation of such cells with their neighbours. Furthermore, it is possible to start with values of Δ smaller than eventually required and then to unite neighbouring classes after the calculation. This method works well if Δ is not chosen too small.

Example 11.4. *Application of the Scheil–Schwartz–Saltykow and Blödner–Mühlig–Nagel methods to synthetic samples of section circle diameters.*

In order to demonstrate the occurrence of numerical and statistical errors, and to indicate the separate ways in which they can arise, the true sphere diameters were assumed to follow a Rayleigh distribution with parameter $a = 3$. This implies $d_V = 3.76$. In this case, the distribution of section circle diameters coincides with that of the sphere diameters, as mentioned in Example 11.3.

The first column of Table 11.2 shows the exact cell-expectations N_i of section circle diameters for $\Delta = 1$. The unfolding procedures of Scheil–Schwartz–Saltykov and Blödner–Mühlig–Nagel produce the values displayed in columns 2 and 3, starting from the cell-expectations in column 1. As noted above, column 1 can also be interpreted as giving the cell-expectations for sphere diameters, so the differences between this and the next two columns show the *numerical* errors that can arise in the unfolding process. They are not small, at least in the higher rows (which correspond to smaller diameters).

Statistical errors are demonstrated in columns 4, 5, and 6. Column 4 records cell-expectations for a sample of 1000 circle diameters taken from the true section circle diameter distribution. The resulting sphere diameter distributions given by Scheil–Schwartz–Saltykov and by Blödner–Mühlig–Nagel are indicated in columns 5 and 6 respectively. Comparison with column 1 indicates the extent of error now resulting both from numerical and statistical causes.

Formula (11.4.9) gives for d_V the estimates

$\hat{d}_V = 4.02$ Scheil–Schwartz–Saltykov

$\hat{d}_V = 3.83$ Blödner–Mühlig–Nagel

while (11.4.7) yields

$$d_V = 3.89.$$

Table 11.2. Data for the example of a solution of Wicksell's problem

i, j	Rayleigh exact cell-expectations	*SSS* from exact *c–e*	*BMN* from exact *c–e*	Sample $n = 1000$ *c–e*	*SSS* from sample *c–e*	*BMN* from sample *c–e*
1	0.054	0.082	0.155	0.050	0.063	0.134
2	0.145	0.166	0.170	0.149	0.172	0.176
3	0.194	0.201	0.187	0.201	0.214	0.200
4	0.195	0.189	0.170	0.197	0.195	0.176
5	0.162	0.149	0.132	0.161	0.153	0.136
6	0.114	0.101	0.088	0.107	0.095	0.084
7	0.070	0.059	0.052	0.066	0.058	0.051
8	0.037	0.031	0.027	0.029	0.019	0.016
9	0.017	0.014	0.012	0.026	0.022	0.019
10	0.007	0.006	0.005	0.008	0.005	0.005
11	0.003	0.002	0.002	0.006	0.005	0.005
12	0.001	0.001	0.001	0.000	0.000	0.000
13	0.000	0.000	0.000	0.000	0.000	0.000

11.4.2 Linear sections

The following quantities are available from a linear section:

N_L mean number of chords per unit length,

D_L distribution of chord lengths.

The distribution function D_L always has a density δ_L.

The aims of statistical analysis in the case of linear unfolding are as in the case of planar unfolding. The mathematical problem of establishing relationships between N_L and D_L on the one hand and N_V and D_V on the other has been solved by SPEKTOR (1950) and LORD and WILLIS (1951). The following formulae can be proved/in a way similar to that of the planar case

$$N_V = 4N_L/(\pi d_V^{(2)}) \tag{11.4.10}$$

$$D_V(r) = 1 - (d_V^{(2)}/2)\, \delta_L(r)/r \tag{11.4.11}$$

$$D_L(r) = \frac{2}{d_V^{(2)}} \int_0^r x[1 - D_V(x)]\, dx \quad \text{for } r \geqq 0. \tag{11.4.12}$$

Here $d_V^{(2)}$ denotes the second moment of D_V.

Example 11.5. *Particular forms of D_V.*

(a) If the sphere diameters are constant then by (11.4.12)

$$\begin{aligned} D_L(r) &= r^2/d_V^2 \\ \delta_L(r) &= 2r/d_V^2 \quad \text{for } 0 \leqq r \leqq d_V. \end{aligned} \tag{11.4.13}$$

The mean chord length satisfies

$$d_L = 2d_V/3.$$

(b) Let the sphere diameters have an exponential distribution with parameter μ. Then

$$D_L(r) = 1 - (1 - \mu r) \exp(-\mu r)$$

$$\delta_L(r) = \mu^2 r \exp(-\mu r) \quad \text{for } r \geqq 0.$$

Relation (11.4.11) implies that

$$d_V^{(2)} = 2 \lim_{r \downarrow 0} (r/\delta_L(r)).$$

Consequently, a density function f is a chord length density function only if

$$0 < \lim_{r \downarrow 0} (f(r)/r) = f'(0) < \infty. \tag{11.4.14}$$

Among the family of gamma distributions with densities

$$f(r) = \mu^\alpha r^{\alpha-1} \exp(-\mu r)/\Gamma(\alpha), \qquad r \geqq 0,$$

this condition (11.4.14) is fulfilled only for $\alpha = 2$. As case (6) shows this case arises when the sphere diameters have an exponential distribution.

The linear section method is even more unstable than that of planar sections; tests suggest that 'it should not be used whenever planar sectioning is possible' (RIPLEY, 1981 p. 212; see also FLOOK, 1983). Errors in measurement and statistical errors play a greater role. Also the numerical errors in the computation are greater as indicated in Example 11.6. GERLACH and OHSER (1985) found condition numbers much greater in the Spektor–Lord–Willis method compared with methods based on planar sections.

The method of Spektor, Lord, and Willis was derived by BOCKSTIEGEL (1966), using 'intensities' as

$$N_V(r) = N_V\delta_V(r)$$

(tacitly assuming that D_V has a density δ_V),

$$N_L(r) = N_L\delta_L(r).$$

Thus $N_V(r)\,\mathrm{d}r$ can be interpreted as the mean number of spheres with a diameter between r and $r + \mathrm{d}r$ per unit volume and $N_L(r)$ can be interpreted analogously. The two 'intensities' are related by

$$N_V(r) = -\frac{2}{\pi}\frac{\mathrm{d}}{\mathrm{d}r}\left[\frac{N_L(r)}{r}\right] \quad \text{for } r \geqq 0$$

eliminating the quantity $d_V^{(2)}$.

Let n_i be the number of chords per unit length with chord length in the cell $C_i = \big((i-1)\,\Delta, i\Delta\big]$ with Δ suitably chosen. Then n_i approximates the integral

$$\int\limits_{(i-1)\Delta}^{i\Delta} N_L(x)\,\mathrm{d}x = \frac{\pi}{2}\int\limits_{(i-1)\Delta}^{i\Delta} x \int\limits_{x}^{\infty} N_V(u)\,\mathrm{d}u\,\mathrm{d}x$$

$$= \frac{\pi}{4}\big((i\Delta)^2 - ((i-1)\,\Delta)^2\big)\int\limits_{\zeta}^{\infty} N_V(u)\,\mathrm{d}u$$

where ζ is a mean-value interpolation for the outer integral

$$(i-1)\,\Delta \leqq \zeta \leqq i\Delta.$$

Analogously n_{i+1} approximates

$$\frac{\pi}{4}\big(((i+1)\,\Delta)^2 - (i\Delta)^2\big)\int\limits_{\eta}^{\infty} N_V(u)\,\mathrm{d}u$$

where $i\Delta \leqq \eta \leqq (i+1)\,\Delta$. Thus

$$\hat{N}_{i+1/2} \simeq \int\limits_{\zeta}^{\eta} N_V(u)\,\mathrm{d}u$$

$$\simeq \frac{4}{\pi}\left[\frac{n_i}{(i\Delta)^2 - ((i-1)\,\Delta)^2} - \frac{n_{i+1}}{((i+1)\,\Delta)^2 - (i\Delta)^2}\right]$$

which approximately equals $N_{i+1/2}$, the mean number of spheres per unit volume with diameters between $i - 1/2$ and $i + 1/2$.

The intensity N_V can be estimated by $\sum_i \hat{N}_{i+1/2}$, since

$$\hat{N}_V = \sum_i \hat{N}_{i+1/2}.$$

Estimation of $d_V^{(2)}$ follows from (11.4.10) and a histogram for δ_V can be constructed by using the values of the $\hat{N}_{i+1/2}$.

Accurate results can only be obtained by using very many chords; it is useful to use the nonparametric method described here only as a pilot study, and then to change to a parametric method.

Note that N_V can be estimated directly if it is known that there is a minimal sphere diameter $d_0 > 0$. Formula (11.4.12) then shows

$$D_L(d_0) = (2/d_V^{(2)})\, d_0^2/2 = d_0^2/d_V^{(2)}$$

Consequently, it is plausible to take

$$N_V = 2p_0 N_L/(\pi d_0)^2$$

where p_0 is the proportion of chord lengths smaller than d_0.

Table 11.3. Results of the applications of the Spector–Lord–Willis method and the Barthel–Ohser method to Example 11.4

i	Rayleigh spheres (exact)	Rayleigh chords (exact)	*SLW* for exact histogram	Sample ($n = 1000$)	*SLW* for sample	*BO* for sample
	0.015					
1		0.054		0.050		0.092
	0.104		0.104		0.007	
2		0.145		0.149		0.202
	0.176		0.177		0.191	
3		0.194		0.201		0.223
	0.200		0.202		0.243	
4		0.195		0.197		0.183
	0.182		0.184		0.207	
5		0.162		0.161		0.138
	0.138		0.141		0.165	
6		0.114		0.107		0.079
	0.091		0.092		0.094	
7		0.070		0.066		0.049
	0.052		0.053		0.064	
8		0.037		0.029		0.009
	0.026		0.027		0.008	
9		0.017		0.026		0.017
	0.011		0.012		0.013	
10		0.007		0.008		0.003
	0.004		0.005		0.002	
11		0.003		0.006		0.004
	0.001		0.002		0.006	
12		0.001		0.000		0.000
	0.000		0.000		0.000	
13		0.000		0.000		0.000

Example 11.6. *Application of the Spektor–Lord–Willis method to a synthetic sample of chord lengths.*

The data of Example 11.4 are used here in order to have an instructive comparison. Because of the assumed Rayleigh distribution for sphere diameter the chord length distribution is also Rayleigh, as is the section circle distribution.

Table 11.3 shows the exact cell-expectations for chord length for $\Delta = 1$; this is placed in column 2. The Spektor-Lord-Willis unfolding procedure produced the values in column 3, which are obtained from the $N_{i+1/2}$. As these are 'intermediate' values they are written between the lines of the other data. Column 1 gives the exact histogram of sphere diameters for the intervals (0, 0.5), (0.5, 1.5), ... The differences between columns 1 and 3 are small; this shows that for this example the numerical errors are small, with the exception of the first class for which the Spektor–Lord–Willis method gives no number.

To demonstrate statistical errors and their interaction with numerical errors, a sample of 1000 chords was taken from the true distribution. Cell-expectations for this are given in column 4. Column 5 gives the resulting sphere diameter distribution unfolded by the Spektor–Lord–Willis method. The errors are not small, especially in the interval (0.5, 1.5). This shows that the method is very sensitive to fluctuations of the n_i.

Barthel, see Barthel, Klimanek and Stoyan (1985), suggested a two-fold application of the Saltykov method in order to unfold the chord length distribution. The first step is to unfold the chord length distribution to a circle diameter distribution. The second step is to unfold the circle diameter distribution to a sphere diameter distribution. This method has a condition number that is smaller than that of the Spektor–Lord–Willis method and it avoids difficulties with the number in the first cell. The last column in Table 11.3 shows the results of this two-step method applied to Example 11.6.

11.4.3 Thick sections

In this section, a system of opaque spheres embedded in a transparent medium is considered. The system is intersected by two parallel planes separated by distance t. Only the projection of the part of the system lying between the two planes is observed, as in Section 11.2.3 (see Figure 11.2). The contrary case, studied by Coleman (1981c, 1983), is when the spheres are transparent and the medium is opaque (as would be the case for holes in Swiss cheese). Just as in the case of planar sections one obtains relationships between N_A and

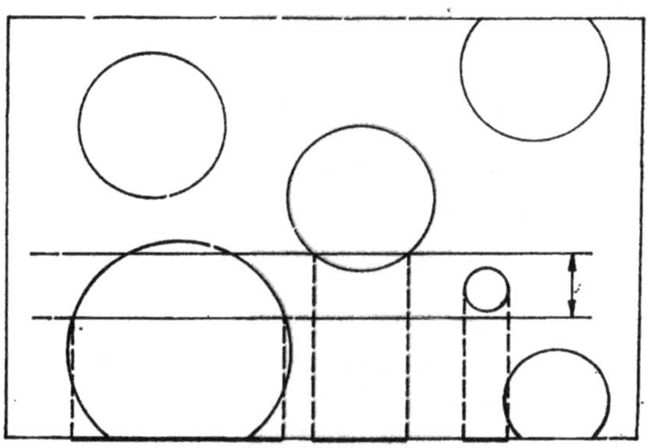

Fig. 11.2. Diagram of a thick section through a system of opaque spheres lying in a transparent medium

D_A on the one hand and N_V and D_V on the other. However the planar quantities must be interpreted more carefully D_A is the *projection* circle diameter distribution, while the method requires the assumption that the *projections* of the system of spheres do not intersect (since N_A is the number density).

Bach (1965a, 1967, 1976) and Mecke and Stoyan (1980b) studied this problem. Bach showed that the solution for D_V is

$$D_V(r) = 1 - \frac{t[1 - D_A(r)] - 2\int\limits_r^\infty p\left(\frac{\sqrt{(2\pi)}}{t}\sqrt{(t^2 - r^2)}\right)[1 - D_A(t)]\frac{t\,dt}{\sqrt{(t^2 - r^2)}}}{t - 2\int\limits_0^\infty p\left(\frac{\sqrt{(2\pi)}}{t}\,t\right)[1 - D_A(t)]\,dt}$$

for $r \geqq 0$.

Here $p(x)$ is given by

$$p(x) = 1 - x\exp(x^2/2)\int\limits_x^\infty \exp(-t^2/2)\,dt \quad \text{for } x \geqq 0.$$

The intensity N_V satisfies

$$N_V = N_A/(d_V + t).$$

Bach (1959, 1967) found some nice relationships between the moments of D_V and D_A. In the case $t = 0$ they reduce of course to those for planar sections. Let

$$M_1 = d_V$$

$$M_k = \int\limits_0^\infty x^k\,dD_V(x), \quad m_k = \int\limits_0^\infty x^k\,dD_A(x) \quad \text{for } k = 0, 1, 2, \ldots.$$

Then the moment relationships are

$$(t + d_V)\,m_k = J_{k+1}M_{k+1} + tM_k \tag{11.4.15}$$

where

$$J_r = \left(\sqrt{\pi}/2\right)\cdot\Gamma(r/2 + 1/2)/\Gamma(r/2 + 1) \quad \text{for } r = 1, 2, \ldots.$$

Bach (1967) suggested a numerical method for the determination of D_V that is similar to that of Saltykov; see also Rose (1980).

Various modifications of the above treatment consider the case where small projections of spheres are ignored, or where the preparation of the projection causes measurement errors; see Bach (1963), Horálek (1978, 1981), Coleman (1980), and the important paper of Cruz-Orive (1983a). The estimation of linear functionals of D_V has been studied, as in Nicholson (1970), Cruz-Orive (1983a), Clarke and Moore (1983), and Piefke (1976).

11.4.4 Sieving distributions for spheres

Sieving distributions for spheres are a particular case of the volume-weighted distributions discussed in Section 6.4. Such distributions can often be more interesting than the number-weighted distributions such as D_V. GUNDERSEN and JENSEN (1983) showed that a direct stereological method allows for their determination in a relatively simple manner.

Let $D_V{}^S$ be the volume-weighted or *sieving diameter distribution*; $D_V{}^S(r)$ is the total volume of spheres of diameter less than r (per unit volume) divided by the total volume of all spheres (again per unit volume). This is related to D_V itself in a manner familiar of Section 6.4; the corresponding densities satisfy

$$\delta_V{}^S(r) = \left[\frac{r}{2}\right]^3 \delta_V(r) \Big/ \int\limits_0^\infty x^3 \delta_V(x) \, \mathrm{d}x. \tag{11.4.16}$$

The stereological measurement is carried out by sampling via a system of test points, lying, for example, on a test grid. For every sample point that falls in a sphere Gundersen and Jensen prescribe that 'the length of a linear intercept, which contains the test point, is measured in a predetermined random direction'. These lengths form a sample with the theoretical probability density δ_0. The formula for total probability shows that

$$\delta_0(l) = \int\limits_l^\infty \delta_0(l \mid x) \, \delta_V{}^S(x) \, \mathrm{d}x \tag{11.4.17}$$

where $\delta_0(l \mid x)$ is the probability density of the length of a point-sampled intercept length through a sphere of diameter x. The function $\delta_0(l \mid x)$ can be shown by geómetric probability, to be

$$\delta_0(l \mid x) = 3l^2/x^3 \quad \text{for } 0 \leqq l \leqq x.$$

This leads to the integral equation

$$\delta_0(l) = \int\limits_l^\infty (3l^2/x^3) \, \delta_V{}^S(x) \, \mathrm{d}x \tag{11.4.18}$$

which has the solution

$$\delta_V{}^S(r) = [2\delta_0(r) - r\delta_0'(r)]/3. \tag{11.4.19}$$

The numerical determination of $\delta_V{}^S$ is straightforward. Note also the mean-value formula (11.3.7).

11.5 Stereological problems for non-spherical grains (Shape and size problems)

11.5.1 General Remarks

In Section 11.4, it is demonstrated that stereology gives rise to hard problems if more is required than mean values; and these difficulties arise even in the simple case of spherical grains. The case of non-spherical grains is even more complicated. Here the section figures (*profiles*) have varying shapes and it is often very difficult to make inferences about the true three-dimensional shape. The size distribution cannot be investigated at all without some information about the shape of the grains. Very delicate problems arise.

If a practical problem requires information about both size and shape then a *trial-and-error method,* such as SMOLEJ and FISCHMEISTER (1982), is probably the only sensible possibility. A trial shape assumption is made, the corresponding size distribution is determined, and the section process is simulated for a germ-grain process with grains of the required size and shape. If the simulated distribution of section figures fits the observed distribution then the problem is taken to be solved. Otherwise another trial assumption is made about size and/or shape and the process is repeated.

Many authors discuss the use of form-factors to reduce the number of unknown parameters. This is particularly helpful in the determination of mean-values as mentioned in Section 11.3.

A popular procedure is to approximate the general grains of the actual process by spherical grains matched in some way so as to be 'equivalent'. For example, the areas of the section profiles can be measured and thus diameters calculated for section circles with the same areas. This method may sometimes produce fairly accurate results but can yield nonsense. WICKSELL (1926) suggested this method which is discussed further in MORAN (1972), ANDERSEN and JAKEMAN (1975a) and CRUZ-ORIVE (1983b); see also Section 11.5.2.

NICHOLSON and MERCKX (1969) give a thorough study of the errors involved in unfolding operations for spherical and non-spherical grains in Boolean models.

In this section, two important special cases are studied which are often considered in the literature; those of ellipsoidal and polyhedral grains. Under shape assumptions such as these it is formally possible to write down integral, equations for the size distributions; see for example, SANTALÓ (1976, p. 286) CRUZ-ORIVE (1983b), and COLEMAN (1979, 1981b). Papers considering particular shapes are listed in COLEMAN (1981b) and HORÁLEK (1982).

It is generally assumed that the structure considered is a motion-invariant germ-grain model. (Note that this need not only be a system of isolated and almost independent particles. For example, the cells of a tessellation produce a germ-grain model.) Theorem 11.1 can be generalized to cover such models.

Theorem 11.3. *Let Ξ be a germ-grain model with 'typical' grain Ξ_0; and let this 'typical' grain have mark distribution M as on page 186. Consider the intersection of Ξ with an arbitrary plane or line independent of Ξ. Let $\Xi_0{}^S$ be the 'typical' grain of the planar or linear germ-grain model resulting from the section. Then $\Xi_0{}^S$ has a distribution that is independent of any assumptions concerning the distribution of the point process of germs and the correlations between germs and grains.*

Consequently, in order to obtain the distribution of $\Xi_0{}^S$ there is no loss of generality in assuming that the germs form a Poisson process, that the grains are independent, and that sections of different grains are distinguishable when they overlap. Furthermore, it is sufficient to determine the distribution of the section figure of a single 'typical' grain intersected by a Poisson line or plane process.

STOYAN (1982) gave a proof of the theorem; the particular case of spheres is discussed in detail in Section 11.4. KOSCHITZKI (1980) used the same method in discussing a germ-grain model using planar discs as grains in space. AMBARTZUMIAN (1982, Prop. 9.2) considered the case of *polygon processes*.

Some distributions for section figures have been obtained by simulation. Methods of simulation are based on the section of Ξ_0 by Poisson line or plane processes (equivalent to sectioning by the so-called IUR-planes; isotropic uniform random planes); both the 'typical' grain itself and the random lines or planes must be simulated. The method is described in MILES (1978).

The ideas described at the end of Section 11.4 concerning sieving distributions can also be applied to non-spherical particles; for example to families of triaxial ellipsoids, cylinders, and so forth. GUNDERSEN and JENSEN (1985), JENSEN (1985), and JENSEN and GUNDERSEN (1985) discuss such problems.

11.5.2 Ellipsoids

The case of non-spherical particles that is most studied is that of ellipsoids. CRUZ-ORIVE (1976, 1978) gave an excellent introduction to work done up to 1976 (dating back to WICKSELL, 1926). As Cruz-Orive points out, it is not possible to solve completely the general problem of reconstructing from planar sections the shape and size distribution of ellipsoids. One must assume either restrictions on the form of the ellipsoids (for example, that the ellipsoids are all spheroids; which is to say ellipsoids of revolution) or restrictions on the orientation of the ellipsoids relative to the plane of intersection. See WICKSELL (1926), BACH (1965), WEIBEL (1980), and KLEINWÄCHTER and ZÄHLE (1986). CRUZ-ORIVE (1976) gave an example employing on the one hand oblate and on the other hand prolate ellipsoids (ellipsoids of revolution generated by ellipses revolving around their major or minor axes respectively); in either case the same collection of planar elliptical section figures is produced. This settles a conjecture of MORAN (1972) that the general ellipsoidal problem is undetermined.

In the case of spheroids, CRUZ-ORIVE (1976) derived and solved an integral equation connecting the bivariate densities of principal semi-axes and eccentricity parameters of ellipses. As in the case of spheres (Section 11.4.1) this equation is of Abel type. KLEINWÄCHTER and ZÄHLE (1986) obtained relationships corresponding to (11.4.3) and (11.4.4) for probability distributions rather than probability densities.

CRUZ-ORIVE (1978) suggested estimation of the size-shape bivariate histogram of spheroids from the corresponding histogram of elliptical parameters measured in the plane of intersection. The case of thick sections of spheroids was studied in CRUZ-ORIVE (1983b); equations being obtained for which numerical unfolding methods were suggested. Using numerical experiments CRUZ-ORIVE concluded that the spherical approximation is acceptable when the ratio of minor axis divided by major axis is greater than 0.7.

11.5.3 Convex polyhedra

The case of polyhedral grains is very complicated. This will be appreciated immediately if the reader considers that the shape of an intersection figure varies considerably depending on the plane of intersection. For example for a cube the intersection profiles can be polygons with three, four, five, or six sides; see Figure 9.5.

The literature is devoted mainly to two problems:

(a) determination of the area and length distributions of random sections and intercepts;

(b) determination of the probabilities p_n that the 'typical' intersection figure has n sides.

The second can be thought of as a shape problem. Simulation methods have often been applied to both problems.

The general formulae for planar and linear sections in Theorem 11.1 and formulae (11.3.4) to (11.3.7) can also be used for the particular case of polyhedra. MILES (1972a) showed that the mean number $\bar{n}_{20}$ of sides of the 'typical'

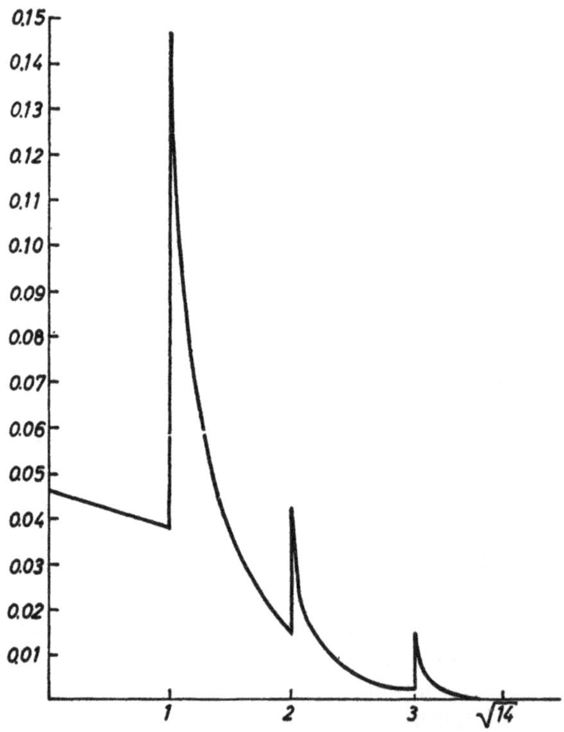

Fig. 11.3. Probability density of length of random intercepts for a parallepiped with side lengths 1, 2, 3. The curve was computed by a program written by GILLE (1985, personal communication).

planar section polygon satisfies

$$\bar{n}_{20} = \bar{l}_3/(2\bar{b}_3) \tag{11.5.1}$$

using the notation of Chapter 10. Determination of distributions is very complicated.

If the form of the polygon is known then the intercept length or section area distribution may help to determine the size distribution. SALTYKOV (1974) studied the case of a cube. In that case ITOH (1970) determined the length distribution of random intercepts. Further references are SANTALÓ (1976), UNDERWOOD (1970), MYERS (1967), and COLEMAN (1981a), who considered parallepipeds as in Figure 11.3, and NAUMOVICH and KRISKOVETS (1982) who considered various prisms. PAUL (1981) determined the distributions of profile areas for some convex polyhedra.

The shape problem (b) has been treated analytically by AMBARTZUMIAN (1973, 1982), SUKIASIAN (1982), and VOSS (1982a). Ambartzumian's method (also used by Sukiasian) is based on his solution of the Buffon-Sylvester problem in $\mathbb{R}^3$, and it studies random plane sections through systems of needles (where the needles correspond to the edges of the polyhedron). His 1973 paper gave the probabilities for the cube. Voss and Sukiasian tabulated p_n for various polyhedra. SUKIASIAN (1982) found some equations and inequalities that the p_n must satisfy. Table 11.4 displays some of their results.

Table 11.4. The probability p_n that the number of sides of a random section of a polyhedron is n

Polyhedron	Number of sides of section					
	3	4	5	6	7	8
Regular tetrahedron	0.711	0.289	–	–	–	–
Regular triangular pyramid with height: side length ratio of						
1	0.715	0.285	–	–	–	–
2	0.774	0.226	–	–	–	–
Cube	0.280	0.487	0.187	0.046	–	–
Parallelepiped with square base and height: side length ratio of						
2	0.248	0.536	0.183	0.033	–	–
4	0.180	0.653	0.152	0.014	–	–
Regular hexagonal straight prism with height: side length ratio of						
1	0.145	0.525	0.235	0.088	0.005	0.001
8	0.109	0.206	0.193	0.421	0.066	0.005

Other candidates for polyhedral grains are the 'typical' cells of the Poisson plane and Poisson-Voronoi tessellations. Note that because of Theorem 11.3 the 'typical' planar profile of the Poisson polyhedron is the Poisson polygon. Consequently the formulae for this polygon, such as the p_n tabulated in Table 10.2, can be employed when the grain is the Poisson polyhedron. The 'typical' planar profile of the Poisson-Voronoi tessellation is much more complicated. The mean-value formulae (11.3.1) to (11.3.3) can be used in connection with the formulae in Section 10.6.2. The mean number of sides is six. Some moments for the section profiles are given by Miles (1972a).

11.5.4 Stereology for spatial tessellations

The intersection of a spatial tessellation by a plane is a planar tessellation so the standard stereological formulae give the relationships

$$\lambda_2^{(2)} = N_A(\text{cells}) = N_V\bar{b} = \lambda_3\bar{b}_3,$$
$$L_A = L_A(\text{edges}) = \pi S_V/4,$$
$$\lambda_0^{(2)} = P_A(\text{nodes}) = L_V/2.$$

If $\bar{n}_{13} = 3$ for the spatial tessellation then $\bar{n}_{02} = 3$ for the planar tessellation (using the notation of Chapter 10).

If the spatial tessellation is a Poisson-Voronoi tessellation with parameter $\tau = \lambda_3$ (the mean number N_V of cells per unit volume) then for the resulting planar tessellation

$$\bar{n}_{02} = 3 \quad \text{and} \quad \bar{n}_{20} = 6$$

and

$$\lambda_2^{(2)} = N_A(\text{cells}) = 1.45\lambda_3^{2/3},$$
$$L_A = 2.29\lambda_3^{1/3},$$
$$\lambda_0^{(2)} = P_A(\text{nodes}) = 2.91\lambda_3^{2/3}$$

and

$$\lambda_1^{(2)} = 4.37\lambda_3^{2/3}$$
$$\bar{l}_1 = 0.52\lambda_3^{-1/3}$$
$$\bar{l}_0 = 1.57\lambda_3^{-1/3}$$
$$\bar{a}_2 = 0.69\lambda_3^{-2/3}.$$

These formulae can be used to test the hypothesis that a given planar tessellation was actually derived as the section of a Poisson-Voronoi tessellation. Note that such a section will not be a planar Poisson-Voronoi tessellation; the formulae above give $L_A^2/P_A = 1.79$ for the section tessellation while $L_A^2/P_A = 2$ for the planar Poisson Voronoi tessellation. It is an open question whether there is a planar point process whose Voronoi tessellation is identical in distribution to the section tessellation.

If the spatial tessellation is a *Poisson plane tessellation* with parameter $\tau = P_L$ (the mean number of planes hitting a test line segment of unit length) then the planar section is a Poisson line tessellation with the same parameter τ

(which in the planar context is the mean number of lines hitting a test line segment of unit length). The quantity N_A(cells) can be used to estimate τ by

$$N_A(\text{cells}) = \pi\tau^2/4.$$

Example 11.7. *Grain boundaries of Austenite in steel.*

This is a continuation of Example 10.1. Figure 10.7b shows the boundaries of Austenite grains in a section through some steel. The cells in the figure are not convex but it seems not inappropriate to approximate them by convex polygons as in Figure 10.7a. The following values were obtained using the methods discussed in Section 10.7

$$N_A = 0.37$$
$$L_A = 0.96$$
$$P_A = 0.69$$

and consequently

$$S_V = 1.22$$
$$J_V = 1.38.$$

Under the assumption that the planar tessellation is a planar section of a Poisson-Voronoi tessellation the corresponding parameter λ_3 can be estimated using N_A, L_A and P_A. The formulae above give the values 0.12, 0.27, 0.12 as estimates for λ_3, and the discrepancy throws doubt on whether the pattern really corresponds to a section of a Poisson-Voronoi tessellation. Conceivably, a simulation test could be conducted to decide whether the fluctuations are significant or whether they are consistent with the Poisson-Voronoi tessellation assumption.

Comparison between Figures 10.7a and 10.7b suggests that a variant of the Johnson-Mehl model might be appropriate; note in particular the curved boundaries, their concavity for large grains, and the clusters of small grains. This might be the result of two stages of generation of Austenite-grain germs. Saltykov (1974, Ch. 2) gives an interesting and relevant discussion.

A further problem for tessellations in $\mathbb{R}^3$ concerns the determination of the distribution of the angles between the faces at a typical edge-point. Under 'equilibrium conditions' one would expect an angle of 120°; see Saltykov (1974) and Harker and Parker (1945). If the angles are constant at 120° then the distribution of the angle at a 'typical' node in the planar section tessellation can be calculated (under the assumption of stationarity and isotropy); the mean is again 120° and the standard deviation is 22°. Duvalian (1971) calculated the standard deviations for other spatial angles, both larger and smaller than 120°. The inverse problem of estimating the spatial angle distribution given the angle distribution in a planar section is an interesting problem of stereology; see Miles (1985b) and Schwandtke (1985).

11.6 Second-order characteristics and directional distributions

11.6.1 Introduction

Since about 1980, workers have investigated methods for the stereological determination of second-order characteristics and directional distributions. This is a step beyond the determination of intensities and the unfolding of particle-size and form distributions. Here a survey is given of attempts on the second-order problem and a brief sketch is given of the ideas and methods used. The case of point systems (such as the centres of spheres) is treated in some detail but the second-order analysis of fibre processes and the determination of directional distributions are only discussed briefly. At the time of writing many of the statistical problems that arise in this field have not been solved.

The second-order analysis of random closed sets of positive volume fraction reduces to the study of the covariance $C(r)$ and this can be measured directly from planar sections. However for thick sections the measurement of the covariance is a complicated problem discussed in GERLACH and STOYAN (1986).

11.6.2 Stereological determination of the pair-correlation function of a system of sphere centres

Consider Φ_V a stationary and isotropic point process of sphere centres with pair-correlation function g_V. Taking a thick section, of thickness t, and projecting onto the plane, the planar point process Φ_A is obtained by considering the centres of projected section circles. Let Φ_A have pair-correlation function g_A. The function g_A can be estimated statistically using the methods of Section 4.6. Under some simplifying assumptions on the diameter distribution g_A can be related to g_V.

If the sphere diameters are constant, equal to d_V, then

$$g_A(r) = \frac{2}{(d_V + t)^2} \int_0^{d_V+t} (d_V + t - x)\, g_V\big(\sqrt{(r^2 + x^2)}\big)\, dx \quad \text{for } r \geqq 0. \tag{11.6.1}$$

If the spheres have independent random diameters with distribution function D_V and if overlapping section circles can be identified then the relation between g_A and g_V is given by

$$g_A(r) = \int_0^\infty f_V(x, t)\, g_V\big(\sqrt{(x^2 + r^2)}\big)\, dx \quad \text{for } r \geqq 0 \tag{11.6.2}$$

where

$$f_V(x, t) = \frac{2}{(d_V + t)^2} \int_u^\infty \big[1 - D_V\big((|x - u| - t)_+\big)\big]\big[1 - D_V\big((u - t)_+\big)\big]\, du$$

for $x \geqq 0$

as given by HANISCH (1983b). Formula (11.6.2) is given for the special case $t = 0$ in HANISCH and STOYAN (1980). That is the case of planar sections and a simple approximation is

$$g_V \simeq g_A.$$

This is exact in the case of a Poisson process and in general is fairly accurate for large r.

Difficulties arise with the above formulae in general because D_V and d_V are unknown and, as seen in Section 11.4, their determination is not easy. This is a powerful recommendation for the integral equation below, found by HANISCH (1984c), which expresses g_V in terms of quantities that can be estimated directly from planar sections

$$\mathfrak{g}_A(r) = \int_0^\infty f_A(x)\, g_V\left(\sqrt{(x^2 + r^2)}\right) dx \quad \text{for } r \geqq 0 \tag{11.6.3}$$

where

$$f_A(t) = \frac{2}{d_A^2} \int_x^\infty [1 - D_A(|u - 2x|)]\,[1 - D_A(u)]\, du$$

for $x \geqq 0$ and d_A is the mean section circle diameter. Furthermore, $\mathfrak{g}_A$ is the pair-correlation function of the mark-sum measure Ψ_A, given by

$$\Psi_A(B) = \sum \{\text{all diameters of section circles with centres in } B\}$$

for B a planar Borel set. The estimation of $\mathfrak{g}_A$ follows the procedure for estimating g_A in 4.6. See also STOYAN (1984b).

The case of linear sections can be treated in a similar way; see HANISCH (1983b).

Remark on the proof of (11.6.1). The second-order factorial moment measure $\alpha_A^{(2)}$ of Φ_A can be expressed in terms of that of Φ_V (namely $\alpha_V^{(2)}$) by

$$\alpha_A^{(2)}(B_1 \times B_2) = \alpha_V^{(2)}(B_1 \times [-\Delta, \Delta] \times B_2 \times [-\Delta, \Delta])$$
$$= \int_{B_1} \int_{-\Delta}^{\Delta} \int_{B_2} \int_{-\Delta}^{\Delta} \alpha_V^{(2)}\left(d\left((x_1, x_2, x_3), (y_1, y_2, y_3)\right)\right)$$

where $\Delta = (d_V + t)/2$ and B_1, B_2 are both planar Borel sets. This equation holds because all points of Φ_A in B_1 and B_2 result from points of Φ_V in $B_1 \times [-\Delta, \Delta]$ and $B_2 \times [-\Delta, \Delta]$. This equation is true for all planar Borel sets B_1 and B_2, and the pair correlation functions are densities of the corresponding second-order factorial moment measures (up to constant factors $N_A^2 = N_V^2(d_V + t)^2$ and N_V^2). Therefore, a similar equation holds for the pair-correlation functions

$$g_A(r) = \frac{1}{(d_V + t)^2} \int_{-\Delta}^{\Delta} \int_{-\Delta}^{\Delta} g_V\left(\sqrt{(r^2 + x_3^2 + y_3^2)}\right) dx_3\, dy_3 .$$

This relationship can be transformed into (11.6.1). The proof of the other formulae is similar.

None of the integral equations occurring in the above admit of easy solutions. Once again numerical problems arise from their ill-posed nature. These numerical problems are similar to those of the Wicksell problem discussed in Section 11.4. HANISCH (1984c) suggested the use of Laplace transform methods, and calculated g_A for some spatial models with known g_V. HANISCH (1983b) proposed estimators for the hard-core distance in hard-core sphere centre processes.

Example 11.8. Sinter metal spheres (Hanisch, 1984c).

Figure 11.4a shows a planar section through a specimen of sinter metal. Figure 11.4b displays the corresponding estimated pair correlation function g_A of the section circle centre point process. Hanisch assumed that the sphere diameters were constant at 1.4; an assumption supported by (11.4.6b) and the measurements for mean diameter and standard deviation of diameter for the sample of section circles,

$$\bar{x} = 0.99 \quad \text{and} \quad s = 0.32.$$

The solution of the integral equation (11.3.1) gives the graph of g_V displayed in Figure 11.4b. The differences between g_A and g_V are typical; the hard-core distance corresponding to g_V is greater than that for g_A while the peaks of g_V are higher and appear at greater values of r. Comparison of g_V with Figure 4.9 suggests the possibility that the spheres in the probe form a random dense packing of spheres.

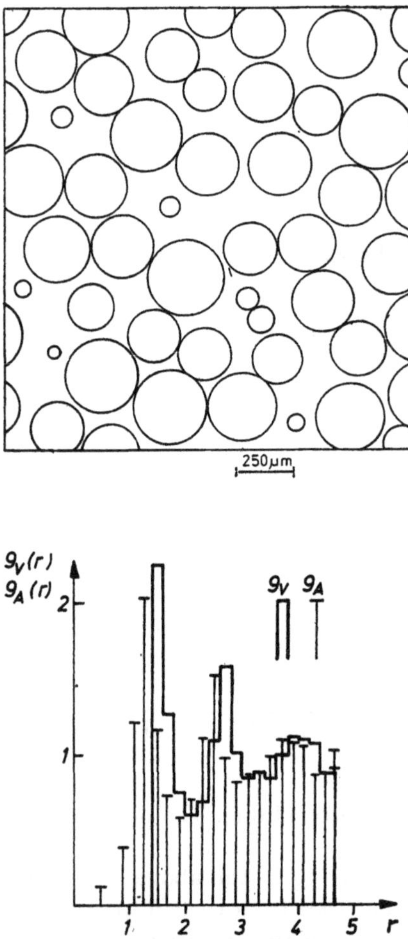

Fig. 11.4(a). Planar section through a specimen of sinter metal. The length unit is approx. 250 µm; (b) The pair correlation functions g_A and g_V for the sinter metal probe. The form of g_V suggests the assumption that the spheres form a dense random packing of spheres; compare with Figure 4.9.

11.6.3 Second-order analysis for spatial fibre systems

Stoyan (1984c, 1985a, 1985c) and Hanisch, König, and Stoyan (1985) suggested an approximate stereological method for the determination of the pair-correlation function for fibre systems. The method applies both to planar and to thick sections. Under a simplifying assumption that one can ignore the correlation of fibre directions at different sampled fibre points, the case of thick sections yields an integral equation similar to (11.6.1). The equation connects the pair correlation functions of the spatial and the projected fibre processes.

Figure 11.5a shows a thick section of a spatial fibre system (a network of dislocation lines induced by diffusion in a single silicon crystal). Figure 11.5b gives the corresponding spatial pair correlation function. See Hanisch, Klimanek, and Stoyan (1985).

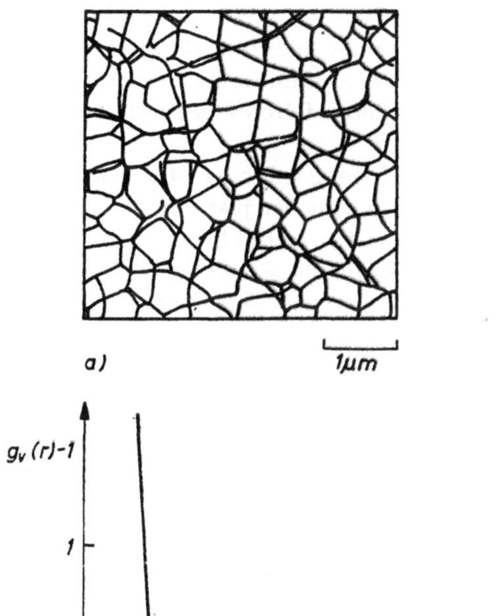

Fig. 11.5(a). Network of dislocation lines induced by diffusion in a silicon crystal (taken from a photograph of a transmission electronmicroscopic image in Hanisch, Klimanek, and Stoyan, 1985). The thickness of the foil was about 0.2 µm; (b) Spatial pair correlation function g_V of the dislocation lines. The limiting value $g_V(0+)$ diverges to $+\infty$; this is typical for fibre processes as discussed in Section 9.3.1.

11.6.4 Determination of directional distributions

Stereological problems increase in complexity when the structures under study are anisotropic. Anisotropy occurs for example for capillaries in skeletal muscles or for grains in rolled steel. In such cases it is common to make sections of different orientations; generally in transverse and longitudinal directions to the main axis of anisotropy if this is known. If the structure can be interpreted as a 'pressed' structure then the formulae of Section 9.3 can be employed.

In this section, a brief discussion will be given of the case of anisotropic fibre systems; similar considerations for surface systems can be found in Weibel (1980).

Cruz-Orive suggested a parametric approach (see Weibel, 1980, and Cruz-Orive *et al.*, 1983). Every fibre direction corresponds to a direction on a hemisphere (or more properly the projective plane) which allows the use of standard distributions on the hemisphere as considered in the theory of directional data (see Mardia, 1972). In particular the *axial Fisher* or *Dimroth-Watson* distribution (Mardia, 1972, 8.5.5) may be employed. It depends on a concentration parameter K and this can be determined together with the intensity J_V by using a nonlinear regression procedure. The procedure uses as data the mean number of intersection points of fibres per unit area in planes of various directions.

This approach also enables tests of the goodness-of-fit of the distributional assumption. Given K it is sufficient to consider just one plane of intersection. Cruz-Orive *et al.* (1983) demonstrated the method by an example concerning the study of orientations of capillaries in the skeletal muscle of cats.

Commencing from the observation that fibres are often tubes a non-parametric method has been developed; see Stoyan (1984a, b) and also papers on the stereology of tubes such as Baddeley and Averback (1983). Assuming that the fibres are of circular cross section with slowly varying diameter the angles at which the fibres intersect the plane may be determined by observation of the intersection figures in planar or thick sections. Figure 11.6 illustrates this point.

Figure 11.6 shows that the intersection angle β is given by

$$\cos \beta = \left(bl + t\sqrt{(l^2 + t^2 - b^2)}\right)/(l^2 + t^2).$$

Experience has shown that it is possible to measure β automatically by exploiting this formula. An alternative means of measuring β exploits the possibility of moving the focal plane up and down when observing thick sections of biological tissue by microscope. The line density J_V can be estimated by

$$J_V = \sum_{i=1}^{N} \left(\nu_2(W) \cos \beta_i\right)^{-1}$$

where β_i is the angle corresponding to the ith intersection figure, and N is the number of intersection figures in the window of observation W.

Stoyan (1985a) demonstrates how the directional distribution can be determined. Figure 11.7b shows the directional distribution, displayed in Schmidt's

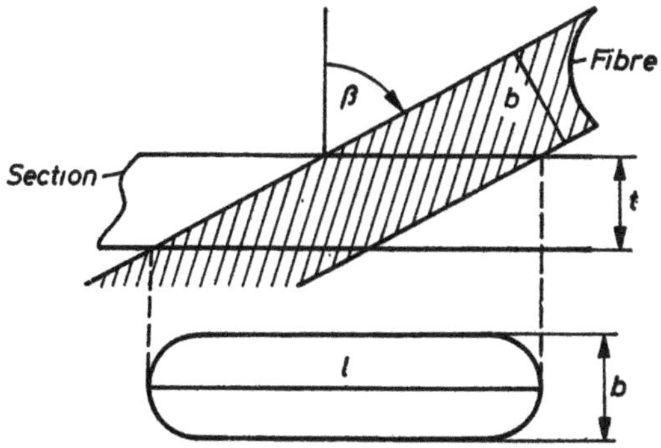

Fig. 11.6. Intersection figure for a cylindrical fibre in a thick section. The same figure would occur if the intersection angle was still β and the orientation of the fibre was reversed (so that it would come from left above to right below).

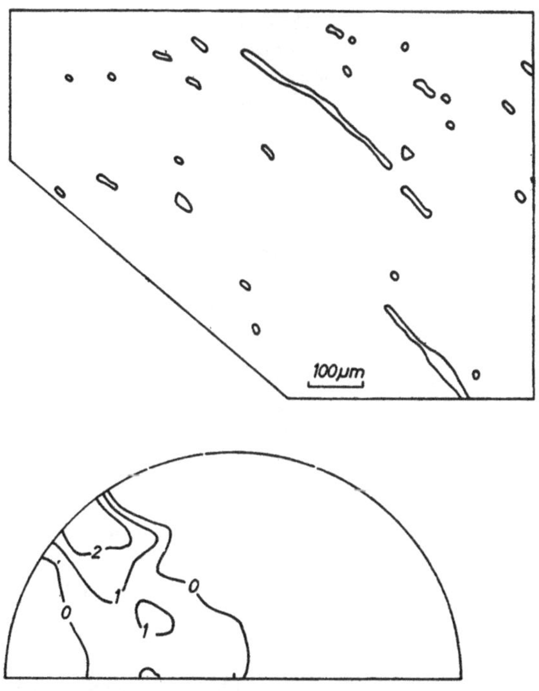

Fig. 11.7 (a). Intersections of capillaries with a section from the medullary zone of the human brain; (b) Orientation distribution derived from the section illustrated in Figure 11.7a, using Schmidt's net representation. The density is given up to a factor of proportionality by contour lines.

net, for the specimen of capillaries in the medullary zone of the human brain in Figure 11.7a (a description of Schmidt's net is in MARDIA, 1972, p. 215). The intersection figures only give fibre directions up to reflection in the normal to the plane of intersection (see Figure 11.6) and so only a half-circle is used.

References

AAP	=	*Adv. Appl. Prob.*
AMS	=	*Ann. Math. Statist.*
BJ	=	*Biometrical J.* (formerly *Biometr. Z.*)
JAP	=	*J. Appl. Prob.*
JRRS	=	*J. Royal Statist. Soc.*
MOS	=	*Math. Operationsf. Statist., ser. statist.* (since 1985: *statistics*)
MOSO	=	*Math. Operationsf. Statist., ser. optimiz.* (since 1985: *optimization*)
MN	=	*Math. Nachr.*
SLNB	=	*Springer Lecture Notes in Biomathematics*
SLNM	=	*Springer Lecture Notes in Mathematics*
SLNS	=	*Springer Lecture Notes in Statistics*
ZW	=	*Z. Wahrscheinlichkeitsth. verw. Geb.* (since 1986: Probab. Th. Rel. Fields)

Abbé, E. (1878) Über Blutkörperzählung. *Sitzungsberichte Jenaische Ges. Medizin Naturwiss.* 98–105.

Agterberg, F. P. (1981) Cell value distribution models in spatial pattern analysis. In: *Future Trends in Geomathematics.* (eds. R. G. Craig and M. L. Labowitz). Pion, London, 5–28.

Ambartzumian, R. V. (1966) On an equation for stationary point processes (Russ.) *Dokl. Akad. Nauk Armjanskoi SSR* **42**, 141–147.

Ambartzumian, R. V. (1973) On random fields of segments and random mosaics in the plane. (Russ.) *Teor. Veroyatn. Prim.* **18**, 515–526 (corrections: **19**, 600).

Ambartzumian, R. V. (1974a) The solution to the Buffon-Sylvester problem in R^3. *ZW* **27**, 53–74.

Ambartzumian, R. V. (1974b) Convex polygons and random tessellations. In: Harding and Kendall (eds) (1974), 176–191.

Ambartzumian, R. V. (1976) Homogeneous and isotropic random point fields in the plane (Russ.) *MN* **70**, 365–385.

Ambartzumian, R. V. (1977) Stochastic Geometry from the standpoint of integral geometry. *AAP* **9**, 792–823.

Ambartzumian, R. V. (1981) Stereology of random planar segment processes. *Rend. Sem. Mat. Torino* **39**, 147–159.

Ambartzumian, R. V. (1982a) *Combinatorial Integral Geometry.* J. Wiley & Sons, Chichester.

Ambartzumian, R. V. (1982b) Random shapes by factorization. In: Ranneby (ed.) (1982), 35–42.

Ambartzumian, R. V. (1984) Factorization in integral and stochastic geometry. In: Ambartzumian and Weil (eds) (1984) 14–33.

Ambartzumian, R. V. and V. K. Oganian (1975) Homogeneous and isotropic fibre fields in the plane. (Russ.) *Izv. AN Armen. SSR Ser. Math.* **10**, 509–528.

Ambartzumian, R. V. and W. Weil (eds.) (1984) Stochastic Geometry, Geometric Statistics, Stereology, Teubner-Texte zur Mathematik 65, Teubner-Verlag, Leipzig.

ANDERSSEN, R. S. and A. J. JAKEMAN (1975a) Abel-type integral equations in stereology. *JM* ***105***, 121–153.

ANDERSSEN, R. S. and A. J. JAKEMAN (1975b) Product integration for functionals of particle size distributions. *Utilitas Math.* ***8***, 111–126.

ARMITAGE, P. (1949) An overlap problem arising in particle counting. *Biometrika* ***36***, 257–266.

ARTSTEIN, Z. (1984) Convergence of sums of random sets. In: AMBARTZUMIAN and WEIL (eds) (1984), 34–42.

ARTSTEIN, Z. and S. HART (1981) Law of large numbers for random sets and allocation processes. *Math. Oper. Res.* ***6***, 482–492.

ARTSTEIN, Z. and R. A. VITALE (1982) A strong law of large numbers for random compact sets. *Ann. Prob.* ***5***, 879–882.

ATKINSON, A. C. (1979a) The computer generation of Poisson random variables. *Appl. Statist.* ***28***, 29–35.

ATKINSON, A. C. (1979b) Recent developments in the computer generation of Poisson random variables. *Appl. Statist.* ***28***, 260–263.

BACH, G. (1959) Über die Größenverteilung von Kugelschnitten in durchsichtigen Schnitten endlicher Dicke. *Z. wiss. Mikrosk.* ***64***, 265–286.

BACH, G. (1963) Über die Bestimmung von charakteristischen Größen einer Kugelverteilung aus der Verteilung der Schnittkreise. *Z. wiss. Mikrosk.* ***65***, 285–296.

BACH, G. (1965a) Über die Bestimmung von charakteristischen Größen einer Kugelverteilung aus der unvollständigen Verteilung der Schnittkreise. *Metrika* ***9***, 228 bis 233.

BACH, G. (1965b) Zufallsschnitte durch ein Haufwerk von Rotationsellipsoiden mit konstantem Achsenverhältnis. *Z. angew. Math. Phys.* ***16***, 224–232.

BACH, G. (1967) Kugelgrößenverteilung und Verteilung der Schnittkreise; ihre wechselseitigen Beziehungen und Verfahren zur Bestimmung der einen aus den anderen. In: *Quantitative Methods in Morphology* (eds. E. R. WEIBEL and H. ELIAS) Springer-Verlag, Berlin, 23–45.

BACH, G. (1976) Über die Auswertung von Schnittflächenverteilungen. *BJ* ***18***, 407–412.

BADDELEY, A. (1977) A fourth note on recent research in geometrical probability. *AAP* ***9***, 824–860.

BADDELEY, A. (1980a) A limit theorem for statistics of spatial data. *AAP* ***12***, 447–461.

BADDELEY, A. (1980b) Absolute curvatures in integral geometry. *Proc. Camb. Phil. Soc.* ***88***, 45–58.

BADDELEY, A. (1981) Combinatorial foundations of Stochastic Geometry. *Proc. London Math. Soc.* ***42***, 151–177.

BADDELEY, A. (1982a) Appendix A to AMBARTZUMIAN (1982a).

BADDELEY, A. (1982b) Stochastic geometry: An introduction and reading-list. *Int. Stat. Review* ***50***, 179–193.

BADDELEY, A. (1983) Applications of the co-area formula to stereology. In: JENSEN and GUNDERSEN (eds.) (1985), 1–17.

BADDELEY, A. (1984a) Vertical sections. In: AMBARTZUMIAN and WEIL (eds.) (1984), 43–52.

BADDELEY, A. (1984b) Stochastic geometry and image analysis. *CWI Newsletter* (Centrum for Wiskunde en Informatica, Amsterdam) September 1984, 2–20.

BADDELEY, A. and P. AVERBACK (1983) Stereology of tubular structures. *JM.* ***131***, 323–340.

BADDELEY, A. and B. W. SILVERMAN (1984) A cautionary example for the use of second-order methods for analyzing point patterns. *Biometrics* ***40***, 1089–1094.

BAECHER, G. B. (1983) Statistical analysis of rock mass fracturing. *Math. Geology* ***15***, 329–348.

BARNARD, G. A. (1963) Discussion of Professor Bartlett's paper, *JRSS* ***B25***, p. 294.

BARTHEL, M., P. KLIMANEK, and D. STOYAN (1985) Stereological substructure analysis in hot-deformed metals from TEM-images. *Czech. J. Phys.* ***B35***, 265–268.

BARTLETT, M. S. (1954) Processus stochastiques ponctuels. *Ann. Inst. H. Poincare'* ***B14***, 35–60.

BARTLETT, M. S. (1963) The spectral analysis of point processes. *JRSS* **B25**, 264–296.
BARTLETT, M. S. (1964) The spectral analysis of two-dimensional point processes. *Biometrika* **51**, 299–311.
BARTLETT, M. S. (1967) The spectral analysis of line processes. *Proc. 5th Berkeley Symp. Math. Statist. Prob.* **3**, 135–153.
BARTLETT, M. S. (1975) *The Statistical Analysis of Spatial Pattern.* Chapman and Hall, London.
BAUDIN, M. (1979) Nearest Neighbour Distance Statistics and Model Fitting in 2-dimensional Poisson Cluster Processes. *Report No.* **266** *Centre de Geostatistique et de Morphologie Mathematique,* Fontainebleau.
BAUDIN, M. (1981) Likelihood and nearest neighbour distance properties of multidimensional Poisson cluster processes. *JAP* **18**, 879–888.
BAUDIN, M. (1983) Note on the determination of cluster centres from a realisation of a multidimensional Poisson cluster process. *JAP* **20**, 136–143.
BAUDIN, M. (1984) Multidimensional point processes and random closed sets. *JAP* **21**, 173–178 and acknowledgement *JAP* **21**, 675.
BERNAL, J. D. (1959) A geometrical approach to the structure of liquids. *Nature* **183**, 141–147.
BESAG, J. and P. J. DIGGLE (1977) Simple Monte Carlo tests for spatial pattern. *Appl. Statist.* **26**, 327–333.
BIGGINS, J. D. (1977) Martingale convergence in the branching random walk. *JAP* **14**, 25–37.
BLÖDNER, R., P. MÜHLIG, and W. NAGEL (1984) The comparison by simulation of solutions of Wicksell's corpuscle problem. *JM* **135**, 61–74.
BOCKSTIEGEL, G. (1966) Eine einfache Formel zur Berechnung räumlicher Größenverteilungen aus durch Linearanalyse erhaltenen Daten. *Z. Metallkunde* **57**, 647–656.
BODZIONY, J. (1965) On certain indices characterizing the geometric structure of rocks. *Bull. Acad. Polon. Science Ser. Science. Techn.* **13**, 469–475.
BOOTS, B. N. (1982) The arrangements of cells in "random" networks. *Metallography* **15**, 53–62.
BOOTS, B. N. (1984) Comments on 'Aboav's rule' for the arrangement of cells in a network. *Metallography* **17**, 411–418.
BOS, A. (1979) Sphere-packings in Euclidean space. In: *Packing and Covering in Combinatorics* (ed. A. SCHRIJVER) Mathematical Centre Tracts **106**, Amsterdam, 161–177.
BOWYER, A. (1981) Computing Dirichlet tessellations. *Computing J.* **24**, 162–166.
BRADLEY, A. and C. G. SMALL (1985) Looking for circular structure in post hole distributions: quantitative analysis of two settlements from Bronze Age England. *J. Archaeological Sci.* **12**, 285–297.
BRILLINGER, D. R. (1975) Statistical inferences for stationary point processes. In: *Stochastic Processes and Related Topics* (ed. M. PURI) Academic Press, New York, 55–59.
BRILLINGER, D. R. (1976) Measuring the association of point processes: a case history. *Amer. Math. Monthly* **83**, 16–22.
BRILLINGER, D. R. (1978) Comparative aspects of the study of ordinary time series and point processes. In: *Developments in Statistics I.* Academic Press, New York, 227–320.
BROADBENT, S. R. (1980) Simulating the ley-hunter. *JPRSS* **A143**, 109–140.
BRUMBERGER, H. and J. GOODISMAN (1983) Voronoi cells: An interesting and potentially useful cell model for interpreting the small-angle scattering of catalysts. *J. Appl. Cryst.* **16**, 83–88.
BYTH, K. (1982) On robust distance-basedinten sity estimators. *Biometrics* **38**, 127–135.
BYTH, K. and B. D. RIPLEY (1980) On sampling spatial patterns by distance methods. *Biometrics* **36**, 279–284.
CAHN, J. W. (1967) The significance of average mean curvature and its determination by quantitative metallography. *Trans. AIME* **239**, 610–616.
CAHN, J. W. and J. NUTTING (1959) Transmission quantitative metallography. *Trans. AIME* **215**, 526–528.

CARGILL, G. S. (1975) Structure of metallic alloy glasses. In: *Solid State Physics*, vol. ***30***. Academic Press, New York, 227–314.
CARGILL, G. S. (1984) Radial distribution functions and microgeometry of dense random packing of hard spheres. In: *Proc. Int. Symp. Physics and Chemistry of Porous Media*, Ridgefield (ed. D. L. JOHNSON), Amer. Inst. Physics, New York 20–36.
CHERNYAVSKIJ, K. S. (1977) *Stereology in metal research.* (Russ.) Metallurgia, Moscow.
CHOQUET, G. (1953/1954) Theory of capacities. *Ann. Inst. Fourier* ***V***, 131–295.
CHOQUET, G. (1960) Le theoreme de representation integrale dans les ensembles convexes compacts. *Math. Hung.* ***5***, 369–394.
CLARKE, K. R. (1981) Statistical techniques in stereology. *Comm. Statist. Theor. Meth.* ***A10*** (***15***), 1459–1478.
CLARKE, K. R. and M. N. MOORE (1983) Unbiased linear property estimation for spherical particles from thin sections. *JM* ***131***, 311–322.
CLIFF, A. D. and J. K. ORD (1981) *Spatial Processes: Models and Applications.* Pion, London.
COLEMAN, R. (1979) *An Introduction to Mathematical Stereology.* Memoirs No. 3, Dept. of Theoretical Statistics, University of Aarhus, Denmark.
COLEMAN, R. (1980) The distribution of the sizes of spheres from observations through a thin slice. *Mikroskopie* ***37*** (Suppl.), 68–73.
COLEMAN, R. (1981a) Intercept length of random probes through boxes. *JAP* ***18***, 276–282.
COLEMAN, R. (1981b) The stereological determination of particle size distributions: theoretical considerations. In: KALIŠNIK (1981), 67–77.
COLEMAN, R. (1981c) Size determination of transparent spheres in an opaque specimen from a slice. *JM.* ***123***, 343–345.
COLEMAN, R. (1983) The sizes of spheres from profiles in a thin slice. I: Opaque spheres. II: Transparent spheres. *BJ* ***24***, 273–286; ***25***, 745–756.
COWAN, R. (1978) The use of the ergodic theorems in random geometry. *Suppl. AAP* ***10***, 47–57.
COWAN, R. (1979) Homogeneous line-segment processes. *Math. Proc. Camb. Phil. Soc.* ***86***, 481–489.
COWAN, R. (1980) Properties of ergodic random mosaic processes. *MN* ***97***, 89–102.
COWAN, R. (1984) A collection of problems in random geometry. In: AMBARTZUMIAN and WEIL (eds) (1984), 64–68.
COX, D. R. (1953) Some simple approximate tests for Poisson variates. *Biometrika* ***40***, 354–360.
COX, D. R. (1955) Some statistical models connected with series of events. *JRSS* ***B17***, 129–164.
COX, D. R. and D. V. HINCKLEY (1974) *Theoretical Statistics*, Chapman and Hall, London.
COX, D. R. and V. ISHAM (1980) *Point Processes.* Chapman and Hall, London.
COX, D. R. and P. A. W. LEWIS (1966) *The Statistical Analysis of Series of Events.* Methuen/Wiley, London/New York.
CRAIN, I. K. (1976) Statistical analysis of Geotectonics. In: *Random Processes in Geology.* (ed. D. F. MERRIAM) Springer-Verlag, Berlin/New York, 3–15.
CRAIN, I. K. and R. E. MILES (1976) Monte Carlo estimates of the distribution of the random polygons determined by random lines in a plane. *J. Statist. Comput. Simul.* ***4***, 293–325.
CRESSLE, N. and G. M. LASLETT (1986) Random set theory and problems of modelling. *SIAM Review.*
CROFTON, M. W. (1885) Probability. *Encyclopaedia Britannica*, IXth edition.
CROW, E. L. and R. S. GARDNER (1959) Confidence intervals for the expectation of a Poisson variable. *Biometrika* ***46***, 441–453.
CRUZ-ORIVE, L.-M. (1976, 1978) Particle size-shape distributions: the general spheroid problem. I, II. *JM.* ***107***, 235–253, ***112***, 153–167.
CRUZ-ORIVE, L.-M. (1979) Estimation of sheet-thickness distributions from linear and plane sections. *BJ* ***21***, 717–730.

CRUZ-ORIVE, L.-M. (1980a) Best linear unbiased estimators for stereology. *Biometrics* **36**, 595–605.
CRUZ-ORIVE, L.-M. (1980b) On the estimation of particle number. *JM* **120**, 15–27.
CRUZ-ORIVE, L.-M. (1983a) Distribution-free estimation of sphere size distribution from slabs showing overprojection and truncation, with a review of previous methods. *JM* **131**, 265–290.
CRUZ-ORIVE, L.-M. (1983b) *The stereology of spheroids. Part I: Analytical results for plane sections and projections. Part II: Dictribution-free unfolding procedures.* University of Berne, Anatom. Inst., Internal Report 81.
CRUZ-ORIVE, L.-M. (1985) Estimating volumes from systematic hyperplane sections. *JAP* **22**, 518–530.
CRUZ-ORIVE, L.-M., H. HOPPELER, O. MATHIEU, and E. R. WEIBEL (1985) Stereological analysis of anisotropic structures using directional statistics. *Appl. Statist.* **34**, 14–32.
CRUZ-ORIVE, L. M. and A. O. MYKING (1981) Stereological estimation of volume ratios by systematic sections. *JM* **122**, 143–157.
CRUZ-ORIVE, L.-M. and E. R. WEIBEL (1981) Sampling designs for stereology. *JM* **122**, 235–258.
DALEY, D. J. and D. VERE-JONES (1972) A summary of the theory of point processes. In: LEWIS (ed.) (1972), 299–383.
DALEY, D. J. and D. VERE-JONES (1986) *An Introduction to the Theory of Point Processes.* (In preparation).
DALEY, D. J. and R. K. MILNE (1973) The theory of point processes: a bibliography. *Int. Stat. Rev.* **41**, 183–201.
DAVIDSON, R. (1974a) Stochastic processes of flats and exchangeability. In: HARDING and KENDALL (eds.) (1974), 13–45.
DAVIDSON, R. (1974b) Construction of Line-processes: Second-order properties. In: HARDING and KENDALL (eds) (1974), 55–75.
DAVIDSON, R. (1974c) Line-processes, Roads, and Fibres. In: HARDING and KENDALL (eds) (1974), 248–251.
DAVY, P. J. (1978) *Stereology: a statistical viewpoint.* Ph. D. thesis, Australian National University.
DAVY, P. J. and R. E. MILES (1977) Sampling theory for opaque spatial specimens. *JRSS* **B39**, 56–65.
DE HOFF, R. T. (1965) The estimation of particle distributions from simple counting measurements made on random plane sections. *Trans. AIME* **233**, 25–29.
DE HOFF, R. T. (1968) Curvature and topological properties of interconnected phases. In: *Quantitative Microscopy* (eds. R. T. DE HOFF and F. N. RHINES) McGraw-Hill, New York, 291–324.
DE HOFF, R. T. (1967) The quantitative estimation of mean surface curvature. *Trans. AIME* **239**, 617–621.
DE HOFF, R. T. (1977) The geometrical meaning of the integral mean curvature. In: *Microstructural Science*, Vol. **5**, 331–348. Elsivier, Amsterdam.
DE HOFF, R. T. (1978) Stereological use of the area tangent count. In: *Geometrical Probability and Biological Structures* (eds. R. E. MILES and J. SERRA) *SLNB* **23**, 99–113.
DE HOFF, R. T. (1983) Quantitative serial sectioning analysis: preview. *JM* **131**. 259–263.
DE HOFF, R. T. and F. N. RHINES (1968) *Quantitative Micsroscopy.* McGraw-Hill, New York.
DELESSE, M. A. (1947) Procede mecanique pour determiner la composition des roches. *C. R. Acad. Sci.* (Paris) **25**, 544–545.
DELFINER, P. (1972) A generalization of the concept of size. *JM* **95**, 203–216.
DIGGLE, P. J. (1975) Robust density estimation using distance methods. *Biometrika* **62**, 39–48.
DIGGLE, P. J. (1977) A note on robust density estimation for spatial point patterns. *Biometrika* **64**, 91–95.
DIGGLE, P. J. (1979) On parameter estimation and goodness-of-fit testing for spatial point patterns. *Biometrics* **35**, 87–101.

DIGGLE, P. J. (1981) Binary mosaics and the spatial pattern of heather. *Biometrics* **37**, 531–539.

DIGGLE, P. J. (1983) *Statistical Analysis of Point Processes*. Academic Press, London.

DIGGLE, P. J. and C. J. F. TER BRAAK (1982) Point sampling of binary mosaics in ecology. In: B. RANNEBY) (ed.) (1982), 107–122.

DIGGLE, P. J., J. BESAG, and J. T. GLEAVES (1976) Statistical analysis of spatial point patterns by means of distance methods. *Biometrics* **32**, 659–667.

DIGGLE, P. J. and T. F. COX (1983) Some distance-based tests of independence for sparsely-sampled multivariate spatial point patterns. *Int. Stat. Rev.* **51**, 11–23.

DIGGLE, P. J. and R. J. GRATTON (1984) Monte Carlo methods of inference for implicit statistical models (with Discussion). *JRSS* **B46**, 193–227.

DIGGLE, P. J. and R. K. MILNE (1983) Bivariate Cox processes: Some models for bivariate spatial point patterns. *JRSS* **B45**, 11–21.

DIRICHLET, G. L. (1850) Über die Reduction der positiven quadratischen Formen mit drei unbestimmten ganzen Zahlen. *J. Reine und Angew. Math.* **40**, 209–227.

DOLEŽAL, F. (1982) The use of stereology for the evaluation of crack patterns in agricultural soils. In: *SAXL* (ed.) (1982), 305–307.

DUPAČ, V. (1980) Parameter estimation in the Poisson field of discs. *Biometrics* **67**, 187–190.

DUVALIAN, A. V. (1971) A method for the approximate determination of the variance of dihedral angles in alloys (Russ.) *Zavods. Lab.* **37**, 939–941.

DWASS, M. (1957) Modified randomization tests for non-parametric hypotheses. *AMS.* **28**, 181–187.

EBBESON, S. D. and D. TANG (1965) A method for estimating the number of cells in histological sections. *J. Roy. Micr. Soc.* **84**, 449–464.

ECONOMO, C. von and G. N. KOSKINAS (1925) *Die Cytoarchitektonik der Hirnrinde des erwachsenen Menschen*. Springer, Wien/Berlin.

EDMUNDS, M. G. and G. H. GEORGE (1981) Random Alignments of quasars. *Nature* **290**, 481–482.

ELIAS, H. (ed.) (1967) *Stereology*. Springer-Verlag, Berlin.

ENNS, E. G. and P. F. EHLERS (1978) Random paths through a convex region. *JAP* **15**, 144–152.

FEDERER, H. (1959) Curvature measures. *Trans. Amer. Math. Soc.* **93**, 418–491.

FEDERER, H. (1969) *Geometric Measure Theory*. Springer-Verlag, Berlin/New York.

FELLOUS, A. and J. GRANARA (1981) Statistique des processus de Poisson stationnaires de sous-varietes lineaires affines. *AAP* **13**, 84–92.

FELLOUS, A., J. GRANARA, and K. KRICKEBERG (1978) Statistics of stationary oriented line Poisson processes in the plane. In: *Geometrical Probability and Biological Structures: Buffon's 200th Anniversary* (eds. R. E. MILES and J. SERRA). *SLNB* **23**, 295–299.

FIKSEL, T. (1984a) Estimation of parametrized pair potentials of marked and nonmarked Gibbsian point processes. *EIK* **20**, 270–278.

FIKSEL, T. (1984b) Simple spatial-temporal models for sequences of geological events. *EIK* **20**, 480–487.

FIKSEL, T. (1987a) Edge-corrected density estimators for point processes. *MOS* **18**.

FIKSEL, T. (1987b) Estimation of interaction potentials of Gibbsian point processes. *MOS* **18**.

FIKSEL, T. and D. STOYAN (1983) Mathematisch-statistische Bestimmung von Gefährdungsgebieten bei Erdfallprozessen. *Z. angew. Geologie* **29**, 455–459.

FINNEY, J. L. (1979) A procedure for the construction of Voronoi polyhedra. *J. Comput. Phys.* **32**, 137–143.

FINNEY, J. L. (1981) The past, present, and future of statistical geometry. In: *Int. Roundtable Congress 50th Anniversary Jap. Statist. Soc.* 1981, 199–209.

FISCHER, R. A. and R. E. MILES (1973) The role of spatial pattern in the competition between crop plants and weeds: A theoretical analysis. *Math. Biosciences* **18**, 335–350.

FLEISCHER, R. L. and A. MOGRO-CAMPERO (1978) Mapping of integrated Radon emanation for detection of long-range migration of gases within the Earth: techniques and principles. *J. Geophys. Research.* **83**, 3539–3549.

FLOOK, A. G. (1983) A comparative study of methods for unfolding size distributions.

FRANKEN, P., D. KÖNIG, U. ARNDT, and V. SCHMIDT (1981) *Queues and Point Processes*, Akademie-Verlag, Berlin and J. Wiley & Sons, Chichester.

FRANKLIN, J. N. (1977) Some stereological principles in morphometric cytology. *SIAM J. Appl. Math.* **33**, 267–278.

FRANKLIN, J. N. (1981) Confidence intervals for stereological estimators with infinite variance. *SIAM J. Appl. Math.* **40**, 179–190.

FRIEDEN, B. R. (1983) *Probability, Statistical Optics, and Data testing*. Springer-Verlag, Berlin/New York.

FRITZ, J. (1970) Generalization of McMillan's theorem to random functions. *Studia Sci. Math. Hung.* **5**, 369–374.

FULLMAN, R. L. (1953) Measurement of particle sizes in opaque bodies. *J. Metals* **5**, 447–452.

GATES, D. J. and M. WESTCOTT (1986) Clustering estimates in spatial point distributions with instable potentials. *Ann. Inst. Statist. Math.* **38A**, 55–67.

GEORGE, E. I. (1982) *Sequential stochastic construction of random polygons*. Tech. Report **320**, Dept. Statistics, Stanford University.

GERLACH, W. and J. OHSER (1986) On the accuracy of numerical solution for some stereological problems such as the Wicksell corpuscle problem. *BJ* **28**.

GERLACH, W. and D. STOYAN (1986) Determination of the covariance of random sets from thin sections. *BJ* **28**, 427–432.

GIGER, H. (1967) Ermittlung der mittleren Maßzahlen von Partikeln eines Körpersystems durch Messungen auf dem Rand eines Schnittbereiches. *Z. angew. Math. Phys.* **18**, 883–888.

GIGER, H. and H. RIEDWYL (1970) Bestimmung der Größenverteilung von Kugeln aus Schnittkreisradien. *BJ* **12**, 156–168.

GILBERT, E. N. (1961) Random plane networks. *J. SIAM* **9**, 533–543.

GILBERT, E. N. (1962) Random subdivisions of space into crystals. *AMS* **33**, 958–972.

GILBERT, E. N. (1967) Random plane networks and needle-shaped crystals. In: *Applications of Undergraduate Mathematics in Engineering* (ed B. NOBLE) Macmillan, New York, Chapter 16.

GIRLING, A. J. (1982) Approximate variances associated with random configurations of hard spheres. *JAP* **19**, 588–596.

GLAGOLEV, A. A. (1933) On the geometrical methods of quantitative mineralogic analysis of rocks. *Trans. Inst. Econ. Min. Moscow* **59**, 1–47.

GLATTER, O. (1979) The interpretation of real-space information from small-angle scattering experiments. *J. Appl. Cryst.* **12**, 166–175.

GLATTER, O. and O. KRATKY (1982) *Small Angle X-ray Scattering*. Academic Press, London.

GLÖTZL, E. (1980) Lokale Energien und Potentiale für Punktprozesse. *MN* **96**, 196–206.

GLÖTZL, E. (1981) Time reversible and Gibbsian point processes I. Markovian spatial birth and death processes on a general phase space. *MN* **102**, 217–222.

GLÖTZL, E. and B. RAUSCHENSCHWANDTNER (1981) On the statistics of Gibbsian processes. In: *The first Pannonian Symposium on Mathematical Statistics* (eds. P. REVESZ, L. SCHMETTERER, V. M. ZOLOTAREV) *SLNS* **8**, 83–93.

GOLDMAN, A. and W. VISSCHER (1978) Applications of integral equations in particle-size statistics. In: *Solution Methods for Integral Equations — Theory and Applications* (ed. M. A. GOLDBERG) Plenum Press, New York/London, 169–182.

GOUDSMIT, S. (1945) Random distribution of lines in a plane. *Rev. Mod. Phys.* **17**, 321–322.

GRANDELL, J. (1976) *Doubly Stochastic Poisson Processes. SLNM* **529**.

GRANDELL, J. (1981) Some recent developments of theory and application of Cox processes. In: *Sixth Conf. Prob. Theory, Brasov 1979*, (eds. B. BEREANU, S. GRIGORESCU, M. IOSIFESCU, T. POSTELNICU) Bucuresti, 288–289.

GRAY, N. H., J. B. ANDERSON, J. D. DEVINE, and J. M. KWASNIK (1976) Topological properties of random crack networks. *Math. Geology* **8**, 617–626.

GRECO, A., D. JEULIN, and J. SERRA (1979) The use of the texture analyser to study sinter structure: application to the morphology of calcium ferrites encountered in basic sinters of rich iron ores. *JM* **116**, 199–211.

GREEN, P. J. and R. SIBSON (1978) Computing Dirichlet tessellations in the plane. *Computer J.* **21**, 168–173.

GREIG-SMITH, P. (1964) *Quantitative Plant Ecology*. Butterworths, London.

GRENANDER, U. (1976, 1978, 1981) *Lectures in Pattern Theory*, vol. 1–3. Springer-Verlag, Berlin/New York.

GUNDERSEN, H. J. G. (1978) Estimators of the numbers of objects pee area unbiased by edge effects. *Micros. Acta* **81**, 107–117.

GUNDERSEN, H. J. G. (1979) Estimation of tubule or cylinder L_V, S_V, and V_V on thick sections. *JM* **117**, 333–346.

GUNDERSEN, H. J. G. and E. B. JENSEN (1983) Particle sizes and their distributions estimated from line- and point-sampled intercepts. Including graphical unfolding. *JM* **131**, 291–310.

GUNDERSEN, H. J. G. and E. B. JENSEN (1985) Stereological estimation of the mean volume of arbitrary particles observed on random sections. *JM* **138**, 127–142.

HAAS, A., G. MATHERON, and J. SERRA (1967) Morphologie mathematique et granulometries en place. *Ann. Mines* **11**, 736–753 and **12**, 767–782.

HADWIGER, H. (1957) *Vorlesungen über Inhalt, Oberfläche und Isoperimetrie*. Springer-Verlag, Berlin/Göttingen/Heidelberg.

HADWIGER, H. and H. GIGER (1968) Über Treffzahlwahrscheinlichkeiten im Eikörperfeld. *ZW* **10**, 329–334.

HALMOS, P. R. (1974) *Measure Theory*. Springer-Verlag, Berlin/New York.

HAMILTON, W. D. (1971) Geometry for the selfish herd. *J. Theor. Biol.* **31**, 295–311.

HANISCH, K.-H. (1981) On classes of random sets and point processes. *Serdica* **7**, 160–166.

HANISCH, K.-H. (1982) On inversion formulae for n-fold Palm distributions of point processes in LCS-spaces. *MN* **106**, 171–179 (see also M. ZÄHLE, 1983b).

HANISCH, K.-H. (1983a) Reduction of n-th moment measures and the special case of the third moment measure of stationary and isotropic planar point processes. *MOS* **14**, 421–436.

HANISCH, K.-H. (1983b) On stereological estimation of second-order characteristics and of hard-core distances of systems of sphere-centres. *BJ* **25**, 731–743.

HANISCH, K.-H. (1983c) *Beiträge zur Stereologie und Geometriestatistik unter besonderer Berücksichtigung von Momentenmassen*. Dissertation, Bergakademie, Freiberg.

HANISCH, K.-H. (1984a) Some remarks on estimators of the distribution function of nearest neighbour distance in stationary spatial point patterns. *MOS* **15**, 409–412.

HANISCH, K.-H. (1984b) Scattering analysis of point processes and random measures. *MN* **117**, 235–245.

ANISCH, K.-H. (1984c) On Palm and second-order quantities of point processes and germ-grain models. *Wissenschaftl. Sitzungen Stochastik WSS* – 01/84, Akademie der Wissenschaften der DDR, Berlin.

HANISCH, K.-H. (1984d) Pair-correlation functions of stationary and isotropic random k-dimensional manifolds in R^d. Bergakademie Freiberg, Internal Report.

HANISCH, K.-H. (1985) On the second-order analysis of stationary and isotropic fibre processes by line-intercept methods. *Geobild* '85, 141–146.

HANISCH, K.-H., P. KLIMANEK, and D. STOYAN (1985) Stereological analysis of dislocation arrangements in crystals from TEM images. *Crystal Res. Techn.* **20**, 921–930.

HANISCH, K.-H., D. KÖNIG and D. STOYAN (1985) The pair correlation function for point and fibre systems and its stereological determination by planar sections. *JM* **140**, 361–370.

HANISCH, K.-H., K.-D. OBERSTEDT, V. REINSCH, and D. STOYAN (1978) Genauigkeitsfragen bei der Vorratsberechnung von Lagerstätten. *Neue Bergbautechnik* **77**, 618–620.

HANISCH, K.-H. and D. STOYAN (1979) Formulas for the second-order analysis of marked point processes. *MOS* **10**, 555–560.

HANISCH, K.-H. and D. STOYAN (1980) Stereological estimation of the radial distribution function of centres of spheres. *JM* **122**, 131–141.

HANISCH, K.-H. and D. STOYAN (1983) Remarks on statistical inference and prediction for a hard-core clustering model. ***MOS 14***, 559–567.

HANISCH, K.-H. and D. STOYAN, (1984) Once more on orientations in point processes. ***EIK 20***, 279–284.

HARDING, E. F. and D. G. KENDALL (eds) (1974) *Stochastic Geometry*, Wiley and Sons, Chichester.

HARKER, D. and E. R. PARKER (1945) Grain shape and grain growth. *Trans Amer. Soc. Metals* ***34***, 156–173.

HASEGAWA, M. and M. TANEMURA (1976) On the pattern of space division by territories. *Ann. Inst. Statist. Math.* ***28B***, 509–519.

HASEGAWA, M. and M. TANEMURA (1980) Spatial patterns of territories. In: *Recent Developments in Statistical Inference and Data Analysis*. North Holland, Amsterdam/New York/Oxford, 73–78.

HASEGAWA, M., M. TANEMURA, and S. TAKIGUCHI (1981) Spatial patterns in ecology. In: *Int. Roundtable Congress 50th Anniversary Jap. Statist. Soc. 1981*, 146–161.

HEINRICH, L. (1984a) On a test of randomness of spatial point patterns. ***MOS 15***, 413–420.

HEINRICH, L. (1984b) Some estimators for the first-order n-fold Palm moment measure and their application to the investigation of dependencies in point patterns. ***EIK 20***, 244–251.

HEINRICH, L. (1986) Asymptotic normality of a random point field characteristic. ***MOS 17***, 453–460.

HERCZYNSKI, R. (1975) Distribution function for random distribution of spheres. *Nature* ***255***, 540–541.

HERMAN, G. T. and F. NATTERER (1981) (eds.) *Mathematical Aspects of Computerized Tomography*. Lecture Notes in Medial Informatics, No. 8. Springer-Verlag, Berlin/New York.

HERMANS, M. and J. P. RASSON (1981) *Stratified sampling of n-dimensional specimens*. Tech. Report, University of Namur.

HILLIARD, J. E. (1962) Specification and measurement of microstructural anisotropy. *Trans. Metall. Soc. AIME* ***224***, 1201–1211.

HILLIARD, J. E. (1967) Determination of structural anisotropy. In: ELIAS (1967), 219–227.

HINDE, A. L. and R. E. MILES (1980) Monte Carlo estimates of the distributions of the random polygons of the Voronoi tessellation with respect to a Poisson process. *J. Statist. Comput. Simul.* ***10***, 205–223.

HINES, W. G. S. and R. J. HINES (1979) The Eberhardt statistic and the detection of non randomness of spatial point distributions. *Biometrika* ***66***, 73–79.

HODDER, I. and C. ORTON (1976) *Spatial Analysis in Archaeology*. Cambridge University Press, Cambridge/London.

HOEFFDING, W. (1948) A class of statistics with asymptotically normal distribution. ***AMS 19***, 293–325.

HONDA, H. (1983) Geometric models for cells in tissues. *Int. Rev. Cytology* ***81***, 191–248.

HORALEK, V. (1978) Influence of resolution power of an instrument on estimation of basic material spatial structure parameters. *Kybernetika* ***14***, 447–460.

HORALEK, V. (1981) On the decomposition of particle size distributions in the extraction replica method. *Apl. Mat.* ***26***, 401–407.

HORALEK, V. (1982) sampling methods for studying spatial structure of specimen. In: SAXL (ed) (1982), 201–214.

HOSEMANN, R. and S. N. BAGCHI (1962) *Direct Analysis of Diffraction by Matter*. North Holland, Amsterdam/New York.

HOUSEHOLDER, A. S. (1964) *The Theory of Matrices in Numerical Analysis*. Blaisdell Publ. Comp., New York/Toronto/London.

HUANG, I. S. (ed.) (1975) *Picture Processing and Digital Filtering*. Springer-Verlag, Berlin/New York.

ISHAM, V. (1981) An introduction to spatial point processes and Markov random fields. *Int. Stat. Rev.* ***49***, 21–43.

ITOH, H. (1970) An analytical expression of the intercept length distribution of cubic particles. *Metallography* **3**, 407–417.

JACOD, J. and P. JOATHON (1971) Use of random genetic models in the study of sedimentary processes. *Math. Geol.* **3**, 265–279.

JAGERS, P. (1973) On Palm probabilities. *ZW* **26**, 17–32.

JAKEMAN, A. J. and R. L. SCHAEFFER (1978) On the properties of product integration estimators for linear functionals of particle size distributions. *Utilitas Math.* **14**, 117–128.

JANSON, S. and O. KALLENBERG, (1981) Maximising the intersection density of fibre processes. *JAP* **18**, 820–828.

JENSEN, E. B. (1984) A design-based proof of Wicksell's integral equation. *JM* **136**, 345–348.

JENSEN, E. B. (1985) On the use of point-sampling in the stereological analysis of arbitrarily shaped particles. In: *Geobild '85*, 105–111.

JENSEN, E. B., A. J. BADDELEY, H. J. G. GUNDERSEN, and R. SUNDBERG (1985) Recent trends in stereology. *Int. Stat. Rev.* **53**, 99–108.

JENSEN, E. B. and H. J. G. GUNDERSEN (eds) (1983) *Proceedings Second Int. Workshop Stereology and Stochastic Geometry*, Aarhus.

JENSEN, E. B. and H. J. G. GUNDERSEN (1985) The stereological estimation of moments of particle volume. *JAP* **22**, 82–98.

JENSEN, E. B., H. J. G. GUNDERSEN, and R. ØSTERBY (1979) Determination of membrane thickness from orthogonal intercepts. *JM*, **115**, 19–33.

JODREY, W. S. and E. M. TORY (1979) Simulation of random packings of spheres. *J. Simulation* **32**, 1–12.

JOLIVET, E. (1980) Central limit theorem and convergence of empirical processes for stationary point processes. In: *Point Processes and Queuing Problems* (eds. P. BARTFAI and J. TOMKO) North-Holland, Amsterdam/New York, 117–161.

JOLIVET, E. (1983) Etude de la vraisemblance d'un processus de Gauss-Poisson et deux applications. *ZW* **64**, 425–443.

JOLIVET, E. (1984) Upper bound of the speed of convergence of moment density estimators for stationary point processes. *Metrika* **31**, 349–360.

JOURNEL, A. and C. HUIJBREGTS (1978) *Mining Geostatistics*. Academic Press, London/ New York.

KALIŠNIK, M. (ed.) (1981) *Proc. Third European Symp. for Stereology*, Ljubljana.

KALLENBERG, O. (1973) Characterization and convergence of random measures and point processes. *ZW* **27**, 9–21.

KALLENBERG, O. (1976) *Random Measures*. Akademie Verlag, Berlin.

KALLENBERG, O. (1976, 1980, 1981) On the structure of stationary flat processes I, II, III. *ZW* **37**, 157–174, **52**, 127–147, 239–253.

KALLENBERG, O. (1977) A counterexample to R. Davidson's conjecture on line processes. *Math. Proc. Camb. Phil. Soc.* **82**, 301–307.

KALLENBERG, O. (1978) On the asymptotic behaviour of line processes and systems of noninteracting particles. *ZW* **43**, 65–95.

KALLENBERG, O. (1983a) The invariance problem for stationary line and flat processes. In: JENSEN and GUNDERSEN (eds) (1983), 105–114.

KALLENBERG, O. (1983b) *Random Measures* (Third Edition, revised and enlarged) Akademie Verlag, Berlin, and Academic Press, London.

KALLENBERG, O. (1984) An informal guide to the theory of conditioning in point processes. *Int. Stat. Rev.* **52**, 151–164.

KANATANI, K. (1984) Stereological determination of structural anisotropy. *Int. J. Eng. Sci.* **22**, 531–546.

KANATANI, K. and O. ISHIKAWA (1985) Error analysis for the stereological estimation of sphere size distributions: Abel type integral equation. *J. Comp. Physics* **57**, 229 bis 250.

KARR, A. F. (1978) Derived random measures. *Stoch. Proc. Appl.* **8**, 159–169.

KARR, A. F. (1985) *Point Processes and Their Statistical Inference*. Dekker, New York/ Zürich.

KEIDING, N., S. T. JENSEN, and L. RANEK (1972) Maximum likelihood estimation of the size distribution of linear cell nuclei from the observed distribution in a plane section. *Biometrics* **28**, 813–830.
KELLERER, A. M. (1971) Considerations on the random transversal of convex bodies and solution for general cylinders. *Radiation Research* **47**, 359–376.
KELLERER, A. M. (1983) On the number of clumps resulting from the overlap of randomly placed figures in a plane. *JAP* **20**, 126–135.
KELLERER, A. M. (1985) Counting figures in planar random configurations. *JAP* **22**, 68–81.
KELLERER, A. M. (1986) The variance of a Poisson process of domains. *JAP* **23**, 307–321.
KELLY, F. P. and B. D. RIPLEY (1976) A note on Strauss's model for clustering. *Biometrika* **63**, 357–360.
KENDALL, D. G. (1974) Foundations of a theory of random sets. In: HARDING and KENDALL (eds.) (1974), 322–376.
KENDALL, D. G. (1983) The shape of Poisson-Delaunay triangles. In: *Studies in Probability and Related Topics in Honour of Octav Onicescu.* (M. C. DEMETRESCU, M. IOSIFESCU eds.) Nagard, Montreal, 321–330.
KENDALL, D. G. (1984) Shape manifolds, procrustean metrics, and complex projective spaces. *Bull. London Math. Soc.* **16**, 81–121.
KENDALL, D. G. (1985) The exact distribution for the shape of a random triangle with IID vertices uniformly distributed inside a circle. *AAP* **17**, 308–329.
KENDALL, D. G. (1986) Further developments and applications of the statistical theory of shape. *Teor. Verojatn. Primen.* **31**, 467–473.
KENDALL, D. G. and W. S. KENDALL, (1980) Alignments in two-dimensional random sets of points. *AAP* **12**, 380–424.
KENDALL, W. S. (1979) *Some problems in probability theory.* D. Phil. Thesis, University of Oxford.
KERSHAW, K. A. (1964) *Quantitative and Dynamic Ecology.* Edward Arnold, London.
KERSTAN, J., K. MATTHES, and J. MECKE (1982) *Infinitely Divisible Point Processes* (Russ.) Nauka, Moscow.
KHINTCHINE, A. YA. (1955) *Mathematical Methods of Queueing Theory* (Russ.). Papers of the Mathematical Institute V. A. Steklov, vol. **49**, Moscow.
KIANG, T. (1966) Random fragmentation in 2 and 3 dimensions. *Z. f. Astrophysik* **64**, 433–439.
KINGMAN, J. F. C. (1967) Completely random measures. *Pacific J. Math.* **21**, 59 bis 78.
KINGMAN, J. F. C. (1969) Random secants of a convex body. *JAP* **6**, 660–672.
KINGMAN, J. F. C. and S. J. TAYLOR (1966) *Introduction to Measure and Probability.* Cambridge University Press, Cambridge/London.
KLEINWÄCHTER, A. and M. ZÄHLE (1986) Size distribution stereology for quasi-ellipsoids in R^n. *MOS* **17**, 435–460.
KLIER, G. (1969) Mathematisch-statistische Untersuchungen zur Verteilung der Bäume im Bestand. *Wiss. Z. TU Dresden* **18**, 1061–1065.
KOCH, R. A., L. PFEIFFER, and L. STAMMLER (1983) *Der Basalt von Stolpen in der Lausitz.* Deutscher Verlag für Grundstoffindustrie, Leipzig.
KOLMOGOROFF, A. N. (1933) *Grundbegriffe der Wahrscheinlichkeitsrechnung.* Springer-Verlag, Berlin.
KÖNIG, D., K. MATTHES and K. NAWROTZKI (1967) *Verallgemeinerungen der Erlangschen und Engsetschen Formeln.* Akademie-Verlag, Berlin.
KÖNIG, D., K.-H. HANISCH, and D. STOYAN (1983) Stereological determination of the pair-correlation function. *Acta Stereologica* **2**, Suppl. I, 19–22.
KOSCHITZKI, S. (1980) Some stereological problems for random discs in R^3. *MOS* **11**, 75–83.
KRATKY, O. (1983) Die Welt der vernachlässigten Dimensionen und die Kleinwinkelstreuung der Röntgen-Strahlen und Neutronen an biologischen Molekülen. *Nova Acta Leopoldina* (Neue Folge) **256**, vol. 55.

KRICKEBERG, K. (1972a) The Cox process. *Instituto Nazionale di Alta Matematicam Symposia Matematica* **9**, 151–167.

KRICKEBERG, K. (1972b) Theory of hyperplane processes. In: LEWIS (ed.) (1972), 514–521.

KRICKEBERG, K. (1974) Moments of point processes. In: HARDING and KENDALL (eds.) (1974), 89–113.

KRICKEBERG, K. (1975) Hyperplane processes. *Bull. Int. Inst. Statist.* **46**, 623–630.

KRICKEBERG, K. (1982) Processus Ponctuels en Statistique. In: *Ecole d'Ete de Probabilities de Saint-Flour X*, 1980 (ed. P. L. HENNEQUIN). *SLNM* **529**.

KROUSTRUP, A. and H. J. G. GUNDERSEN (1983) Sampling problems in a heterogeneous organ: quantification of relative and total volume of pancreatic islets by light microscopy. *JM* **132**, 43–55.

KRYSCIO, R. J. and R. SAUNDERS (1983) On interpoint distances for planar Poisson cluster processes. *JAP* **20**, 513–528.

KUTOYANTS, YU. A. (1984) *Parameter estimation for stochastic processes*. Heldermann, Berlin.

LASLETT, G. M. (1982) The survival curve under monotone density constraints with application to two-dimensional line segment processes. *Biometrika* **69**, 153–160.

LASLETT, G. M., W. S. KENDALL, A. J. W. GLEADOW, and I. R. DUDDY (1982) Bias in measurement of fission track length distributions. *Nuclear Tracks* **6**, 79–85.

LEDER, D. (1979) Stichprobentechnik in der quantitativen Morphologie. *Micr. Acta Suppl.* **3**, 129–138.

LEE, D. T. and B. J. SCHACHTER (1980) Two algorithms for constructing a Delaunay triangulation. *Int. J. Comput. Inform. Sci.* **9**, 219–242.

LEICHTWEISS, K. (1980) *Konvexe Figuren*. Deutscher Verlag d. Wissenschaften, Berlin.

LEWIS, F. T. (1946) The shape of cells as a mathematical problem. *Amer. Scientist* **34**, 359–369.

LEWIS, H. D., K. L. WALTHERS, and K. A. JOHNSON (1973) Particle size distributions by area analysis: Modifications of the Saltykov method. *Metallography* **6**, 93–101.

LEWIS, P. A. W. (ed) (1972) *Stochastic Point Processes. Statisticae Analysis, Theory and Applications*. Wiley and Sons., Chichester.

LEWIS, P. A. W. and G. S. SHEDLER (1979) Simulation of non-homogeneous Poisson processes by thinning. *Naval Res. Log. Quart.* **26**, 403–413.

LIEMANT, A. (1981) Kritische Verzweigungsprozesses mit allgemeinem Phasenraum, IV. *MN* **102**, 235–254.

LIESENBERG, O., and J. OHSER (1983) Die Beziehung zwischen der Zugfestigkeit und der Graphitausbildung von Gußeisen mit Lamellengraphit. *Gießereitechnik* **29**, 106–108.

LIKEŠ, J. (1963) On the problem of particle number and size determination in opaque bodies. *Acta Techn. Acad. Sci. Hung.* **42**, 325–341.

LIPSKIJ, J. N. et al. (1977) *Catalogue of the Craters of the Mars and Statistics of the Craters of the Mars, Mercury and Moon*. Acad. Sciences USSR, Geol. Inst., Moscow. (The analysis is based on: Shaded relief map of Mars, 1 : 25000000, U. S. Geol. Survey 1972.)

LORD, G. W. and T. F. WILLIS (1951) Calculation of air bubble distribution from results of a Rosiwal traverse of aerated concrete. *A. S. T. M. Bull.* **56**, 177–187.

LOTWICK, H. W. (1982) Simulation of some spatial hard-core models and the complete packing problem. *J. Statist. Comput. Simulation* **15**, 295–314.

LOTWICK, H. W. (1984) Some models for multitype spatial point processes, with remarks on analysing multitype patterns. *JAP* **21**, 575–582.

LOTWICK, H. W. and B. W. SILVERMAN (1981) Convergence of spatial birth-and-death processes. *Math. Proc. Camb. Phil. Soc.* **90**, 155–165.

LOTWICK, H. W. and B. W. SILVERMAN (1982) Methods for analysing spatial processes of several types of points. *JRSS* **B44**, 406–413.

LYASHENKO, N. N. (1982) Limit theorems for sums of independent, compact, random subsets of Euclidean space. *J. Soviet Math.* **20**, 2187–2196.

LYASHENKO, N. N. (1983) Statistics of random compacts in Euclidean space. *J. Soviet Math.* **21**, 76–92.

MACK, C. (1954) The expected number of clumps formed when convex laminae or bodies are placed at random and with random orientation on a plane area. *Proc. Camb. Phil. Soc.* **50**, 581–585.

MACK, C. (1956) On clumbs formed when convex laminae or bodies are placed at random in two or three dimensions. *Proc. Camb. Phil. Soc.* **52**, 246–250.

MARCHANT, J. C. and P. L. P. DILLON (1961) Correlation between randomdot samples and the photographic emulsion. *J. Optical Soc. America* **51**, 641–644.

MARCUS, A. (1972) Some point processes models of lunar and planetary sufaces. In: LEWIS (ed.) (1972), 682–699.

MARDIA, K. V. (1972) *Statistics of Directional Data.* Academic Press, London.

MARDIA, K. (1980) Contribution to the discussion of Broadbent (1980).

MARDIA, K. V., R. EDWARDS, and M. L. PURI (1977) Analysis of central place theory. *Bull. I. S. I.* **47** (2), 93–110.

MARRIOTT, F. H. C. (1979) Barnard's Monte Carlo tests: how many simulations? *Appl. Statist.* **28**, 75–77.

MASE, S. (1982a) Asymptotic properties of stereological estimators of volume fraction for stationary random sets. *JAP* **19**, 111–126.

MASE, S. (1982b) Properties of fourth-order mixing rates and its application to random set theory. *J. Multivariate Analysis* **12**, 549–561.

MASE, S. (1984) Locally asymptotic normality of Gibbs models on a lattice. *AAP* **16**, 585–602.

MATÉRN, B. (1960) *Spatial Variation.* Meddelanden från Statens Skogsforskningsinstitut **49**, 5, 1–144.

MATÉRN, B. (1971) Doubly stochastic Poisson processes in the plane. In: *Statistical Ecology*, vol. I. (ed. G. PATIL) Penn. State Univ. Press, University Park and London, 195–213.

MATHERON, G. (1967) *Eléments pour une Théorie des Mileux Poreux.* Masson, Paris.

MATHERON, G. (1975) *Random Sets and Integral Geometry.* J. Wiley & Sons, New York/ London.

MATHERON, G. (1978) La formule de Steiner pour les erosions. *JAP* **15**, 126–135.

MATTHES, K. (1963) Stationäre zufällige Punktfolgen. *Jahresbericht Deutsche Math. Verein.* **66**, 66–79.

MATTHES, K., J. KERSTAN, and J. MECKE (1978). *Infinitely Divisible Point Processes.* J. Wiley & Sons, Chichester.

MATTHES, K., W. WARMUTH, and J. MECKE (1979) Bemerkungen zu einer Arbeit von Nguyen Xuan Xanh und Hans Zessin. *MN* **88**, 117–127.

MEAD, R. (1974) A test for spatial pattern at several scales using data from a grid of contiguous quadrats. *Biometrics* **30**, 295–307.

MECKE, J. (1967) Stationäre zufällige Maße auf lokalkompakten Abelschen Gruppen. *ZW* **9**, 36–58.

MECKE, J. (1968) Eine characteristische Eigenschaft der doppelt stochastischen Poissonschen Prozesse. *ZW* **11**, 74–81.

MECKE, J. (1976) *Momentenmasse zufälliger Maße.* Unpublished Manuscript, Jena.

MECKE, J. (1977) Eine Charakterisierung des Westcottschen Funktionals. *MN* **80**, 295–313.

MECKE, J. (1979a) *Das n-fache Campbell-Maß.* Unpublished Manuscript, Jena.

MECKE, J. (1979b) An explicit description of Kallenberg's lattice-type point process. *MN* **89**, 185–195.

MECKE, J. (1980) Palm methods for stationary random mosaics. In: *Combinatorial Principles in Stochastic Geometry.* (ed. R. V. AMBARTZUMIAN) Armenian Academy of Sciences Publ. Erevan, 124–132.

MECKE, J. (1981a) Formulas for stationary planar fibre processes III — Intersection with fibre systems. *MOS* **12**, 201–210.

MECKE, J. (1981b) Stereological formulas for manifold processes. *Prob. and Math. Statist.* **2**, 31–35.

MECKE, J. (1983a) Inequalities for intersection densities of superpositions of stationary Poisson hyperplane processes. In: JENSEN and GUNDERSON (eds) (1983), 115–124.
MECKE, J. (1983b) Zufällige Mosaike. In: STOYAN and MECKE (1983), Chapter 8.
MECKE, J. (1984a) Parametric representation of mean values for stationary random mosaics. *MOS* **15**, 437–442.
MECKE, J. (1984b) Isoperimetric properties of stationary random mosaics. *MN* **117**, 75–82.
MECKE, J. (1984c) Random tessellations generated by hyperplanes. In: AMBARTZUMIAN and WEIL (eds) (1984), 104–109.
MECKE, J. and W. NAGEL (1980) Stationäre räumliche Faserprozesse und ihre Schnittzahlrosen. *EIK* **16**, 475–483.
MECKE, J. and D. STOYAN (1980a) Formulas for stationary planar fibre processes I — general theory. *MOS* **12**, 267–279.
MECKE, J. and D. STOYAN (1980b) Stereological problems for spherical particles. *MN* **96**, 311–317.
MEIJERING, J. L. (1953) Interface area, edge length, and number of vertices in crystal aggregates with random nucleation. *Philips Res. Rep.* **8**, 270–290.
MENDELSON, B. (1975) *Introduction to Topology*. Allyn & Bacon, Boston.
MILES, R. E. (1964) Random polygons determined by random lines in a plane. *Proc. Nat. Acad. Sci.* (*USA*) **52**, 901–907, (II), 1157–1160.
MILES, R. E. (1970) On the homogeneous planar Poisson point process. *Math. Biosciences* **6**, 85–127.
MILES, R. E. (1971) Poisson flats in Euclidean spaces. Part II: Homogeneous Poisson flats and the complementary theorem. *AAP*, **3**, 1–43.
MILES, R. E. (1972a) The random division of space. *Suppl. Adv. Appl. Prob.*, 243 bis 266.
MILES, R. E. (1972b) Multidimensional perspectives in stereology. *JM*, **95**, 181 bis 195.
MILES, R. E. (1973) The various aggregates of random polygons determined by random lines in a plane. *Adv. Math.* **10**, 256–290.
MILES, R. E. (1974a) On the elimination of edge-effects in planar sampling. In: HARDING and KENDALL (eds) (1974), 228–247.
MILES, R. E. (1974b) A Synopsis of 'Poisson Flats in Euclidean Spaces'. In: HARDING and KENDALL (eds) (1974), 202–227.
MILES, R. E. (1976) Estimating aggregate and overall characteristics from thick sections by transmission microscopy. *JM* **107**, 227–233
MILES, R. E. (1978) The importance of proper model specification in stereology. In: *Geometrical Probability and Biological Structures: Buffon's 200th Anniversary*. (eds. R. E. MILES and J. SERRA) *SLNB* **23**, 115–136.
MILES, R. E. (1983) Stereology for embedded aggregates of not-necessarily-convex particles. In: JENSEN and GUNDERSON (eds) (1983), 127–147.
MILES, R. E. (1985a) A comprehensive set of stereological formulae for embedded aggregates of not-necessarily-convex particles. *JM* **138**, 127–136.
MILES, R. E. (1985b) Estimating dihedral angle distributions by planar sections. In: *Extended Abstracts of 4th Europ. Symp. Stereology*, Göteborg (eds. B. KARLSSON, P. WARREN, J. WASEN) No. 1.
MILES, R. E. and P. J. DAVY (1977) Particle number can be stereologically estimated by wedge sections. *JM* **113**, 45–51.
MILES, R. E. and R. J. MAILLARDET (1982) The basic structures of Voronoi and generalized Voronoi polygons. *JAP* **19 A**, 97–112.
MILNE, R. K. and W. WESTCOTT (1972) Further results for Gauss-Poisson processes. *AAP* **4**, 151–176.
MINLOS, R. A. (1968) Lectures on statistical physics. (Russ.) *Usp. Mat. Nauk* **23**, 133–190.
MITTELBACH, P. (1964) Zur Röntgenkleinwinkelstreuung verdünnter kolloider Systeme, VIII. *Acta Phys. Austriaca* **19**, 53–102.
MOLEK, H., S. POHLMANN, F. REUTER, and D. STOYAN (1981) Entwicklung eines kom-

plexen Durchtrennungsgrades von Gesteinsverbänden mit Hilfe stereologischer Methoden. *Neue Bergbautechnik* ***11***, 221–224.

MOLLISON, D. (1977) Spatial contact models for ecological and epidemic spread (with discussion). *JRSS* ***B39***, 283–326.

MORAN, P. A. P. (1972) The probabilistic basis of stereology. *Suppl. Adv. Appl. Prob.*, 69–91.

MORAN, P. A. P. (1973) The random volume of interpenetrating spheres in space. *JAP* ***10***, 483–490.

MORAN, P. A. P. (1976) Another quasi-Poisson planar point process. *ZW* ***33***, 269–272.

MORTON, R. R. A. (1966) The expected number and angle of intersection between random curves in a plane. *JAP* ***3***, 559–562.

MÜRMANN, M. G. (1978) Poisson point processes with exclusion. *ZW* ***43***, 23–37.

MYERS, E. J. (1967) Size distributions of cubic particles. In: ELIAS (ed.) (1967), 187–188.

NAGEL, W. (1983) Dünne Schnitte von stationären räumlichen Faserprozessen. *MOS* ***14***, 569–576.

NAGEL, W. (1985) Weighted size distributions for stationary grain models. *Geobild '85*, 118–125.

NAUMOVICH, N. V. and T. I. KRISKOVETS (1982) Influence of the body shape on the profile of chord length distribution. *Acta Stereol.* ***1***, 51–59.

NEYMAN, J. (1939) On a new class of contagious distributions applicable in entomology and bacteriology. *AMS* ***10***, 35–57.

NEYMAN, J. and E. L. SCOTT (1958) Statistical approach to problems of cosmology. *JRSS* ***B20***, 1–43.

NEYMAN, J. and E. L. SCOTT (1972) Processes of clustering and applications. In: LEWIS (ed) (1972), 646–681.

NGUYEN, X. X. (1979) Ergodic theorems for subadditive spatial processes. *ZW* ***48***, 159–176.

NGUYEN, X. X. and H. ZESSIN (1979) Ergodic theorems for spatial processes. *ZW* ***48***, 133–158.

NICHOLSON, W. L. (1970) Estimation of linear properties of particle size distributions. *Biometrika* ***57***, 273–297.

NICHOLSON, W. L. (1976) Estimation of linear functionals by maximum likelihood. *JM* ***107***, 323–336.

NICHOLSON, W. L. (1978) Application of statistical methods in quantitative microscopy. *JM* ***113***, 223–239.

NICHOLSON, W. L. and K. R. MERCKX (1969) Unfolding particle size distributions. *Technometrics* ***11***, 707–723.

NORBERG, T. (1984) Convergence and existence of random set functions. *Ann. Prob.* ***12***, 726–732.

NUTTING, P. G. (1913) On the absorption of light in heterogeneous media. *London, Edinburg & Dublin Phil. Mag. Ser. 6*, ***26***, 423–426.

NYCHKA, D. (1983a) Solving integral equations with noisy data: An application of smoothing splines in pathology. In: *Computer Science and Statistics: Proc. of the 14th Symp. on the Interface.* (eds. K. W. HEINER, R. S. SACHER, J. W. WILKINSON) Springer-Verlag, Berlin/New York, 157–163.

NYCHKA, D. (1983b) Smooth non-parametric estimates of particle size distributions. *Acta Stereologica* ***2***, Suppl. I, 25–28.

OGATA, Y. (1981) On Lewis' simulation method for point processes. *IEEE Trans. Inf. Theor.* ***IT-27***, 23–31.

OGATA, Y. and M. TANEMURA (1981) Estimation of interaction potentials of spatial point patterns through the maximum likelihood procedure. *Ann. Inst. Statist. Math.* ***33B***, 315–338.

OGATA, Y. and M. TANEMURA (1984) Likelihood analysis of spatial point patterns. *JRSS* ***B46***, 496–518.

OGATA, Y. and M. TANEMURA (1985) Estimation of interaction potentials of marked spatial point processes through the maximum-likelihood method. *Biometrics* ***41***, 421–433.

OGAWA, T., and M. TANEMURA (1974) Geometrical considerations on hard core problems. *Progress Theor. Phys.* ***51***, 399–417.

OGANIAN, V. K. (1973) On random Markovian colouring of the plane with two colours (Russ.) *Dokl. Akad. Nauk. Armenian SSR* ***58***, 193–198.

OGANIAN, V. K. (1978) Combinatorial principles in stochastic geometry of random segment processes. In: *Combinatorial Principles in Stochastic Geometry* (ed. R. V. AMBARTZUMIAN) Publ. Akad. Sci. Armen. SSR, 81–106.

OHSER, J. (1980) On statistical analysis of the Boolean model. *EIK* ***16***, 651–653.

OHSER, J. (1981) A remark on the estimation of the rose of directions of fibre processes. *MOS* ***12***, 581–585.

OHSER, J. (1983) On estimators for the reduced second moment measure of point processes. *MOS*, ***14***, 63–71.

OHSER, J. and G. HEINZEL (1985) Bemerkungen zur metallographischen Bestimmung des Volumenanteils von Gefügebestandteilen. *Neue Hütte* ***30***, 229–232.

OHSER, J., B. D. RIPLEY, and D. STOYAN (1985) *On the efficiency of edge corrections for spatial point processes.* Manuscript, University of Strathclyde and Bergakademie Freiberg.

OHSER, J. and D. STOYAN (1980) Zur Beschreibung gewisser zufälliger Muster in der Geologie. *Z. angew. Geol.* ***26***, 209–212.

OHSER, J. and D. STOYAN (1981) On the second-order and orientation analysis of planar stationary point processes. *BJ* ***23***, 523–533.

OHYA, T., M. IRI, and K. MUROTA (1984) Improvement of the incremental method for the Voronoi diagram with computational comparison of various algorithms. *J. Oper. Res. Soc. Japan* ***27***, 306–336.

ORNSTEIN, L. S. and F. ZERNICKE (1914) Accidental deviations of density and opalescence at the critical point of a single substance. *Proc. Akad. Sci.* (Amsterdam) ***17***, 793–806.

PALM, C. (1943) Intensitätsschwankungen im Fernsprechverkehr. *Ericsson Technics* ***44***, 1–189.

PAPANGELOU, F. (1974) On the Palm probabilities of processes of points and processes of lines. In: HARDING and KENDALL (eds) (1974), 114–147.

PAPANGELOU, F. (1976) Point processes on spaces of flats and other homogeneous spaces. *Math. Proc. Camb. Phil. Soc.* ***80***, 297–314.

PAUL, J. L. (1981) Distribution curves of sectional area through some families of convex particles. *JM* ***122***, 165–172.

PENTTINEN, A. (1984) Modelling interactions in spatial point patterns: parameter estimation by the maximum-likelihood method. *Jyväskyla Studies in Comp. Sci., Econ. and Statist.* ***7***, Jyväskyla.

PERSSON, O. (1971) The robustness of estimating density by distance measurements. In: *Statistical Ecology*, vol. II. (eds. G. PATIL, E. C. PIELOU, W. E. WATERS) Penn. State Univ. Press, University Park, 175–187.

PIEFKE, F. (1976) Estimation of linear properties of spherical bodies in thin foils from their projections. *Fourth Intern. Cong. Stereology* 1975, 497–498.

POHLMANN, S., J. MECKE, and D. STOYAN (1981) Stereological formulas for stationary surface processes. *MOS* ***12***, 429–440.

POLYA, G. (1918) Zahlentheoretisches und Wahrscheinlichkeitstheoretisches über die Sichtweite im Walde. *Arch. Math. Phys.* ***27***, 135–142.

POROD, G. (1951) Die Röntgenkleinwinkelstreuung von dichtgepackten kolloiden Systemen I. *Kolloid Zeitschrift* ***124***, 83–114.

POROD, G. (1952) Die Röntgenkleinwinkelstreuung von dichtgepackten kolloiden Systemen II. *Kolloid Zeitschrift* ***125***, 51–57.

PRESTON, C. J. (1974) *Gibbs States on Countable Sets.* Cambridge University Press, Cambridge/London.

PRESTON, C. J. (1975) Spatial birth-and-death processes. *Bull. Int. Statist. Inst.* ***46*** (2), 371–391.

PRESTON, C. J. (1976) Random fields. *SLNM* ***534***, 1–200.

QUINE, M. P. and D. F. WATSON (1984) Radial simulation of n-dimensional Poisson processes. *JAP* **21**, 548–557.

RADECKE, W. (1980) Some mean value relations on stationary random mosaics in the space. *MN* **97**, 203–210.

RAHMAN, A. (1966) Liquid structure and self-diffusion. *J. Chem. Phys.* **45**, 2584–2592.

RANNEBY, B. (ed) (1982) *Statistic in Theory and Practice.* Swedish University of Agricultural Science, Umea.

RENYI, A. (1967) Remarks on the Poisson process. *Studia Sci. Math. Hung.* **2**, 119–123.

RIPLEY, B. D. (1976a) The second-order analysis of stationary point processes. *JAP* **13**, 255–266.

RIPLEY, B. D. (1976b) The foundations of stochastic geometry. *Ann. Prob.* **4**, 995–998.

RIPLEY, B. D. (1976c) On stationarity and superposition of point processes. *Ann. Prob.* **4**, 999–1005.

RIPLEY, B. D. (1976d) Locally finite random sets: Foundations for point process theory. *Ann. Prob.* **4**, 983–994.

RIPLEY, B. D. (1977) Modelling spatial patterns (with discussion). *JRSS* **B39**, 172–212.

RIPLEY, B. D. (1979a) Simulating spatial patterns: dependent samples from a multivariate density. Appl. Statist. **28**, 109–112.

RIPLEY, B. D. (1979b) Test of 'randomness' for spatial point patterns. *JRSS* **B41**, 368–374.

RIPLEY, B. D. (1981) *Spatial Statistics.* J. Wiley & Sons, New York/Chichester.

RIPLEY, B. D. (1982) Edge effects in spatial stochastic processes. In: Ranneby, B. (ed.) (1982), 247–262.

RIPLEY, B. D. (1983) The lattice structure of pseudo-random number generators. *Proc. R. Soc. London* **A389**, 197–204.

RIPLEY, B. D. (1984) Edge correction for spatial processes. In: AMBARTZUMIAN and WEIL (eds) (1984), 144–153.

RIPLEY, B. D. (1985a) Summaries and models for digitized images. Geobild '85, 6–9.

RIPLEY, B. D. (1985b) Analyses of nest spacings. In: *Statistics in Ornithology* (eds. B. J. T. MORGAN and P. M. NORTH), *SLNS* **29**, 151–158.

RIPLEY, B. D. (1986) *Statistical Simulation.* J. Wiley & Sons, Chichester.

RIPLEY, B. D. and F. P. KELLY (1977) Markov point processes. *J. London Math. Soc.* **15**, 188–192.

ROBBINS, H. E. (1944, 1945) On the measure of a random set I, II. *AMS* **15**, 70–74; **16**, 342–347.

ROGERS, C. A. (1964) *Packing and Covering.* Cambridge University Press, Cambridge/London.

ROSE, P. E. (1980) Improved tables for the evaluation of sphere site distributions including the effect of section thicknesses. *JM* **118**, 135–141.

ROSENFELD, A. and A. C. KAK (1982) *Digital Picture Processing.* 2nd edition. Academic Press, New York.

ROSIWAL, A. (1898) Über geometrische Gesteinsanalysen. Ein einfacher Weg zur ziffernmäßigen Feststellung des Quantitätsverhältnissess der Mineralbestandteile gemengter Gesteine. *Verh. K. K. Geol. Reichsanst.* **5/6**, 143–175.

RUELLE, D. (1969) *Statistical Mechanics.* Benjamin, New York.

RUELLE, D. (1970) Superstable interaction in classical mechanics. *Comm. Math. Phys.* **18**, 127–159.

RUSSELL, M. J. (1973) Base-metal mineralisation in Ireland and Scotland and the formation of Rockall Trough. In: *Implications of Continental Drift to the Earth Sciences* I, (eds. D. H. TARLING, S. K. RUNCORN) Academic Press, London, 581–597.

RYSZ, J. and K. WIENCEK (1980) Stereology of spherical carbide particles in steels. *Arch. Nauki o Materialach* **1**, 151–168.

SACHS, L. (1984) *Applied Statistics.* Springer-Verlag, Berlin/New York.

SALTYKOV, S. A. (1950) *Introduction to Sterometric Metallography* (Russ.), Publ. Acad. Sci. Armen. SSR, Erevan.

SALTYKOV, S. A. (1974) *Stereometrische Metallographie.* Deutscher Verlag für Grundstoffindustrie, Leipzig.

SANTALÓ, L. A. (1976) *Integral Geometry and Geometric Probability*. Addison-Wesley, Reading (Mass.).
SANTALÓ, L. A. (1980) Random lines and tessellations in a plane. *Stochastica* **4**, 3–13.
SANTALÓ, L. A. (1984) Mixed random mosaics. *MN* **117**, 129–133.
SAUNDERS, R., R. J. KRYSCIO, and G. M. FUNK (1982) Poisson limits for a hard-core clustering model. *Stoch. Proc. Appl.* **12**, 97–106.
SAXL, I. (1981) Overview of contemporary theoretical approaches to stereology. In: KALIŠNIK (ed) 1981, 19–41.
SAXL, I. (ed) (1982) Colloquium on Math. Morphology, Stereology, and Image, Analysis, ČSAV, Praha.
SAXL, I. (1984) *Stereologie vnitřni strukturu objektů*. Academia, Prague.
SAXL, I. (1986) *Stereology of Objects with Internal Structure*. Elsevier, Amsterdam.
SCALES, L. E., V. HOWARD, J. R. GREEN, and R. LYNCH (1982) *Some observations on the parametric unfolding of size distributions*. Report, University of London.
SCHEIDEGGER, A. E. (1979) Beziehungen zwischen Orientierungsstruktur der Tallagen und der Kluftstellungen in Österreich, *Mitt. Österr. Geograph. Ges.* **121**, II, 187–195.
SCHNEIDER, R. (1979a) Boundary structure and curvature of convex bodies. In: *Contributions to Geometry*. Proc. Geometrie-Symp. Siegen 1978 (eds. J. TÖLKE, J. M. WILLS) Birkhäuser, Basel/Boston/Stuttgart, 13–59.
SCHNEIDER, R. (1979b) Bestimmung konvexer Körper durch Krümmungsmasse. *Comment. Math. Helvet.* **54**, 42–60.
SCHNEIDER, R. (1980) Parallelmengen mit Vielfachheit und Steiner-Formeln. *Geom. Ded.* **9**, 111–127.
SCHNEIDER, R. and W. WEIL (1983) Zonoids and related topics. In: *Convexity and its Applications* (eds. P. GRUBER, J. M. WILLS), Birkhäuser, Basel/Boston/Stuttgart, 360–412.
SCHWANDTKE, A. (1985) Distributional analysis of dihedral angles in single-phase polycrystalline structures — comments on a paper by J. RYSZ and A. KASPRZYK. *Metalurgie i Odlewnictwo* **11**, 171–182.
SCHWANDTKE, A., J. OHSER, and D. STOYAN (1987) Improved estimation in planar sampling. *Acta Stereol.* **6**.
SERRA, J. P. (1980) The Boolean model and random sets. *Computer Graphics and Image Processing* **12**, 99–126.
SERRA, J. P. (1982) *Image Analysis and Mathematical Morphology*. Academic Press, London.
SHANNON, C. E. (1949) Communication in the presence of noise. *Proc. I. R. E.* **37**, 10–21.
SHIYOMI, M. (1978) A rapid graphic estimation of population density by a quasi-sequential method. *Bull. Nat. Inst. Agricult. Sci.* **A25**, 33–57.
SIBSON, R. (1980) The Dirichlet tessellation as an aid in data analysis. *Scand. J. Statist.* **7**, 14–20.
SIBSON, R. (1981) A brief description of natural neighbour interpolation. In: *Interpreting Multivariate Data* (ed. V. BARNETT) J. Wiley & Sons, Chichester, 21–36.
SIEGEL, C. L. (1945) A mean value theorem in geometry of numbers. *Ann. Math.* **46**, 340–347.
SILVERMAN, B. W. and T. C. BROWN (1978) Short distances, flat triangles and Poisson limits. *JAP* **15**, 815–825.
SMALL, C. (1982) Random uniform triangles and the alignment problem. *Math. Proc. Camb. Phil. Soc.* **91**, 315–322.
SMALL, C. (1983) Characterization of distributions from maximal invariant statistics. *ZW* **63**, 517–527.
SMALLEY, I. J. (1966) Contraction crack networks in basalt flows. *Geol. Mag.* **103**, 110–114.
SMOLEJ, V. and H. F. FISCHMEISTER (1982) Digital image analysis in quantitative metallography. In: SAXL (ed) (1982), 189–200.
SNYDER, D. L. (1975) *Random Point Processes*. J. Wiley & Sons, New York/London.
SONNTAG, U., D. STOYAN, and H. HERMANN (1981) Random set models in the interpretation of small-angle scattering data. *Phys. stat. sol.* (*a*) **68**, 281–288.

Spektor, A. G. (1950) Analysis of distribution of spherical particles in non-transparent structures. *Zav. Lab.* ***16***, 173–177.

Stapper, C. H., A. McLaren, and M. Dreckmann (1980) Yield model for productivity optimization of VLSI memory chips with redundancy and partially good product. *IBM J. Res. Devel.* ***24***, 398–409.

Stephan, H. (1975) Allocortex. In: *Handbuch der mikroskopischen Anatomie des Menschen*, Band 4/9, Teil Nervensystem. Springer-Verlag, Berlin/New York.

Sterio, D. G. (1984) The unbiased estimation of numbers and sizes of arbitrary particles using the disector. *JM* ***134***, 127–136.

Stiny, J. (1929) *Technische Gesteinskunde für Bauingenieure, Kulturtechniker, Land- und Forstwirte*. 2nd. ed. Springer-Verlag, Wien.

Stoyan, D. (1979a) On some qualitative properties of the Boolean model of stochastic geometry. *Z. angew. Math. Mech.* ***59***, 447–454.

Stoyan, D. (1979b) On the accuracy of lineal analysis. *BJ* ***21***, 439–449.

Stoyan, D. (1979c) Proof of some fundamental formulas of stereology for non-Poisson grain models. *MOSO* ***10***, 573–581.

Stoyan, D. (1979d) Interrupted point processes. *BJ* ***21***, 607–610.

Stoyan, D. (1981) On the second-order analysis of stationary planar fibre processes. *MN* ***102***, 189–199.

Stoyan, D. (1982) Stereological formulae for size distributions through marked point processes. *Prob. and Math. Statist.* ***2***, 161–166.

Stoyan, D. (1983) Inequalities and bounds for variances of point processes and fibre processes. *MOS* ***14***, 409–419.

Stoyan, D. (1984a) On correlations of marked point processes. *MN* ***116***, 197–207.

Stoyan, D. (1984b) Correlations of the marks of marked point processes — statistical inference and simple models. *EIK* ***20***, 285–294.

Stoyan, D. (1984c) Further stereological formulae for spatial fibre processes. *MOS* ***15***, 421–428.

Stoyan, D. (1984d) Weighted fibres and surfaces in stereology. In: Ambartzumian and Weil (eds) (1984), 188–196.

Stoyan, D. (1984e) Two examples for the use of two-point Palm distributions in the statistics of point processes. In: Ambartzumian and Weil (eds) 1984, 179–187.

Stoyan, D. (1985a) Stereological determination of orientations, second-order quantities and correlations for random fibre systems. *BJ* ***27***, 411–425.

Stoyan, D. (1985b) Estimating the volume density from thin sections. *BJ* ***27***, 427 bis 430.

Stoyan, D. (1985c) Practicable methods for the determination of the pair-correlation function of fibre processes. *Geobild '85*, 131–140.

Stoyan, D. (1986) On generalized random tessellations. *MN* ***128***, 97–101.

Stoyan, D. and H. Hermann (1986) Some methods for statistical analysis of planar random tessellations. *MOS* ***17***, 407–420.

Stoyan, D. and J. Mecke (1983) *Stochastische Geometrie: Eine Einführung*. Akademie-Verlag, Berlin.

Stoyan, D., Mecke, J., and S. Pohlmann (1980) Formulas for stationary planar fibre processes II — Partially oriented fibre systems. *MOS* ***11***, 281–286.

Stoyan, D. and J. Ohser (1982) Correlations between planar random structures (with an ecological example). *BJ* ***24***, 631–647.

Stoyan, D. and J. Ohser (1984) Cross-correlation measure of weighted random measures and their estimation. (Russ.) *Teor. Verojatn. Primen.* ***29***, 338–347.

Stoyan, D. and H.-L. Steyer (1979) Zur Genauigkeit der Linearanalyse. *Neue Hütte* ***24***, 303–307.

Stoyan, D. and H. Stoyan (1980a) On some partial orderings of random closed sets. *MOSO* ***11***, 145–154.

Stoyan, D. and H. Stoyan (1980b) Gedanken zur Entstehung der Säulenformen bei Basalten. *Z. Geol. Wiss.* ***8***, 1529–1537.

Stoyan, D. and H. Stoyan (1984) Über eine Methode zur Quantifierung von Korrelationen zwischen geologischen Liniensystemen. *Z. angew. Geologie* ***29***, 512–517.

STOYAN, D. and H. STOYAN (1985) On one of Matérn's hard-core point process models. *MN* ***122***, 205–214.
STOYAN, H. and D. STOYAN (1986) Simple stochastic models for analysis of dislocation distributions. *Phys. stat. sol.* (a) ***97***.
STRAUSS, D. J. (1975) A model for clustering. *Biometrika* ***63***, 467–475.
SUKIASIAN, G. S. (1978) Random triangles on the plane. (Russ.) *Dokl. Akad. Nauk Armenian SSR* ***66***, 150–155.
SUKIASIAN, G. S. (1980) On chord processes on lines intersecting random fields of circles in the plane. (Russ.) *Dokl. Akad. Nauk Armenian SSR* ***70***, 297–300.
SUKIASIAN, G. S. (1982) On random sections of polyhedra. *Dokl. Akad. Nauk SSR* ***263***, 809–812.
SUNDBERG, R. and E. B. JENSEN (1983) Modelling stereological data. In: JENSEN and GUNDERSEN (eds) (1983), 149–171.
SUWA, N., T. TAKAHASHI, K. SAITO, and T. SAWAI (1976) Morphometrical method to estimate the parameters of distribution functions assumed for spherical bodies from measurements on a random section. *Tohoku J. Exp. Med.* ***118***, 101–118.
SUZUKI, A. and M. IRI (1986) Approximation of a tessellation of the plane by a Voronoi diagram. *J. Opns. Res. Japan* ***29***, 69–96.
TAKACS, R. (1983) Estimator for the pair-potential of a Gibbsian point process. *Johannes Kepler Univ. Linz. Inst. für Math. Inst. Ber.* ***238***.
TAKACS, R. (1986) Estimator for the pair-potential of a Gibbsian point process. *MOS* ***17***, 429–433.
TANNER, J. C. (1983a) The proportion of quadrilaterals formed by random lines ir a plane. *JAP* ***20***, 400–404.
TANNER, J. C. (1983b) Polygons formed by random lines in a plane: some further results. *JAP* ***20***, 778–787.
TAYLOR, C. C. (1983) A new method for unfolding sphere-size distributions. *JM* ***132***, 57–66.
THIELE, B. (1979) Eine Umkehrformel für einfache stationäre Punktprozesse endlicher Intensität im R^s. *MN* ***91***, 171–179.
THIESSEN, A. H. (1911) Precipitation averages for large areas. *Monthly Weather Review* ***39***, 1082–1084.
THOMAS, C. (1984) Extremum properties of the intersection densities of stationary Poisson hyperplane processes. *MOS*, ***15***, 443–450.
THOMPSON, E. (1930) Quantitative microscopic analysis. *J. Geol.* ***38***, 193–222.
UNDERWOOD, E. E. (1970) *Quantitative Stereology.* Addison-Wesley, Reading, Mass.
UNDERWOOD, E. E. (1976a) Appendix: Basic Stereology. In: *Proc. of the Fourth Int. Cong. for Stereology 1975*, National Bureau of Standards Special Publ. ***431***, 509 bis 513.
UNDERWOOD, E. E. (1976b) Three-dimensional shape parameters from planar sections. In: *Proc. Fourth Int. Cong. Stereology 1975*, 91–92.
UNDERWOOD, E. E. (1981) A standardized system of notation for stereologists. In: KALIŠNIK (ed.) (1981), 715–728.
UPTON, G. and B. FINGLETON (1985) *Spatial Data Analysis by Example*, vol. 1. J. Wiley & Sons, Chichester.
VITALE, R. A. (1983) Some developments in the theory of random sets. *Bull. Int. Statist. Inst.* ***50***, No. 2, 863–871.
VORONOI, G. (1908) Nouvelles applications des parametres continus a la theorie des formes quadratiques. *J. Reine Angew. Math.* ***134***, 198–287.
VOSS, K. (1980) Exakte stereologische Formeln und Näherungslösungen für konvexe Körper. *EIK* ***16***, 485–491.
VOSS, K. (1982a) Frequencies of n-polygons in planar sections of polyhedra. *JM* ***128***, 111–120.
VOSS, K. (1982b) Morphometry and stereology of cellular tissues. In: SAXL (ed) (1982), 137–146.
VOSS, K. and W. SCHUBERT (1975) Zur numerischen Auswertung von Schnittflächenverteilungen. *BJ* ***17***, 189–195.

Voss, K. and D. Stoyan (1985) On the stereological estimation of numerical density of particle systems by an object counting method. *BJ* **27**, 919–924.

Warren, W. G. (1962) Contributions to the study of spatial point processes. *University of North Carolina, Institute of Statistics Series* no. **337** (mimeo).

Warren, W. G. (1971) The centre-satellite concept as a basis for ecological sampling. In: *Statistical Ecology*, vol. II. (eds. G. Patil, E. C. Pielou, W. E. Waters) Penn. State Univ. Press, University Park, 87–118.

Wassilew, G. and K. Voss (1982) Integral geometric methods in stereology and their applications in light microscopy. *Acta Stereologica* **1**, 67–71.

Watson, D. F. (1981) Computing the n-dimensional Delaunay tessellation with application to Voronoi polytopes. *Comput. J.* **24**, 167–172.

Watson, G. S. (1971) Estimating functionals of particle size distributions. *Biometrika* **58**, 483–490.

Watson, G. S. (1974) Geometrical statistics: a report of the Satellite Symposium. *Int. Stat. Rev.* **42**, 211–212.

Watson, G. S. (1978) Characteristic statistical problems of stochastic geometry. In: *Geometrical Probability and Biological Structures: Buffon's 200th Anniversary.* (eds. R. E. Miles and J. Serra) *SLNB* **23**, 215–234.

Weber, W. (1977) Zur Methode der Lokalisierung und Charakterisierung tiefer Bruchstrukturen für minerogenetische Untersuchungen. In: Weber, W. and V. A. Korcemagin: *Tiefenbruchstrukturen und postmagmatische Mineralisation. Freiberger Forschungshefte* **C329**, 9–52.

Weibel, E. (1980) *Stereological Methods*, vol. 2. Academic Press, London/New York.

Weil, W. (1982a) Inner contact probabilities for convex bodies. *AAP* **14**, 582–599.

Weil, W. (1982b) An application of the central limit theorem for Banach-space-valued random variables to the theory of random sets. *ZW* **60**, 203–208.

Weil, W. (1983) Stereology: A survey for geometers. In: *Convexity and Its Applications* (eds. P. Gruber and J. M. Wills) Birkhäuser, Basel/Boston/Stuttgart, 360–412.

Weil, W. (1984) Densities of quermassintegrals for stationary random sets. In: Ambartzumian and Weil (eds) (1984), 233–247.

Weil, W. and J. A. Wieacker (1983) Densities for stationary random sets and point processes. *AAP* **16**, 324–346.

Weil, W. and J. A. Wieacker (1987) A representation theorem for random sets. *Prob. and Math. Statist.* **6**.

Westcott, M. (1982) Approximations to hard-core models and their application to statistical analysis. *JAP* **19 A**, 281–292.

Wicksell, S. D. (1925, 1926) The corpuscle problem I, II. *Biometrika* **17**, 84–89, **18**, 152–172.

Widom, J. S. and B. Rowlinson (1970) New model for the study of liquid-vapour phase transitions. *J. Chem. Phys.*, **52**, 1670–1684.

Wieacker, J. A. (1986) Intersections of random hypersurfaces and visibility. *ZW* **71**, 405–433.

Yadin, M. and S. Zacks (1985) The visibility of stationary and moving targets in the plane subject to a Poisson field of shadowing elements. *JAP* **22**, 776–786.

Zähle, M. (1982) Random processes of Hausdorff rectifiable closed sets. *MN* **108**, 49–72.

Zähle, M. (1983a) Random set processes in homogeneous Riemannian spaces. *MN* **110**, 179–193.

Zähle, M. (1983b) Review of Hanisch (1982); *Math. Reviews* **84b**: 60069, p. 699.

Zähle, M. (1984a) Curvature measures and random sets I. *MN* **119**, 327–339.

Zähle, M. (1984b) Thick section stereology for random fibres. *MOS* **15**, 429–436.

Zähle, M. (1984c) Properties of signed curvature measures. In: Ambartzumian and Weil (eds) (1984), 256–266.

Zähle, M. (1985) Curvature measures and random sets II. *ZW* **71**, 37–58

Zessin, H. (1983) The method of moments for random measures. *ZW* **62**, 395–409.

Zuiderwijk, E. J. (1982) Alignment of randomly distributed objects. *Nature* **295**, 577–578.

Author Index

Subject Index

Bei Fragen zur Produktsicherheit wenden Sie sich bitte an:
If you have any questions regarding product safety,
please contact:

Walter de Gruyter GmbH
Genthiner Straße 13
10785 Berlin
productsafety@degruyterbrill.com